Agricultural Land Use

This book presents contemporary trends and challenges in agricultural land use, drawing on an array of global case studies.

This volume examines agricultural land use through a three-pronged approach: structural transformations, environmental challenges, and planning and policy. Reflecting the global relevance of contemporary challenges to agricultural land use, the book presents a wide range of novel case studies from across the world, including Poland, Hungary, Denmark, Slovenia, Czechia, Tanzania, Uganda, South Africa, Algeria, Northeast African countries, China, India, Australia, New Zealand, USA, and Brazil. Each chapter focuses on a particular issue related to agricultural land use in that country or region, including land fragmentation, reduction of cropland, agricultural intensification, desertification of soils, climate change, biodiversity loss, traditional agricultural systems, urbanization and rural development, and planning challenges for agricultural land. Together the chapters present a global view of the challenges facing the agricultural industry and offer solutions for developing sustainable agricultural practices to ensure food security and environmental and biodiversity conservation.

This book will be of great interest to students and scholars of food and agriculture, land use and land management, rural studies, and sustainable development.

Jerzy Bański is Professor in the Institute of Geography and Spatial Organization at the Polish Academy of Sciences, Poland. From 2018 to 2021, he was the director of this institute, and from 2006 to 2012, he was the president of the Polish Geographical Society. He is the editor of *The Routledge Handbook of Small Towns* (Routledge, 2021) and *Dilemmas of Regional and Local Development* (Routledge, 2020).

Earthscan Food and Agriculture

Digital Communication for Agricultural and Rural Development
Participatory Practices in a Post-COVID Age
Edited by Ataharul Chowdhury and Gordon Gow

Regenerative Farming and Sustainable Diets
Human, Animal and Planetary Health
Edited by Joyce D'Silva and Carol McKenna

Sustainable Cacao Cultivation in Latin America
Edited by Luz Cecilia García, Naga Raju Maddela, Freddy Zambrano Gavilanes and Carolina Aguilar Duarte

Food Policy in the United States
An Introduction, 3rd edition
Parke Wilde

Agricultural and Climate Change Adaptation Law in Africa
Reflections from Kenya, Nigeria and South Africa
Habib Sani Usman

Pathways to African Food Security
Challenges, Threats and Opportunities towards 2050
Edited by Michiel de Haas and Ken Giller

Multiple Capitals Approach for Upland Agricultural Systems
Lois Mansfield

Agricultural Land Use
Structural Transformations, Environment Challenges, Planning and Policy
Edited by Jerzy Bański

For more information about this series, please visit: www.routledge.com/Earthscan-Food-and-Agriculture/book-series/ECEFA

Agricultural Land Use

Structural Transformations, Environment Challenges, Planning and Policy

Edited by Jerzy Bański

Routledge
Taylor & Francis Group
LONDON AND NEW YORK

earthscan
from Routledge

First published 2025
by Routledge
4 Park Square, Milton Park, Abingdon, Oxon OX14 4RN

and by Routledge
605 Third Avenue, New York, NY 10158

Routledge is an imprint of the Taylor & Francis Group, an informa business

British Library Cataloguing-in-Publication Data
A catalogue record for this book is available from the British Library

ISBN: 978-1-032-67275-5 (hbk)
ISBN: 978-1-032-67279-3 (pbk)
ISBN: 978-1-032-67277-9 (ebk)

DOI: 10.4324/9781032672779

Typeset in Times New Roman
by Apex CoVantage, LLC

Contents

Contributors

Tayeb Addoun obtained a doctorate in Science in Geography and Regional Planning in 2021 from the University of Oran 2 Mohammed Ben Ahmed under the title 'Spatial dynamics of the agglomerations of the North-Central Algerian Sahara, (wilaya of Ghardaïa)'. He has worked in the Faculty of Earth and Universe Sciences, Oran 2 University, Department of Geography. Currently, he is a doctor at the Higher College of Teachers 'ENS' Ouargla, Department of Geography. He is a member of the team of the Geographic Space and Regional Planning (EGEAT) research laboratory, working on the dynamics of steppe and Saharan spaces. In addition to his tasks as a teacher-researcher, he has published several scientific articles in indexed journals and participated in several international scientific events.

Georges A. Agonvonon holds a master's degree in agricultural sciences with a focus on natural resources management and has a strong background in forest ecology and biodiversity conservation. Currently pursuing a PhD in the Land Systems and Sustainable Land Management Unit, Institute of Geography, University of Bern, his research investigates the ecology and conservation of forest patches in West and Central Africa. By examining tree community composition, natural regeneration, and functional ecology, he aims to enhance understanding of tree community dynamics and contribute to sustainable land management practices.

Phydias Agossou is pursuing a PhD in landscape connectivity and the influence of spatial patterns on ecological processes and species movement in the Land Systems and Sustainable Land Management Unit, Institute of Geography, University of Bern. His experience includes dispersal processes, ecosystem dynamics, land use planning, and sustainable resources management. Passionate about biodiversity conservation and ecosystem resilience, he focuses on the sustainable management of protected areas and the interplay between human activities and natural landscapes.

Felicia O. Akinyemi is Associate Professor in Geomatics at Karlstad University, Sweden. Certified as a Geographic Information System Professional, she examines how land system changes relate to degrading processes in human-natural systems. She leads the efforts to develop remote sensing-based metrics for operationalizing the Land Degradation Neutrality target (SDG 15.3) and its indicator on the proportion of degraded land over the total land area. She contributes to the Biodiversity Task Force of the European Geosciences Union and engages in science-policy dialogue in the United Nations Convention to Combat Desertification and the Intergovernmental Science-Policy Platform on Biodiversity and Ecosystem Services.

Jerzy Bański is Professor of Human Geography in the Institute of Geography and Spatial Organization, Polish Academy of Sciences (IGSO PAS). His main research interests include rural and agricultural geography, land use, regional policy, spatial organization, and local development. He has been the president of the Polish Geographical Society from 2006 to 2012, the director of the IGSO PAS (2018–2021), and the chairman of the Commission of Local and Regional Development under the International Geographical Union (2017–2022). He is the author of 440 publications, including 29 books and more than 210 papers with review processes. He has been coordinating over 52 research projects and is a member of 49 other national and international undertakings (such as FP6, FP7, Horizon, and ESPON).

Anne Gravsholt Busck's research addresses planning processes, policy analysis, and governance regimes – primarily in a Danish, European, and Scandinavian context but also in Global South contexts such as Tanzania and Latin America. Her current thematic focus is on rural development, rural–urban linkages including urbanization processes in peri-urban areas, human adaptation to climate change, and temporary activities as a strategic planning tool. She often studies decision-making processes, visions, policies, and regulation at local, national, and international scale, relationships between the planning systems, citizens, and other private actors.

Zhi Cao is an associate professor and master's supervisor in the Institute of Geographic Sciences and Natural Resources Research, Chinese Academy of Sciences.

Sean Connelly is a senior lecturer in the School of Geography at Ōtākou Whakaihu Waka/University of Otago and co-director of the Otago Energy Research Centre. He teaches environmental management, and his research interests are in community responses to sustainability challenges in the context of agrifood and energy systems and in rural and regional development.

Dhanjit Deka is an assistant professor in the Department of Geography at Gauhati University, Guwahati, India, and has over 14 years of teaching experience. He is presently the steering committee member of the International Geographical Union (IGU) Geomorphology Commission and a representative member of India in the Commonwealth Geographical Bureau (CGB). His research interests lie in watershed management, biodiversity conservation, land use change, urban mapping, remote sensing, and GIS applications. He has to his credit a few research and consultancy projects funded by national and state organizations. He has published many articles and research papers in international and national journals of repute.

Danièllé A. du Plooy is a postgraduate researcher at Ōtākou Whakaihu Waka/ University of Otago, School of Geography in New Zealand. Her PhD research explores the future of Aotearoa New Zealand's agri-food system through understanding the diverse values, narratives, and relationships of stakeholders with 'place'. With a professional and academic background focused on place-based agri-food systems, she has consulted on several projects aimed at enhancing food resilience and fostering a circular economy in the Queenstown-Lakes District of Central Otago, New Zealand.

Jenő Zsolt Farkas is a senior research fellow at the HUN-REN Centre for Economic and Regional Studies, Institute of Regional Studies, Hungary. He holds an MSc in Geography, an MA in History, and a PhD in Earth Sciences. His main research areas are agricultural and rural geography. His work primarily involves applied research, focusing on land use, land use change modelling, assessing the impact of agricultural production and subsidies on ecosystem services, and landscape management.

Giulia F. Curatola Fernández is a research associate in the Land Systems and Sustainable Land Management Unit, University of Bern, Switzerland. She holds a PhD in Natural Sciences from the University of Marburg, Germany. In her doctoral thesis she studied the relation between deforestation, human-induced fires, and the spread of an invasive species in Ecuador. Her work integrates remote sensing, geographic information systems, and mixed methods to analyse land cover/use changes particularly in forests, wetlands, and agricultural landscapes. Through transdisciplinary approaches, her research bridges science and practice to promote sustainable landscape management.

Juri Fitz is a research associate in the Land Systems and Sustainable Land Management Unit, Institute of Geography, as well as in the Impact Area Transformative Education and Science Impact at the Centre for Development and Environment (CDE), both at the University of Bern, Switzerland. He holds a master of science degree in Geography from the University of Bern. In his master's thesis, he applied a mixed-methods approach, combining remote sensing, geographic information systems, and qualitative data to explore farmers' motivation to conserve biodiversity-rich dry grasslands in the Swiss Alps.

Diana B. Frimpong is a faculty specialist at the University of Maryland/NASA Harvest. She holds an MSc in Urban Forestry and Natural Resource Management from Southern University and A&M College and a bachelor's degree in Geography and Resource Development from the University of Ghana. She specializes in GIS and remote sensing applications and works under Dr Catherine Nakalembe to address environmental challenges through geospatial techniques. Her notable projects include developing a scalable yield estimation methodology in Kenya and conducting a 40-year remote sensing analysis of climate patterns in Senegal. Also, she supports crop monitoring training across Africa.

Matej Gabrovec is a principal research associate at the Anton Melik Geographical Institute, Research Centre of the Slovenian Academy of Sciences and Arts, and assistant professor at the University of Primorska. His main research interests are transport geography, land use and land cover change, and regional geography.

Uelson Serra Garcia holds a PhD in Regional Development and Agribusiness from Western Parana State University, Brazil (UNIOESTE). He is a professor in the Department of Rural Economy at the State University of Maranhão, Brazil. He is a researcher of the Research Group of Regional Development and Agribusiness of the UNIOESTE. His main research interests include regional development, sustainable development, and crop–livestock–forest integrated systems. He has participated in national and international projects (Embrapa and Ministry of Family Agriculture of the State of Maranhão).

Ashley Gunter is a professor in geography at the University of South Africa (UNISA) with expertise in urban geography. His research integrates innovative methodologies and emphasizes community engagement, particularly in African development and spatial analysis. Professor Gunter has led large-scale international grants, fostering partnerships across Africa and Europe. He is committed to decolonizing education, promoting inclusive research practices, and addressing spatial disparities in education and development contexts. His scholarly contributions have advanced discourse on development, distance education, and migration. An advocate for practical, impactful research, he develops multimedia content to enhance learning experiences.

Liu Hao is a master's student in Land Resource Management at Sichuan University, focusing on the studies of land use and management, integrated land consolidation, cultivated land abandonment restoration, and intensive land use.

Robin Hartmann is currently pursuing a master's degree in Geography at the University of Bern, Switzerland. He works as a research assistant in the Land Systems and Sustainable Land Management Research Unit and in the SUSTAINFORESTS research project, supporting the team in different administrative and scientific tasks.

Samuel Hepner holds a master's degree in Biogeoscience from the University of Lausanne and is currently pursuing PhD in the Land Systems and Sustainable Land Management Unit, Institute of Geography University of Bern. As part of the SUSTAINFORESTS project, his research focuses on the ecology of tropical forest patches in West Africa. He employs terrestrial laser scanning, botanical inventories, satellite data, geographic information systems, and statistics to study forest structure, quantify aboveground biomass and carbon, and assess the impact of management on biodiversity and forest integrity. Outside research, Samuel enjoys juggling clubs on a slackline and skiing in the Alps.

Chima Iheaturu is a PhD researcher at the Institute of Geography, University of Bern, Switzerland, where he specializes in forest dynamics and land use interactions in West African landscapes. He holds an MSc in Geospatial and Mapping Sciences from the University of Glasgow, UK. His interdisciplinary research integrates advanced remote sensing technologies, including UAVs and satellite imagery, with stakeholder narratives to model the spatiotemporal dynamics of forests across multiple spatial scales. His work addresses critical challenges at the intersection of deforestation and ecosystem conservation, contributing to sustainable land management and policy strategies across Africa and beyond.

Bimal Kumar Kar is a professor in the Department of Geography, Gauhati University, Assam, India. His fields of research interest include population, social, urban and health geography, and development studies. He has to his credit more than a hundred research articles published in national and international journals and edited books and a few books on human geography. Also, he has authored a number of school textbooks on geography and social science. Currently, he is involved in a collaborative research project on small tea plantations of Assam with Freie University Berlin funded by German Research Council. Also, he is Vice-President of the Institute of Indian Geographers (IIG), Secretary General of the CGB, and Vice-President of the Indian National Cartographic Association (INCA). Moreover, he is the editor of *North Eastern Geographer* and *Indian Cartographer*.

Ingrid Kjelsen is currently pursuing a master's degree in Geography at the University of Bern, Switzerland. She works as a research assistant in the Land Systems and Sustainable Land Management Research Unit and in the SUSTAINFORESTS research project, supporting the team in various administrative and research tasks.

Michael Helt Knudsen's research has primarily focused on rural livelihoods, mobility, and settlement patterns within an African context. More specifically, he has explored the role of small towns in driving rural development and the deep integration of rural–urban interactions into livelihoods in Ghana and Tanzania. This work was carried out in close collaboration with universities in both Europe and Africa. In Denmark, his work centred on the structural transformation of the agricultural sector and its impact on rural areas. Additionally, he has examined how Danish municipalities are adapting to the challenges of rising water levels resulting from climate change.

Ana Cecília Kreter is the project coordinator of Agroforestry Living Lab (Agroforst Reallabor) Rhine-Waal University of Applied Sciences – HSRW, Germany. From 2018 to 2022, she was Chief Editor of Agricultural Bulletins of the IPEA, Minister of Economy. Her main research interests include agricultural policy, sustainable rural development, crop–livestock–forest integrated, and gender. She has been a consultant for Conservation International, ITAIPU Binational, and others. She has participated in several projects and been the coordinator of other national and international projects. She has published books, book chapters, and papers.

Søren Bech Pilgaard Kristensen's research focuses on rural development in a European and African context. In a European context, he studies the changing role of agriculture in rural and peri-urban landscapes, the environmental impact of agriculture, and the decision-making process of farmers. In addition, he studies the uptake level of different agri-environmental schemes by landowners in different locations. His fieldwork has focused on Denmark but has also included European perspectives through joint EU projects. In an African context, he has studied the challenges facing rural youth engaging in agriculture or non-farm economic activities with a focus on young entrepreneurs in rural areas in Uganda. In addition, he has studied the impacts of climate change on rural communities in Tanzania and how rural–urban linkages affect the climate resilience of these communities.

József Lennert is a senior research fellow at the HUN-REN Centre for Economic and Regional Studies, Institute of Regional Studies, Hungary. Holding an MSc in Geography and a PhD in Earth Sciences from the University of Szeged (2017), his research explores the post-socialist transformation of rural areas, with additional interests in spatial modelling and forecasting.

Chandra Kala Magar has 13 years of teaching experience in the Post Graduate Department of Geography, B. Borooah College, Gauhati, Assam, India. She has completed her doctoral thesis titled 'Place of tea in the culture, economy and polity of Assam: a geographical analysis' under the guidance of Professor Bimal Kumar Kar. She has published a handful of research papers and book chapters and reviewed articles in reputed journals and edited books at regional level. Her major research interest includes plantation economy, forest conservation, climate change, and traditional conservation practices by different tribes in north-eastern India.

Frank Mintah is a PhD researcher at the Institute of Geography, University of Bern, Switzerland. His research, which is part of the ERC-funded SUSTAINFORESTS research project, analyses sustainable governance pathways for forest patches in the agricultural landscapes of West Africa. This entails engaging with different stakeholders to understand the complex factors, including institutional structures and relational values that influence forest use and management, and, particularly, identifying pathways for sustainable and just social–ecological transformation. He, therefore, utilizes participatory systems dynamics modelling, scenario planning, focus group discussions, and in-depth interviews to understand forest social-ecological systems and facilitate the co-production of knowledge.

Francisco M. P. Mugizi is a senior lecturer in Economics at the University of Dar es Salaam, Mkwawa University College of Education, Tanzania. He holds a PhD in Development Economics and an MA in Public Economics from the

National Graduate Institute for Policy Studies, Japan. He obtained his BA in Political Science and Economics and MA in Economics from the University of Dar es Salaam, Tanzania. His expertise in research and teaching is development economics and applied microeconomics. His research interests are in the areas of development economics, and his geographical area of focus is mainly on sub-Saharan Africa. He has researched and published widely. His research results have been published in renowned international academic journals such as *Ecological Economics*, *Land Use Policy*, *Environmental Management*, *Journal of Social and Economic Development*, *Development Southern Africa*, *GeoJournal*, and *Africa Spectrum*.

Gordon Y. Mwesigwa holds a master's degree in Environment and Natural Resources and a first-class honours bachelor's degree from Makerere University, Uganda. His research focuses on climate change, natural and human-induced hazards, and disaster risk impacts on community livelihoods. He specializes in Earth Observation and data science, integrating community knowledge for local applications. With five years of experience in GIS and remote sensing technologies, including Google Earth Engine, he is dedicated to advancing research innovation and best practices in the field.

Catherine Nakalembe is an assistant professor at the University of Maryland and Africa Program Director for NASA Harvest. She serves as Agriculture and Food Security Thematic Lead on the NASA SERVIR Applied Sciences team. Her research focuses on applying satellite remote sensing and machine learning to agriculture, food security, and climate change. Notable awards include the 2020 Africa Food Prize and 2022 AL Sumait Prize for African Development, for improving smallholder farmers' lives through satellite technology. Her work has established numerous food security monitoring systems across Africa. Dr Nakalembe holds a PhD in Geographical Sciences from the University of Maryland, an MS from Johns Hopkins University, and a BS from Makerere University.

Sivapuram Venkata Rama Krishna Prabhakar works in the field of climate change adaptation and disaster risk reduction. He is particularly focused on understanding the impacts of climate change on various sectors and for promoting adaptive strategies to mitigate these impacts. Prabhakar has been associated with the Institute for Global Environmental Strategies (IGES) in Japan, where he plays a role in various climate change projects as a principal policy researcher. His research focuses on the intersections of climate change policy, disaster risk reduction, and community resilience. He is involved in policy advocacy and capacity-building efforts for enhancing the resilience of vulnerable communities to climate change. He has published several research papers and reports, contributing significantly to the discourse on climate adaptation and disaster risk reduction.

Zhao Qianyu is an associate professor in the Department of Land Resource Management at Sichuan University. She has been a postdoctoral fellow in the

Department of Land Economy at the University of Cambridge and holds a PhD from the Department of Land Resource Management at Renmin University of China and a joint PhD from the Department of Real Estate at the National University of Singapore. Her main research interests are land policy and land use analysis. She has presided over more than ten projects such as the National Natural Science Foundation and the National Social Science Foundation (Youth Project). She has published more than 30 papers in top journals such as *Land Use Policy* and has published two monographs.

Daniela Ribeiro is a research associate at the Anton Melik Geographical Institute, Research Centre of the Slovenian Academy of Sciences and Arts, and assistant professor in the doctoral programme in Environmental and Regional Studies in the Postgraduate School of the Research Centre of the Slovenian Academy of Sciences and Arts. Her main research interests are landscape ecology and management, sustainability of landscapes, land use and land cover change, and human–nature interactions.

Laudelina Alves Ribeiro holds a PhD in Regional Development and Agribusiness from Western Parana State University, Brazil (UNIOESTE). She is a researcher at the Development, Labor and Gender Research Group/Department of Economics of the UNIOESTE. Her main research interests include regional development, sustainable development, labor market development, and gender. She holds an interchange experience in Evora, Portugal, and Del Valle University, Colombia.

Guy M. Robinson is an adjunct professor of Geography at the University of Adelaide and a fellow in the Laboratory for Interdisciplinary Spatial Analysis in the Department of Land Economy, University of Cambridge. He is Editor-in-Chief of the Elsevier journal, *Research in Globalization*; Editor Emeritus of *Land Use Policy*; Chair of the IGU's Commission on Agricultural Geography and Land Engineering (AGLE); and Vice-President of the Royal Geographical Society of South Australia. His books include *Transforming Rural China* (Elgar), *Handbook on the Globalization of Agriculture* (Elgar), *Sustainable Rural Systems* (Ashgate), and *Geographies of Agriculture* (Pearson). The author of 14 books, 152 refereed papers, and 88 book chapters, his recent research has focused on multifunctional agriculture in Australia and China, land use competition in the peri-urban fringe, and various aspects of globalization.

Bingjie Song is a senior postdoctoral research fellow at Central Queensland University, having obtained her doctorate at the University of Adelaide and both her bachelor's and master's degrees at Shaanxi Normal University in China. She is a human geographer, specializing in rural and regional development and community engagement in both China and Australia. To date, she has published 14 papers in refereed high-quality journals. These journals include several ranked as Q1: *Land Use Policy*, *Ecological Indicators*, *Habitat International*, *Journal of Rural Studies* and *Journal of Geographical Sciences*. She is currently conducting research on the social impacts of mining and agriculture industries on local communities in central Queensland in the context of Industry 4.0.

Chinwe Ifejika Speranza is a geographer and professor at the Institute of Geography, University of Bern, Bern, Switzerland. She heads the Land Systems and Sustainable Land Management Research Unit. Her research focuses on understanding land management practices and how to improve their positive and reduce their negative effects on natural ecosystems and agricultural lands. She applies an integrative research approach that draws from physical and human geography methods and engages in science-policy dialogue. She conducts research in Switzerland and in African, European, and South American countries.

Milada Šťastná graduated from Mendel University in Brno, majoring in agronomy, and received her doctorate in applied and landscape ecology. She is the head of the Department of Applied and Landscape Ecology at the Faculty of AgriSciences and the editor-in-chief of the journal *European Countryside*.

Jefferson Andronio Ramundo Staduto is a professor in the Department of Economics at the State University of Western Paraná, Brazil (UNIOESTE). He is a CNPq research productivity fellow. His main research interests include sustainable rural development, gender, family agriculture, and labor market. From 2015 to 2017, he was the chief editor of *Brazilian Journal of Rural Economics and Sociology*. He has participated in several projects and been the coordinator of other national and international projects. He has published books, book chapters, and papers.

Paule Pamela Tabi Eckebil is a PhD researcher in the Land Systems and Sustainable Land Management Unit, Institute of Geography, University of Bern, Switzerland. She holds a master's degree in Auditing and Certification of Forests from the University of Yaoundé I in Cameroon. She gained experience in forest sciences and natural resources management as a researcher at the Center for International Forestry Research (CIFOR). Her current research focuses on assessing ecosystem services and livelihoods as well as modelling land system dynamics. She employs qualitative and quantitative methods such as interviews, surveys, participatory workshops, and modelling to study social–ecological systems across Central and West Africa.

Antonín Vaishar graduated in social and economic geography from Comenius University in Bratislava and received a doctorate from the Czechoslovak Academy of Sciences. He is an employee of the Department of Applied and Landscape Ecology at Mendel University in Brno and the executive editor of *European Countryside*.

Gregory Veeck is an emeritus professor in the School of Environment, Geography and Sustainability at Western Michigan University located in Kalamazoo, Michigan. He completed his doctorate in geography in 1988 at the University of Georgia and taught at Louisiana State University and Western Michigan University. Over the course of his career, he conducted extensive field research on agriculture and rural development in China, East Asia, and the Midwest of the United States.

Shiqi Wang is a master's degree student in Institute of Geographic Sciences and Natural Resources and Jiangxi Normal University.

Vladimir R. Wingate is a postdoctorate in the Land Systems and Sustainable Land Management Unit, Institute of Geography, University of Bern, Switzerland. He gained his PhD in Geography from the University of Basel's Department of Environmental Sciences. Subsequently, he completed a postdoctorate with the University of Zurich's Remote Sensing group in the Department of Geography (2019–2021). He holds an MSc degree in Environmental Management from the University of Stirling (2011) and a BSc in Environmental Science from the University of Aberdeen (2010). His research focuses on mapping ecological changes across biomes using satellite remote sensing to help inform natural resources management.

Simon Measho Yhdego is a GIS lecturer and thesis supervisor at Nanjing University of Information Science and Technology, NUIST Reading Academy. He did his PhD in Cartography and GIS from the University of the Chinese Academy of Sciences. His research assesses land use/land cover changes, vegetation dynamics, ecosystem resilience to drought, and agroecosystem changes. He received the NSFC Research Fund for International Young Scientists, 2020 Excellent International PhD Student award, and the 2021 Excellent International PhD Graduate award from UCAS. He is the author of 22 international journal papers and a book chapter, and his Google Scholar citations are 328, h-index 10, and i10-index 11.

Ruan Yizhu is a master's degree student admitted by IGSNRR.

Introduction to the book

Jerzy Bański

Land used by the farmer in the agricultural production process is the main land use category in most countries around the world. Agricultural land is the basis for crop and livestock production and thus guarantees food security in every region. Increasingly, such land is also used to produce green energy. Urbanization, the development of settlement networks, communications, and other technical infrastructure are causing a permanent loss of land used for agriculture and shaping serious spatial conflicts in land use. The response to these changes is the intensification of agricultural production, the introduction of more productive crops, searching of modern cultivation methods, and attempts by the food sector to develop new land. However, the problem of farmland loss and the accompanying food shortages in some countries is increasing. Therefore, it seems very important to diagnose the contemporary state of agricultural land use in various regions of the world and identify the dilemmas and challenges associated with the processes outlined earlier. It is equally important to identify agricultural policies aimed at protecting and/or optimizing agricultural land use that guarantee sufficient and sustainable food production.

The aforementioned objectives are addressed in the chapters of the book, which have been prepared by scholars from different regions of the world (Europe, Africa, Asia, America, and Australia and Oceania) specializing in the topic of agricultural land use and land management issues. The themes of the proposed book will be covered in three main parts: *Part I Structural transformation of agricultural land*, *Part II Environmental and social challenges in agricultural land use*, and *Part III Farmland in planning and policy*.

Part I Structural transformation of agricultural land consists of six chapters. The purpose of the first chapter 'Contemporary processes of transformation of agricultural land use in Central and Eastern Europe' is to diagnose the contemporary state and identify the trends in the structure of agricultural land use in the countries of Central and Eastern Europe. Particular attention is paid to the spatial variation characterizing this structure and the role of the conditions shaping it. Also, the chapter addresses the transformation of forms of landownership. Chapter 2 'Agricultural transformation in Hungary: trends and challenges since 1990' is devoted to the transformation of Hungarian agriculture from 1990 to 2022, with special attention to the evolution of land use patterns, changes in holdings, and

DOI: 10.4324/9781032672779-1

their production structure. Changes in agricultural land use in North and East Africa form the subject of Chapter 3 titled 'Structural transitions of agricultural land use across Northeast Africa from 2001 to 2020'. It shows that rapid changes in the structure of crop production are primarily due to climate variability and changes in agroecosystems. The analysis uses rich data sources (ESA, LULC, FAOSTAT, Google Earth Engine, Climate Engine) and applications enabling the prediction of changes. The issue of agricultural land fragmentation is the subject of Chapter 4 titled 'Land fragmentation and agricultural productivity in Tanzania'. The findings of the chapter reveal that land fragmentation is negatively correlated with the productivity of both permanent and seasonal crops. In Chapter 5, 'Structural transitions of agricultural land use in China after 1978', the researchers analyse the quantitative changes in agricultural land use and evaluate the main mechanisms of these changes at the regional level. China is a vast country with a complex and diverse geographical environment, and agricultural land use changes in each region show different characteristics and trends. The last chapter in this part, 'Growth of tea plantations and its impact on agricultural land use in Assam, is devoted to trends and spatial patterns of tea plantation growth in Assam, India.

The second part of the book is devoted to environmental issues. The first chapter in this part, Chapter 7 'Agriculture's response to diverse environmental and other challenges in the driest state in the driest continent', examines some of the key environmental challenges affecting agriculture in South Australia, and especially the horticultural sector. It discusses the experiences of two contrasting areas: the Adelaide Hills, where the rural economy benefits from proximity to the state capital, and the Riverland – the most productive horticultural region. Agriculture–environment relationships in neighbouring New Zealand are the subject of Chapter 8 'Shaping agriculture and agri-food system futures in Aotearoa New Zealand: the role of values and relationships with land and self'. In New Zealand, agriculture generates the majority of the country's exports by value and is a significant contributor to the national GDP. Yet the agricultural sector also contributes nearly 50% of GHG emissions, is increasingly dependent on irrigation, and has negatively impacted freshwater systems. These issues have resulted in calls for rethinking the future of agricultural land use. The next chapter 'Recent changes of land use in Southern Moravia's lowland and highland regions identifies the environmental challenges in the Czech Republic. The agricultural land use has changed little in Czechia over the past decades. However, a significant challenge there is climate change, which in the lowlands means the danger of drought and accentuates the problem of irrigation, while in the highlands it is causing a shift to an organic form of agriculture – mainly extensive livestock production combined with agritourism. The transformation of agricultural land in Slovenia is mainly due to urban sprawl and the development of infrastructure systems. Additionally, as the authors of Chapter 10 titled 'Transformation of cultural landscapes: case studies from Slovenia' note, one of the most striking processes of change in rural landscapes in Slovenia is forest overgrowth, which is associated with the migration of rural population to cities and the withdrawal of agriculture from rural and marginal areas. Chapter 11 'Agricultural land use and its climate change vulnerability in Asia:

moving from vulnerability to resilience' offers a concise review of the prevailing knowledge concerning the agricultural land use changes and other land use-related issues in Asia. It evaluates how the use (or misuse) of agricultural land can affect the vulnerabilities and capacities of national agricultural systems. The issue of agricultural activity in extremely dry areas is the subject of the chapter titled 'Dynamics of agriculture and sustainability of the oasis ecosystem: the case of Mansoura (Algerian Sahara)'. The chapter concerns the assessment of the dynamics of agricultural development with particular emphasis on land use and irrigation problems in the Mansoura oasis system.

The third part of the book addresses issues related to agricultural policy and agricultural land use planning and their impact on the economic situation of farms. In Chapter 13 titled 'Targeting agricultural land expansion across Africa: pathways through effective nature protection, sustainable intensification, and strengthening land governance', the authors examine the spatial dynamics of agricultural expansion and its environmental implications. The analysis results in a proposal of three complementary paths to reconcile agricultural productivity with environmental protection. The topic of Chapter 14 'Development policy and its influence on land use: a study of institutional markets in Maranhão, Brazil' is the analysis of the set of municipalities in Brazil covered by the HDI+ programme (High Human Development Index Plan) from a sustainable rural development perspective, particularly assessments of institutional markets supporting land use change and promoting sustainable agriculture. The institutional market programmes, which comply with various regulations, drive land use change and promote sustainable development. Chapter 15, 'Rapid increases in arable land values and long-term implications to the farm sector and rural economy of the United States: 2000–2023', introduces recent changes in arable land costs and the implications of these changes on the future of the agricultural sector of the United States. This chapter presents the spatial-temporal changes in arable land prices in the United States from 2000 to 2023 and the long-term implications of high arable land prices on the agricultural sector and rural communities. The problem of urbanization and the resulting need to employ rural residents outside the agricultural sector is the subject of Chapter 16 titled 'Urbanization and agricultural land use transformation in China: a perspective of off-farm employment'. The basic research question is this: how does the transfer in the employment structure affect the agricultural use of land? Chapter 17 'Transitioning agricultural land to eco-estates in rural Gauteng: farmers' perceptions and policy and planning implications' examines the transition of agricultural land to eco-estates in rural Gauteng (South Africa). The chapter investigates the role of agricultural land use in South African planning and policy, as well as farmers' perceptions of the new estates and how they perceive the change in land use. Political and historical contexts are key elements in the study of agricultural land use in Uganda (Chapter 18 'Long-term agricultural land use change analysis and land use policy in Karamoja, Uganda, since 2000'). The chapter analyses Karamoja's conflict history, past policies, policies implemented during the study period, and the impacts of recent regional developments on agricultural land use changes and the local community. The last chapter, Chapter 19 'Rural idyll or forgotten backwaters: what and who determines the transition of rural areas in Denmark? A case study from Central Jutland, Denmark', is devoted to agriculture in Denmark. The chapter concerns the roles that rural landscapes play in rural development. Key research questions include the following: 1) what is the relationship between agricultural production and landscapes? 2) what are the landscape management strategies of different types of rural residents? 3) what is the role of public policies and instruments in this process?

Jerzy Bański

Part I

Structural transformation of agricultural land

1 Contemporary processes of transformation of agricultural land use in Central Europe

Jerzy Bański

Introduction

Until the beginning of the 1990s, the farming economy of the Central European countries – except for Poland – was unquestionably dominated by the socialized sector, namely the state farms and the production cooperatives. Family farming, on the other hand, has had a rather marginal significance. The years following the collapse of the socialist system, during which the processes of privatization and restitution of the previously sequestered assets took place, brought deep transformations in the agricultural land use, not only in terms of the very ownership but also with respect to the production-related structure of agricultural land (Bański, 2019; Doucha et al., 2005; Petrick & Weingarten, 2004; Rusu & Florian, 2003). The contemporary image of the agricultural land use is not undergoing any more such dynamic changes, which are now mainly the effect of the broadly conceived economic conditions associated with food production and the agricultural policy of the European Union (Bański, 2018; Page & Popa, 2013; Todorová, 2016; Veznik et al., 2013), as well as pressures from other economic functions with respect to the areas used to date for farming purposes (Bański, 2017).

Agricultural land is the fundamental category in the land use structure of all Central European countries. Thus, for instance, in Hungary, Poland, and Romania this is at least half of the entire territory of these countries. A persistent process, which characterizes not less than the last several decades, is the shrinking of agricultural land area to the advantage of other forms of use, particularly so regarding overbuilt areas, technical infrastructure, as well as forest areas (Bański & Mazur, 2021). This process is accompanied by the optimization of agricultural production, with increase of crop yields and livestock production effectiveness. At the same time, the crop structure is being modified, with due account of the contemporary demand from the food market and of the climate change.

The purpose of this chapter is to diagnose the state and to identify the directions of changes in the ownership and structure of agricultural land use in the region of Central Europe after the fall of the socialist economy, that is during the last three decades. Special attention is paid to the differentiation of the spatial structure of agricultural land and the significance of various conditioning factors in shaping it. The chapter concerns the countries, having belonged to the former Eastern Bloc,

DOI: 10.4324/9781032672779-3

including the three former Baltic republics of the Soviet Union (Estonia, Latvia, Lithuania). The source materials originate from the EUROSTAT and FAO, and it should be indicated that the potential divergences, regarding the values of some statistical indicators, are the effect of adopting different definitions of the respective notions and the methodologies of data acquisition used by the national statistical bureaus.

Transformations in the ownership of agricultural land

During the period of socialism, agriculture in the countries of Central Europe was subject to a very strong influence from the side of the state, aiming at the establishment of a new model of a farm. The primary expression of the activities of the authorities at that time was elimination of private ownership to the advantage of the social and national property, and this ended with success in almost all countries of the region. In former Czechoslovakia, virtually all of agricultural land was socialized. On the territory of the present-day Czech Republic, socialist farming cooperatives operated on 65% of agricultural land, while state farms operated on the remaining 35%. Similarly, in Hungary, the process of collectivization of land, which was especially intensive in the years 1948–1962, consisted in taking over of privately owned land by the newly established state farms and socialist production cooperatives. In the peak period, some 75% of agricultural land belonged to production cooperatives. Then, in turn, in Romania, the socialist authorities nationalized all of the church property, royal family properties, as well as all of private land exceeding 50 hectares. In 1989, the essential form of land property was constituted by the production cooperatives, administering 59% of agricultural land, along with the state farms, functioning on close to 30% of agricultural land (Balteanu & Popovici, 2010). It was only in Poland and in former Yugoslavia, that during the entire period, socialism and family farming persisted, disposing of the majority of agricultural land and playing the leading role in the food economy of these two countries (Bański, 2011).

The processes of socialization or nationalization of the agricultural sector in the Eastern Bloc brought appropriation of a vast majority of previously private farming land to the advantage of the socialist farming enterprises, which had, as a rule, the character of large-scale farming establishments. This, in turn, caused changes in the agrarian structure and in the structure of agricultural use of land. The collapse of the socialist system (at the turn of the 1990s) entailed a reverse direction of changes, consisting in the reduction of the socialized property to the advantage of private ownership (Mejstrik, 1997; Smith & Pechota, 1994). The character of farms changed, with family farms playing today the leading role. Most symptomatic of the ownership changes that were taking place was land property itself, first of all of agricultural land (Bański, 2014).

At the beginning of the 1990s, dynamic shifts of the landownership structure were taking place in all countries of Central Europe, leading to the increase in the significance of the private sector (Balteanu & Popovici, 2010; Bański, 2017; Bicik & Jelecek, 2009; De Arriba, 2007; Tisenkopf, 1999; Swinnen, 1996). And so, for instance, in Hungary, the share of agricultural land in use by the private sector increased from 14% in 1990 to 54% in 2000 (Kovacs, 2005). This was accompanied by the process of fragmentation of agricultural land and the increase in the number of farms. On the other hand, in Czechia and Slovakia, where privatization affected, as well, the entirety of farming land, changes had a different character. The socialist cooperatives and the state farms were transformed into private enterprises, but the land, owned by them, was not fragmented, even though this land became again the property of hundreds of thousands of its legal owners. In 2007, in Czech Republic, some 4% of farms had surfaces exceeding 500 hectares each, and these farms functioned on more than 72% of total farming land (Basek & Divila, 2008).

In the entire period, which followed after the collapse of the socialist economy, one can distinguish two main phases which shaped the contemporary image of agricultural land in Central Europe – the phase of systemic transformation and polarization of ownership and the phase of ownership stabilization. The first of these phases took on a different form and had different duration, depending upon the country. Thus, for instance, in Czechia, some 90% of agricultural land was handed over to private persons, while 10% remained at the disposal of the state but was used by private farmers, companies, as well as production cooperatives. Transformation preserved the previous acreage structure of farming estates, which only changed their legal and administrative form of functioning, and so it was possible to maintain large production units. In Hungary, during the final period of socialism, some 86% of agricultural lands were administered by 1,500 farming cooperatives and 124 state farms. In effect of privatization and restitution a complete change of the landownership structure took place. A clear bipolarity of this structure took shape – small farms of just a couple of hectares and large production holdings. Altogether, roughly 2.5 million people took part in Hungary in the process of land re-privatizing, of whom 1.5 million obtained land that had been common property before, while half a million had their former property returned to them (Kovacs, 2005). In 2011, around 80% of agricultural land was in private hands (Toth-Naar et al., 2014). Then, the Romanian model of privatization envisaged the return of up to ten hectares of farmland to the former owners of land and the maximum farm area of 100 hectares. Yet laws introduced later on liberalized landownership regulations. Altogether, in the course of ownership transformations, former owners regained more than nine million hectares of land, which entailed strong fragmentation of property and similarly intensive dispersion of farmland use. The negative consequence of the restitution of land property consisted, first of all, in the division of the land having been in use of the production cooperatives into small plots. The systemic transformation resulted in the increase of the number of agricultural land users up to roughly four million (Benedek, 2000). In the case of Poland, where private farming dominated, privatization of agricultural land concentrated in only those regions where the state farms had functioned. Restitution of the estates did not have, unlike other countries of Central Europe, a common character but, instead, would occur sporadically. Regarding the state farms, the idea of their complete liquidation was favoured, even though some of them were in a good economic shape (Bański, 2011). The very last state farms were liquidated in 1994.

The period of re-privatization in the farming sector of the Central European countries ended in the second half of the 1990s. Yet, over several years after this, dynamic changes in the landownership structure were still taking place, resulting from the development of a new model of commercial farms and gradual taking over of the land having had belonged to the socialized sector. A very strong influence was exerted on the ownership relations in this region of Europe from the side of the processes associated with the preparation and then membership of the countries of the region in the European Union.

In the phase of stabilization of ownership, which has been lasting by now for close to two decades, the land-related transactions concerned solely the private entities. After the countries of Central Europe entered the European Union, advantageous symptoms appeared in agriculture, linked with direct payments and with the support for farming and for rural areas in the framework of the structural programmes. On the one hand, groups of farms took shape at that time featuring specialized production, large areas, and higher competitive effectiveness on the European market. On the other hand, though, small farms coexisted with them, bringing their owners very modest incomes. Thus, for instance, the study of W. Dzun and W. Józwiak (2008) shows that in Poland, in the middle of the 2000s, there were around 99,000 farms which would attain good economic and financial effects, while there were, at the same time, some 1.6 million very small farms which actually did not participate in the food products market. Land would be transferred from the unprofitable farms towards those offering definite development perspectives.

An extreme polarization of agricultural ownership structure existed in Romania. According to Eurostat data, in 2016, 30% of agricultural land belonged to farms of less than five hectares each (there were altogether more than three million such farms), while 48% of land belonged to farms having more than 100 hectares (there were some 12,000 of such farms). A persistent and advantageous phenomenon was the decrease of the total number of farms and the process of transfer of land to the large farms, conducive to the concentration of farming land.

Transformation of the agrarian structure

According to the available Eurostat data, there were in 2016 in the countries of Central Europe almost six million farms using agricultural land. In only three countries of the region (Poland, Romania, Hungary) there were five million farms (Table 1.1), and this number was higher than in all other countries of the European Union taken together. The consequence of adopting various models of privatization and restitution of land is a distinct difference in the ownership structures and magnitudes of farms. Thus, for example, while in Czechia an average farm has at its disposal more than 130 hectares of farmland, Romania has only 3.6 hectares.

The most advantageous farm magnitude structure exists in Czechia and Slovakia. On the other hand, in the remaining bigger countries of the region, resulting

Table 1.1 Number of entities using agricultural land in selected countries of Central Europe

Country/Year	*2005*	*2007*	*2010*	*2013*	*2016*
Czech Republic	41,180	38,490	22,580	25,950	26,530
Hungary	662,370	565,950	534,020	453,090	430,000
Poland	2,465,830	2,380,120	1,498,660	1,421,560	1,410,000
Romania	4,121,250	3,851,790	3,724,330	3,563,770	3,422,030
Slovakia	66,360	66,520	23,720	22,050	25,660

Source: Eurostat.

from privatization and restitution of agricultural land, a dualistic model of the acreage structure of farms developed – there are entities with significant acreage of agricultural land, which coexist with a great number of dwarf farms. So, for instance, in Bulgaria, the number of large farms is limited, but they dispose of the majority of agricultural land in the country, while the number of small farms is high, but they dispose of a relatively limited share of total agricultural land. At the beginning of the 2000s, it was estimated that there were some 8.7 million farming land plots, which were owned by roughly 5.1 million citizens of Bulgaria, that is – 65% of the total population of the country (Bencheva, 2005; Kopeva, 2003).

In Hungary, in the middle of the 1990s, there were more than 1.4 million farms (Hartsa et al., 1998). Over time, their number decreased relatively fast. It is estimated that in the years 1990–2000, the number of farms decreased by 30%, and in the period 2000–2010, by approximately 50%. This phenomenon resulted from several reasons, among which the following ones are mentioned: low profitability of agricultural production, subsistence character of the small farms, simultaneous employment of farm owners in other sectors of economy, as well as the process of ageing of the population (Sadowski & Takacs-György, 2005; Toth-Naar et al., 2014). In the same period, an increase in the number of large farming enterprises has been observed. At present time, though, the ownership structure in Hungarian agriculture has still a bipolar character.

The biggest fragmentation of farms existed in Romania. After the process of re-privatization some 2.8 million farms (around 70% of all farms in Romania) had less than one hectare (Table 1.2). Such farms occupied altogether five million hectares, that is 35% of the total agricultural land in Romania. Then, a bit more than one million farms (27% of all farms) had between one hectare and ten hectares and disposed of somewhat more than 10% of the total agricultural land. It is interesting that only 12,000 farms, that is 0.3% of all farms in Romania, having on the average more than 100 hectares, carried out commercial activity and disposed of 34% of the totality of agricultural land in Romania (Otiman, 2013). This is the best example of the strong acreage polarization of farms, which, however, undergoes reduction with time.

After the termination of privatization in the farming sector of the Central European countries, the short-lived process of abrupt increase in the number of private farms first slowed down and then was replaced by the reverse process, that is the

Table 1.2 Numbers of farms according to their acreage in countries of Central Europe in 2016

Country	*Acreage in hectares*								
	0	*0.1–1.9*	*2.0–4.9*	*5.0–9.9*	*10.0–19.9*	*20.0–29.9*	*30.0–49.9*	*50.0–99.9*	*100 or more*
Bulgaria	15,530	129,910	20,090	9,650	7,100	3,900	4,340	3,600	6,110
Croatia	90	2,710	1,890	5,180	4,470	2,370	2,380	2,450	4,710
Czechia	620	1,410	3,240	3,490	2,700	1,170	1,120	1,050	1,900
Estonia	1,330	50,810	40,840	20,080	9,470	3,160	3,160	3,540	1,620
Hungary	290	12,370	11,950	15,880	14,560	4,990	3,760	2,890	3,250
Latvia	190	22,330	52,680	32,770	19,320	6,880	5,370	5,490	5,290
Lithuania	64,890	243,120	42,110	27,560	20,060	8,450	7,760	7,280	8,760
Poland	5,320	299,340	461,600	306,220	202,350	60,820	40,390	22,440	12,010
Romania	79,840	2, 400,930	6,60,000	194,200	50,210	10,990	7,530	6,010	12,310
Slovakia	80	17,440	24,050	16,060	8,230	2,180	1,250	500	120
Slovenia	1,750	5,660	6,880	3,570	2,540	1,000	920	940	2,400

Source: Eurostat.

decrease in the number of farms. This turn of events was characteristic for all the countries of the region in question, but the dynamics of the respective processes was strongly differentiated. As regards liquidation, it affected first of all the small, economically feeble farms, run by the elderly people, not having successors. This particular process got amplified after the accession of these countries to the European Union, the basic cause being the increase of competitiveness on the food market.

Summing up, with respect to the contemporary acreage structure of farms, one can distinguish in Central Europe three groups of countries. The first group is constituted by Bulgaria, Romania, and Hungary, with a very important share of the dwarf farms, whose land shall presumably quite soon be taken over by the larger farms. A significant part of these farms has, actually, only statistical aspect, for their land is already used by the larger and economically stronger entities. The second group encompasses Croatia, Lithuania, Poland, and Slovenia, where small farms, below ten hectares, dominate in the magnitude structure. The third group is composed of Czechia, Slovakia, Latvia, and Estonia, where there are also a lot of small farms, but medium-sized and large farms constitute, as well, a very important category.

Directions of changes in the areas and structure of agricultural plots

The consequence of changes in ownership and in the agrarian structure of farms was expressed through transformations in the use of farming land. The changes concerned the proportions of the shares taken by the basic components of farming land (arable land, grasslands, and permanent crops) and the directions of crop production on arable land. At the same time, there has been a persistent loss of the area of land used for agricultural purposes. An essential influence was exerted on the dynamics of changes and development of new directions of agricultural land use by the fact that the Central European countries joined the European Union and the possibility opened of taking advantage of direct payments, structural funds, as well as special programmes, aimed at farming in the framework of the Common Agricultural Policy. Transformation of the agricultural use of land was also influenced by a number of other socio-economic conditions, among which one should mention the following: change of the relation between the prices of agricultural production means and farming products, shift in the directions of foreign trade in agricultural production means and in agricultural products, internal agricultural policies of the respective countries, changes in the consumption habits, increase of the ecological awareness of the society, activities associated with the development of green energy, and many others. Also, one cannot neglect the importance of the environmental factors, including changing climate and hydrological conditions.

All countries of the region in question are characterized by the persistent shrinking of the share of agricultural land in the structure of land use (Figure 1.1), the rate of decrease displaying very significant spatial differentiation. The biggest absolute changes in the area of agricultural land during the last 20 years (2000–2020) took place in Poland and Romania, which is primarily due to the magnitude of these two

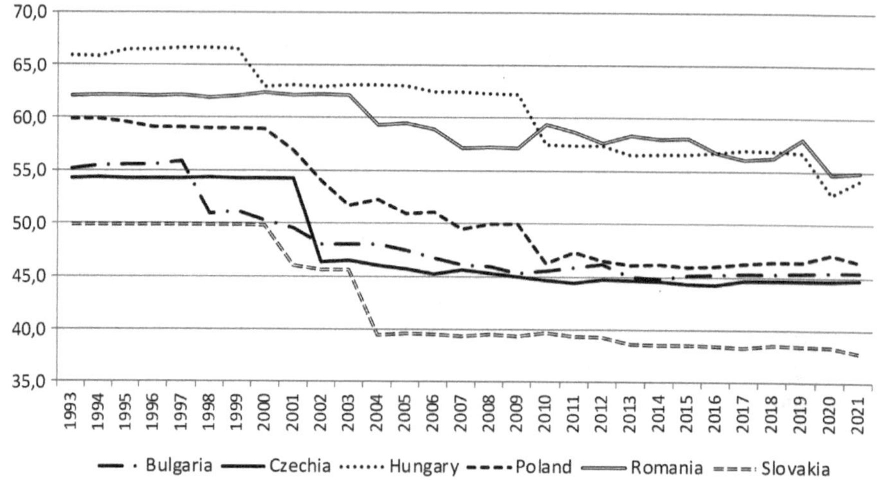

Figure 1.1 Changes in the shares of agricultural land in total surfaces of the countries of Central Europe.

Source: FAOSTAT.

countries and the large surface taken by this category of land in them (agricultural land occupies some 14 million hectares in each of them). Decreases of the area of farming land occurred mainly in the surroundings of large cities, where dynamic development of family housing took place along with the accompanying development of technical infrastructure, primarily of transport networks. The biggest drop in the share of agricultural land was observed in Poland, which resulted from several reasons. One of them was territorial expansion of towns and particularly of the suburban housing areas. This phenomenon distinctly intensified at the turn of the century. Agricultural land was also taken over for new projects in transport networks, mainly for the construction of expressways and modernization of road networks within the metropolitan areas. On the other hand, introduction of the environmental programmes contributed to forestation of the low-quality farmland, where agricultural production was unprofitable.

The decrease in the agricultural land area should also be associated with the disadvantageous demographic processes, including ageing of population in rural areas and migration outflow of the young and active persons. An important proportion of farm owners are older persons, who often give up active functioning in the farming sector or switch to subsistence farming. This was especially clearly visible in Romania (Otiman, 2013). The areas which are affected by the disadvantageous demographic processes do not develop, and the farms there, largely uncompetitive, are being liquidated. Similar processes were observed within the marginal areas of other countries of the region studied (Bicik & Jelecek, 2009; Gajdos, 2005; Balteanu & Popovici, 2010).

The decrease of the surface area of agricultural land resulted, as well, from the privatization processes. New owners of land on the small-acreage farms would abandon cultivation in view of the insufficient profitability of production, lack of investment means, or of availability of an alternative, more profitable occupation. In the literature of the subject, the phenomenon of land fragmentation is considered to constitute a significant problem in the agricultural use of land (Bentley, 1987; Dijk van, 2003). Excessive fragmentation of land generates additional costs and time, associated with accessing individual plots, makes difficult use of larger machines and, in general, the fieldwork itself, adds length to plot boundaries, leading to exclusion of larger boundary zones from cultivation, necessitates construction of a dense network of field roads, complicates the registry of plots, and in some cases leads to conflicts between the landowners.

From among all the countries of Central Europe, the most stable structure of land use is observed in Czechia. This is the effect of the well-established already some time ago high degree of urbanization and good condition of transport infrastructure. Yet, in the regional perspective also in Czechia, one can notice distinct differentiation regarding changes in land use. The biggest loss of agricultural land occurred in the vicinity of towns, where farming land is taken over for construction purposes (Janku et al., 2016). Such a process was particularly strong in the zone of influence of Prague. On the other hand, the fertile Moravian lowland is rather characterized by the stable structure of land use.

The decrease in the area of agricultural land was accompanied by the changes in the structure of its main components (arable land, grasslands, permanent crops), but these changes were mostly limited and had different directions in particular countries (Table 1.3). The biggest structural changes took place in the countries with smaller territories, where even relatively small changes in terms of area would

Table 1.3 Development of the shares of basic categories of agricultural land in the countries of Central-Eastern Europe (in %)

Country	*Arable land*				*Permanent crops*				*Grassland*			
	1995	*2000*	*2015*	*2020*	*1995*	*2000*	*2015*	*2020*	*1995*	*2000*	*2015*	*2020*
Bulgaria	64.9	63.2	70	69.2	3.3	4.5	2.7	3.0	31.8	32.3	27.3	27.8
Croatia	47.9	72	54.9	59.1	5	5.9	4.9	5.2	47.1	22	40.2	36.8
Czechia	77.1	75.7	74.4	70.5	1.8	1.8	1.8	1.4	21.1	22.4	23.8	28.1
Estonia	89.2	85.5	67.4	70.5	1.2	1.2	0.6	0.5	10.6	13.3	32	29.0
Hungary	77.8	78.6	82.5	81.9	3.6	3.4	3.2	3.2	18.6	19.5	14.2	14.9
Latvia	54.7	61.1	65.3	67.7	1.6	0.8	0.4	0.4	43.7	38.2	34.4	31.8
Lithuania	84.5	84.2	72.3	76.4	1.3	1.3	1.1	1.2	14.7	14.5	26.6	22.4
Poland	76.3	76	75.8	75.7	2	1.8	2.7	2.6	21.7	22.2	21.5	21.7
Romania	63.1	63.1	63.3	65.0	3.8	3.5	3.1	3.1	33	33.3	33.7	31.9
Slovakia	63.6	62.6	71.9	71.5	2	1.9	1	0.9	34.3	35.4	27.1	27.6
Slovenia	36.4	33.4	29.8	29.7	6.1	6	8.7	8.7	57.4	60.6	61.5	61.6

Source: FAOSTAT.

bring important relative shifts. Thus, for instance, in Estonia and Lithuania, the share of arable land decreased distinctly to the advantage of grasslands, while the transformation in Latvia had the opposite character. In the bigger countries, the shares of the particular components of agricultural land were rather stable, undergoing quite limited fluctuations.

The structural changes, when perceived in their spatial aspect, had diverse directions and character. Depending upon the particular country they may indicate an increase in the intensity of production or, to the contrary, a more extensive production. There was an increase in permanent cropland, which implies intensified production of fruits and fruit products. They became a very important export item. Besides, after joining the European Union, the farmers started to obtain high subsidies for some categories of fruit orchards, which resulted in new plantings and in the introduction of more productive varieties of fruit trees. Contrary to the phenomenon observed in Poland, in the majority of countries situated in the south of the region considered, the significance of the permanent crops in the structure of farming land decreased. This is a sign of more extensive production structure and of giving up the highly specialized crops (Takacs, 2008). The phenomenon concerns, first of all, Romania, Slovakia, and Bulgaria and, to a lesser extent, Hungary. The factor behind such processes is constituted, most probably, by the ownership changes and the associated giving up by small farm owners the labour-intensive and capital-intensive crops, as well as those requiring modern technologies and appropriate technical infrastructure. And so, for instance, in Romania, in the years 1990–2000, farmers would be abandoning the intensive crops, and in some regions the new owners of agricultural land would entirely abandon cultivation because of incapability to do the work, lack of investment means, and a lack of professional skills (Popovici et al., 2013). It is only in Slovenia, where the economic situation of agriculture, compared to other countries of the region, is better that the significance of permanent crops increased.

In Bulgaria, decrease affected the area not only under permanent crops but also under grasslands, while the share of arable land increased. It appears that this direction of change is justified, for more than half of agricultural land in Bulgaria has very fertile soils. The structure of crops is dominated nowadays by wheat and maize for grain, which are considered to be strategic food crops (Moteva et al., 2014). An increase is also observed in the significance of industrial crops, first of all oil crops for biofuel production. On the other hand, in Czechia, there has been an increase in the share of grasslands, mainly within the mountainous and foothill areas, which are characterized by low quality of agro-ecological conditions. Privatization and reduction of support for farming (socialized in the past) from the side of the state resulted in forestation of the arable land or their transformation into grasslands. Increase of the pasture areas resulted also from the changes in the cattle-raising technologies. During the 1990s, cattle would be kept in the stables, which required production of feedstuff on the arable land, and this has been replaced by pasturing on grasslands (Bicik & Jelecek, 2009).

Crop production, mainly of cereals and industrial crops, plays a very important role in the farming production of the countries of Central Europe (Figure 1.2). That

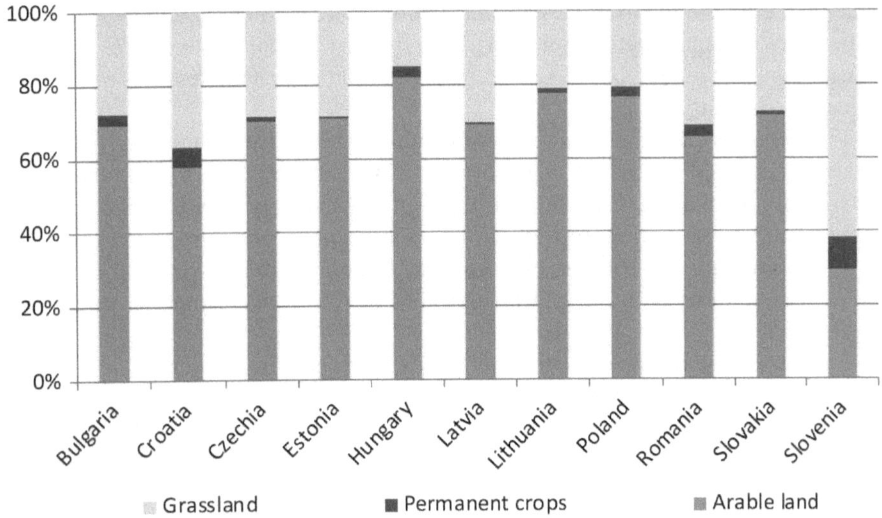

Figure 1.2 Structure of agricultural land use in the countries of Central-Eastern Europe in 2021.

Source: FAOSTAT.

is why the structure of agricultural use of land is dominated decidedly by the arable land. The grasslands, often featuring low production qualities (primarily so in the case of mountainous areas), account for some 20–30% of total farmland. Only in the cases of Croatia and Slovenia, where a significant proportion of territory has mountainous character and is not fit for field cultivation, the shares of meadows and pastures are significantly higher. Simultaneously, in these two countries, the shares of permanent crops are also relatively high, this being largely linked with the climatic capacity of vineyard and orchard cultivation. In the remaining countries, permanent crops occupy rather limited shares of farmland, although there are definite areas where they can play a leading role (e.g. vineyards in the region of Tokay in Hungary or apple orchards in the area of Grójec-Skierniewice in Poland).

The structure of field crops in the particular countries of Central Europe is, first of all, the consequence of geographical conditions, including climatic and hydrological ones. The differentiation of climate from moderately cool to subtropical, as well as differences in the duration of the growing season, soil qualities, and terrain relief, shapes the specificity of crop production. Farming depends upon climate and atmospheric conditions like no other branch of economy. Thus, for instance, in the countries of the northern and central parts of the region considered here, wheat cultivation is of foremost importance, since the most advantageous climatic and hydrographic conditions for cultivation of wheat is found there. The share of wheat is smaller in the south, where the leading role is played by the maize for grain. Climatic conditions distinctly determine the directions of cultivation of oil plants. In the northern part of the region, only rapeseed and field mustard are cultivated, while in the southern part, rapeseed and sunflower are cultivated. Other crops, such as potatoes or sugar beets, do not display any clear spatial dependence, and their distribution, to a higher extent, results from tradition and characteristics of the food processing industry.

The diversity of the crop structures results, as well, from the socio-economic and infrastructural conditions. The collapse of the socialized sector and the strong agrarian fragmentation, especially in the south of the region considered, brought about abandoning of the crops that require high production means inputs. The state of technical infrastructure plays, as well, a definite role, especially regarding the irrigation systems, which, in the cases of Romania and Bulgaria, underwent degradation.

Cereals are the basic crop cultivated, with their share in field crops in all countries of the region attaining the level of around 70%. In view of the diversity of varieties, adapted to different natural conditions, wheat is the most commonly cultivated cereal. The biggest producers of wheat are Romania, Poland, Bulgaria, and Hungary, which together account for more than 70% of total wheat production in the region considered. The second place after wheat is occupied by maize, followed by barley, oats, and rye. The species of cereal that is produced depends primarily upon the environmental conditions and is determined by the geographical location. So, for instance, wheat is the leading crop in the countries situated in the northern part of the region, while maize is the leading crop in the southern part.

In Baltic countries, high shares in the crop structure, side by side with wheat, are taken by barley, oats, and rapeseed. Crop structure in Poland, on the other hand, is characterized by a wide diversity of components, with similar shares of barley, rapeseed, rye, and somewhat smaller shares of maize, oats, and potatoes, with wheat remaining the most important element of this structure. In Czechia, crop structure features high shares of wheat and barley, as well as rapeseed. There is a relatively high proportion of sugar beets, which also play an important role in the crop structures of Croatia, Poland, and Slovakia. A similar structure as in Czechia is observed in Slovakia, with maize and sunflower taking larger shares in this country. Then, in Hungarian agriculture three crops dominate in the area of surfaces occupied by them: maize, wheat, and sunflower. The crop structure in Romania is characterized by high shares of maize for grain and wheat, as well as significant proportions of sunflower, rapeseed, and barley. Finally, the crop structure in Bulgaria is dominated by wheat and sunflower.

The direction of changes during the last three decades in the entire region considered here resulted, first of all, in the increase of shares of rapeseed and wheat and a decrease of those of potatoes and barley. The increase in the area under wheat resulted from the high yield potential, common use in feeding people and animals, as well as export possibilities. On the other hand, the increased interest in cultivating rapeseed, especially after the countries of Central Europe joined the European Union, was the effect of swelling demand for rapeseed grain, including the possibility of its use in biofuel production. Currently, the region in question provides some 40% of the total production of this crop in the European Union, the biggest producers being Poland, Romania, and Czechia. At the same time, the persistent drop in the area of cultivation of potatoes as animal feedstuff was the effect of its replacement by other root crops used for feeding animals, and as an alimentary crop it was replaced by other vegetables.

The changes in the areas under cereals had differentiated character, depending upon the particular countries. One can speak, in general, of a slight decrease or of stabilization as to the areas of their cultivation. Yet one should expect further slow decrease of the areas under cereals in connection with the overall shrinking of agricultural land. It is also highly probable that their share in the area-wise crop structure will decrease in association with the introduction of new varieties, offering higher yields, resulting altogether, though, in increased production. This, however, in turn, can lead to overproduction problems and lower profitability of their cultivation. The biggest annual fluctuations in the areas under cereals during the last two decades have been observed in the south-eastern part of the region, that is in Romania and Bulgaria, this phenomenon having resulted from deep ownership changes, leading to an unstable situation on the food market.

Conclusions

Privatization of farmland in Central Europe was in agreement with the "western" model of organization of agriculture, according to which the foundation for the farming system was to be constituted by family farms. Yet the ultimate effects have not confirmed this direction of action; in Czechia, in Slovakia, and partly also in Hungary, large agricultural enterprises persisted, as the most effective ones in which, though, the manner of managing them changed. On the other hand, in Romania, very strong fragmentation of agricultural property and field plots took place, which lowered the effectiveness and competitiveness of Romanian agriculture. In Poland, the analogous changes had a rather regional character and affected altogether only approximately 20% of the total agricultural land. It can therefore be stated that the countries considered adopted a similar concept of privatization and restitution of the landownership, but the effects of these processes were clearly diversified.

In all of these countries, the process of systemic transformation brought an abrupt fragmentation of agricultural land and the later actions concentrated on consolidation of land. The effect of ownership changes consisted first of all in an increase in the number of small and unprofitable farms. Yet the main successors of the post-communist cooperatives and state farms, capable of competing on the food market in all the countries of the region (except for Poland, where during the socialist period private farmland ownership persisted), are now large agricultural enterprises, characterized by the differentiated ownership forms.

The analysis of changes in the agricultural use of land in Central Europe shows that these changes were multidirectional and spatially diversified. In the biggest countries of the region (Poland, Romania, Bulgaria, Hungary), the share of agricultural land in their total surfaces decreased since the downfall of the Eastern Bloc by some 10% on the average. During the 1990s, though, the analysed changes were rather limited, and they gained momentum only at the turn of the century. It can be generally stated that the intensity of the phenomenon of decrease in the area of agricultural land became more significant since the period of preparation of the countries in question to the membership in the European Union. Shrinking of the area of farmland was the effect of territorial expansion of towns, development of the transport infrastructure, as well as exclusion from cultivation of the land featuring the lowest agro-ecological qualities and its forestation.

All in all, in the entire region, with the exception of Slovenia, field crops dominate in the crop structure. In the south, relatively important shares are occupied by permanent crops, this being a result of advantageous climatic conditions. The directions of changes in the structure of use of agricultural land were spatially differentiated, but a general shrinking of the share of permanent crops (except for Poland) has been observed, the land being transferred to arable land or, less frequently, to grasslands. Side by side with economic factors, an important influence on the transformation of the structure of farmland was exerted by environmental conditions. In the countries where the quality of agro-ecological conditions is low, there is usually an increase in the share of grasslands, while in the countries featuring higher quality of these conditions – mainly regarding soils – the share of arable land increased.

Regarding the proportions of crops, distinct difference between the north and the south of the region considered can be observed. In the northern countries, a clear domination of cereals, primarily wheat, is observed. In the southern countries, the crop structure is more diversified, as more of industrial crops are cultivated there, along with the cereals. While the crop structures in the particular parts of the region depend to a large extent upon the environmental conditions, the contemporary directions of their changes are first of all the effects of the economic and social conditioning. The ownership changes, which entailed a disadvantageous agrarian structure, along with the process of ageing of the rural population, resulted in abandoning of cultivation of crops that require important inputs in terms of production means and labour. On the other hand, joining the European Union by the countries of Central Europe contributed to the increase of significance of the oil-producing crops in connection with production of biofuels. Yet, in a longer-time perspective, one should expect an increasing dependence of agriculture upon the global climate change and the sudden weather phenomena. It is quite probable that under their influence, the reaches of particular crops, the dates of fieldwork, the yields etc. will undergo changes. Actually, farmers are already now forced to look for new solutions concerning methods and directions of agricultural production. It is possible that global warming will increase the production potential of agriculture in the region considered, but it can get reduced by the more frequent extreme phenomena and problems associated with water shortage.

References

Balteanu, D., & Popovici, E. (2010). Land use changes and land degradation in post-socialist Romania. *Romanian Journal of Geography*, *54*(2), 95–105.

Bański, J. (2011). Changes in agricultural land ownership in Poland in the period of the market economy. *Agricultural Economics*, *57*(2), 93–101.

Bański, J. (2014). Structure and ownership of agricultural land in Poland in the first years of new millennium. *Journal of Agricultural Science and Technology B*, *4*, 85–93.
Bański, J. (2017). The consequences of changes of ownership for agricultural land use in Central European countries following the collapse of the Eastern Bloc. *Land Use Policy*, *66*, 120–130.
Bański, J. (2018). Phases to the transformation of agriculture in Central Europe – selected processes and their results. *Agricultural Economics*, *64*, 546–553.
Bański, J. (2019). Land-use and ownership changes of agriculture. In G. Gorzelak (Ed.), *Social and economic development in Central and Eastern Europe: Stability and change after 1990* (pp. 225–243). Routledge.
Bański, J., & Mazur, M. (2021). *Transformation of agricultural sector in the East-Central Europe after 1989*. Springer.
Basek, V., & Divila, E. (2008). *Struktura czeskich gospodarstw rolnych – dziś i jutro* [Structure of the Czech farms – today and tomorrow]. Dziś i jutro gospodarstw rolnych w krajach Centralnej i Wschodniej Europy, Prace IERiGŻ, 98, 53–66, Warszawa.
Bencheva, N. (2005). Transition of Bulgarian agriculture: Present situation, problems and perspectives for development. *Journal of Central European Agriculture*, *6*(4), s. 473–s. 480.
Benedek, J. (2000). Land reform in Romania after 1989: Towards market oriented agriculture? In P. Tillack & E. Schulze (Eds.), *Land ownership, land markets and their influence on the efficiency of agricultural production in Central and Eastern Europe*. IAMO.
Bentley, W. (1987). Economic and ecological approaches to land fragmentation: In defence of a much-maligned phenomenon. *Annual Review of Anthropology*, *16*, 31–67.
Bicik, I., & Jelecek, L. (2009). Land use and landscape changes in Czechia during the period of transition 1990–2007. *Geografie*, *114*(4), 263–281.
De Arriba, R. (2007). Evaluating land reform and market in Bulgaria. *Journal of Economics and Business*, *10*(1), 11–33.
Dijk van, T. (2003). Scenarios of Central European land fragmentation. *Land Use Policy*, *20*, 149–158.
Doucha, T., Divila, E., & Fischer, M. (2005). Land use and ownership and the Czech farm development. *Rural Areas and Development, European Rural Development Network (ERDN)*, *3*, 1–13.
Dzun, W., & Józwiak, W. (2008). *Gospodarstwa rolne w Polsce przed i po wejściu do UE* [Farms in Poland before and after the entry into the European Union]. Dziś i jutro gospodarstw rolnych w krajach Centralnej i Wschodniej Europy, Program Wieloletni 2005–2009, 98, 10–24, IERiGZ, Warszawa.
Gajdos, P. (2005). Marginal regions in Slovakia and their developmental disposabilities. *Agricultural Economics*, *51*(12), 555–563.
Hartsa, I., Kovach, I., & Szelenyi, I. (1998). The price of privatization. The post-communist transformational crisis of the Hungarian agrarian system. In I. Szelenyi (Ed.), *Privatising the land. Rural political economy in post-communist societies*. Routledge.
Janku, J., Sekac, P., Barakova, J., & Kozak, J. (2016). Land use analysis in terms of farmland protection in the Czech Republic. *Soil and Water Resources*, *11*(1), 20–28.
Kopeva, D. (2003). Land market in Bulgaria. *Land Reform*, *3*, 41–58.
Kovacs, T. (2005). Restructuring agriculture. In G. Barta, E. Fekete, I. Szorenyine, & J. Timar (Eds.), *Hungarian spaces and places: Pattern of transition* (pp. 259–271). Centre for Regional Studies, Pecs.
Mejstrik, M. (Ed.). (1997). The privatization process in East-Central Europe. In *International studies in economics and econometrics* (Vol. 36). Springer.
Moteva, M., Mondeshka, M., Stoeva, A., & Yarlovska, N. (2014). Contemporary issues of land use and water management for agriculture in Bulgaria. *Lucrari Stiintifice*, *57*(2), 59–68.
Otiman, P. (2013). *Romania's present agrarian structure*. Romanian Academy.

Page, N., & Popa, R. (2013). *Family farming in Romania*. Fundaţia ADEPT Transilvania.
Petrick, M., & Weingarten, P. (2004). *The role of agriculture in Central and Eastern European rural development: Engine of change or social buffer?* Institute of Agricultural Development in Central and Eastern Europe IAMO.
Popovici, E., Balteanu, D., & Kucsicsa, G. (2013). Assessment of changes in land-use and land-cover pattern in Romania using Corine Land Cover database. *Carpathian Journal of Earth and Environmental Sciences*, *8*(4), 195–208.
Rusu, M., & Florian, V. (2003). Rural space and rural development in Romania. In J. Bański & J. Owsiński (Eds.), *Alternatives for European rural areas* (Vol. 1, pp. 39–56). European Rural Development Network.
Sadowski, A., & Takacs-György, K. (2005). Results of agricultural reform: Land use and land reform in Poland and Hungary. *Studies in Agricultural Economics*, *103*, 53–70.
Smith, H., & Pechota, V. (Eds.). (1994). *Privatization in Eastern Europe: Legal, economic, and social aspects*. Columbia University, Transnational Juris Publications Inc., Martinus Njihoff Publishers.
Swinnen, J. F. M. (1996). *An explanation of land reform choices in Central and Eastern Europe* (Working Papers, 5). Policy Research Group.
Takacs, I. (2008). Longitudinal analysis of changing partial efficiency of assets in the EU agriculture at the beginning of the new 21st century. *Annals of the Polish Association of Agricultural and Agribusiness Economists*, *10*(5), 149–154.
Tisenkopfs, T. (1999). Constructed countryside: Post-socialist and late modern mixture in rural change. *Humanities and Social Sciences Latvia. Latvia in Europe Sociologically*, *1*, 72–111.
Todorová, S. (2016). Bulgarian agriculture in the conditions of the EU common agricultural policy. *Journal of Central European Agriculture*, *17*(1), 107–118.
Toth-Naar, Z., Molnar, M., & Vinogradov, S. (2014). Impact of land use changes on land value in Hungary. *Roczniki Naukowe, Stowarzyszenie Ekonomistów Rolnictwa i Agrobiznesu*, *16*(6), 500–504.
Veznik, A., Kral, M., & Svobodova, H. (2013). Agriculture of the Czech Republic in the 21st century: From productivism to post-productivism. *Quaestiones Geographicae*, *32*(4), 7–14.

2 Agricultural transformation in Hungary

Trends and challenges since 1990

József Lennert and Jenő Zsolt Farkas

Introduction

Over the past century, Hungarian agriculture has undergone three fundamental changes: Soviet-style forced collectivization (from 1948 to the early 1960s), the re-establishment of a market economy after the political transition (1990), and accession to the European Union (2004). The first change falls outside the scope of our analysis. However, some key characteristics of the state socialist agricultural system are discussed to better understand the second transformation after 1990. The third and least dramatic event was Hungary's accession to the European Union, during which the common market and the Common Agricultural Policy (CAP) became the new economic, legal, and social frameworks for production. This process of change is specific to Hungary and certain other post-socialist countries. Its significance is heightened by its close alignment with the transition to a market economy and its interaction with the subsequent aftershocks of this transformation.

In parallel with the successive transformations in Hungary, the global system of agricultural production, its conceptual frameworks, and its socio-economic goals have also evolved. Since the 1950s, industrial agriculture has been established, marked by the Green Revolution, which was characterized by modernization and technological development, farm concentration, the use of chemicals, and high-yield varieties in production (Bowler, 1986; Goodman & Redclift, 2002). This period can be termed the productivist era of agriculture (Ilbery & Bowler, 1998; Lowe et al., 1993). The underlying motive on every continent was to eliminate hunger and achieve self-sufficiency, a goal that was perhaps overachieved in Europe, leading to significant overproduction in Western European agriculture by the 1980s. In Hungary, not only was self-sufficiency achieved by the early 1960s, but intensive industrial agriculture also brought significant foreign currency revenues to the socialist state, which had few marketable export products. The most important cost of this success was generally the degradation of ecosystem services (Bowler, 1986), which increasingly appeared in the criticism of the Common Agricultural Policy (CAP) and led to successive reforms, strengthening sustainability considerations (Mather et al., 2006; Hodge et al., 2015). In the 1990s, agri-environmental management measures emerged, and after 2008, there was increasing attention to mitigating the impacts of climate change.

DOI: 10.4324/9781032672779-4

The European Union adopted the model of multifunctional agriculture, where farmers receive agricultural subsidies not based on the quantity of crops they produce but for non-market outputs that contribute to important societal goals for the European Union, including preserving the environmental state of rural landscapes (Van Huylenbroeck et al., 2007). European agriculture could no longer be described solely by the characteristics of the productivist era because the post-productivist transition (PPT) emerged as an alternative (Ilbery & Bowler, 1998; Lowe et al., 1993). The spread of sustainable agriculture principles (low-input production, organic and ecological farming) significantly depends on geographical and spatial differences in environmental, economic, and social characteristics. The same applies to the non-market outputs of agricultural production, which can hold entirely different values in different regions. Therefore, according to Marsden and his colleagues (1987), the PPT leads to different developmental paths and further differentiation within agriculture and rural areas. While in the productivist era, the main identified issue was the negative environmental impacts of agriculture, during the PPT, the primary concern has increasingly shifted to the reduction of utilized agricultural area on which the European Environmental Agency (2006) issued a comprehensive report. This trend is the amalgamation of two fundamentally different phenomena: urban expansion (sprawl) and the abandonment of distant agricultural lands (Freshwater, 2008).

These developments coincided with Hungary's EU accession. While Hungarian decision-makers and producers familiarized themselves with the CAP's agricultural support system and adapted the required policies, Western environmental and sustainability concerns could not receive true emphasis. At that time, Hungary was still dealing with problems arising from the political and economic transition, such as the loss of markets, the dramatic decline in agricultural employment, and the fate and development of hundreds of thousands of forced agricultural enterprises that emerged after the compensation process.

In the latter half of the 2010s, it became apparent that due to global population growth, the efficiency of agricultural production needed to be further increased but in a way that also met environmental goals (Maffezzoli et al., 2022; Dayıoğlu & Turker, 2021). Precision agricultural practice in crop production has a history of several decades but only became widespread after 2000 with the cessation of GPS signal scrambling and improved accuracy. The second boost could come from advances in sensor technologies (IoT), the application of artificial intelligence, and the digitization of agricultural production, referred to as Agriculture 4.0 or Agri-Food 4.0, following the concept of Industry 4.0. Again, this system of agriculture does not replace the PPT but offers another new alternative, maybe further differentiating agricultural enterprises and rural areas (Wolfert et al., 2017).

Thus, our analysis aims not only to present the changes that have occurred in Hungarian agriculture but also to evaluate them from an agro-geographical perspective. In doing so, we strive to identify the characteristics of production, as well as the economic and social factors behind them, which allows us to determine whether the characteristics of the productivist or PPT era dominate and where the sector is in the adaptation process of Agriculture 4.0.

Methodology

To examine the various aspects of Hungary's agricultural transformation, we conducted statistical and GIS analyses, utilizing diverse data sources:

- Hungarian Central Statistical Office (HCSO) – yearly data tables (www.ksh.hu)
  - Land use structure (1960–2019)
  - Sale of pesticides (2001–2022)
  - Output of crop production and livestock breeding (1960–2023)
  - Yield and cultivated area of main crops (1990–2022)
  - Volume of vegetable, fruit, grape and wine production (1990–2022)
  - Livestock (poultry and pig) (1990–2023)
  - Consumption of poultry and pork (1990–2022)
  - Gross Value Added, agricultural sector (1990–2022)
- HCSO – General Population Census (https://nepszamlalas2022.ksh.hu/)
  - Number of people employed in the agricultural sector (1990–2022)
- HCSO – Agricultural Census (https://map.ksh.hu/timea/?locale=en)
  - Spatial characteristics of crop, fruit, and vegetable production (2020)
- EEA Corine Land Cover (https://land.copernicus.eu)
  - Land cover change, GIS data (1990–2018)
- Eurostat (https://ec.europa.eu/eurostat)
  - Share of high- and low-input farms (2005–2021)
  - Share of organic farming (2004–2021)
  - Harmonized pesticide risk indicator (2011–2022)
  - Common farmland bird index (2000–2020)
  - Number of farms by utilized agricultural area and legal status (2005–2020)
  - Share of utilized agricultural area by farm size and legal status (2005–2020)
- farmsubsidy.org.
  - Value of subsidies per recipient (2022)
  - Total value of subsidies per county (2022)

Changing land use patterns

Like other European countries (Rabbinge & van Diepen, 2000), Hungary has experienced a significant decline in utilized agricultural area (UAA) (from 7.14 million hectares in 1960 to 5.31 million hectares in 2019) as a key trend over the last 60 years. A main component of this is shrinkage in the area of arable land, gardens, and grasslands. Statistics also point to dynamic accretion of land excluded from cultivation. Despite these changes, Hungary still retains an agricultural character in its land use, with the share of UAA at 57.1% in 2019. Table 2.1 contains the changes by reference to the land use categories.

After 1990, the economic transition and the modernization of the country's infrastructure accelerated land use conversions. Before the regime change, approximately 25,000 hectares were converted annually, which increased to 42,000–46,000 hectares per year from 1990 to 2012, then to 31,000 hectares annually between 2012 and 2018 according to Corine Land Cover Change data. Further decreases in agricultural land are expected, with the Hungarian National Forest Plantation Programme aiming to increase forest cover by 750,000 hectares by 2050. Concurrently, the demand for new built-up areas persists, driven by suburbanization and foreign direct investments (recently automotive industry and battery manufacturing). Corine Land Cover Change data indicate that from 1990 to 2018, 1.15 million hectares of land were converted, with 56% occurring in the agriculture-dominated Hungarian Plain (Figure 2.1).

Meso-regions particularly susceptible to conversion are characterized by weak, sandy soils with poor water retention. These areas compensate this disadvantage with labour-intensive agriculture, including vineyards and orchards, and a diverse landscape with scattered farms. These regions are environmentally vulnerable, and one of them (Homokhátság) is identified as increasingly arid in the FAO's forecast, where desertification is a possible outcome (UNCCD, 2006).

The main reason for this spatial pattern is that farmers on poor soils tried to maximize subsidies. Pre-2004 domestic and post-accession CAP subsidies formulated contradictory goals, for example, concerning vineyard plantings. Thus, the producers made land use decisions related to their short-term interests due to their difficult economic (income) situations. Forest plantations also contributed to the changes in these areas because it has been supported since the early 2000s, though concerns about using alien species and increasing aridity remain.

In contrast, fertile regions with chernozem soil have stable land use due to area-based EU payments and higher incomes. In these areas, the typical land use is large-scale industrial arable crop cultivation.

Based on Eurostat data, the intensification of land use changed favourably between 2005 and 2021. The proportion of the area used by high-input farms decreased by 9.8% of the UAA (from 45% to 35.2%, which means -522,172 hectares in absolute terms), while in the case of low-input farms, this value increased by 8.8% (from 21.2% to 30%, which means 464,362 hectares). As a result of these changes, Hungary's data were similar to the EU27 data in 2021. These changes can be considered as steps towards sustainability, which happened despite the fact that the enthusiasm for agri-environmental measures or organic farming (OF) experienced at the beginning of the 2000s waned after joining the EU, mainly due to the accompanying bureaucratic administration processes. However, after a long period of stagnation (between 2004 and 2015), a significant increase can be seen in the size of the area covered by OF. From the UAA, its share was only 2.43% in 2015, which increased to 6.31% in 2021 according to the Eurostat Agri-environmental Indicators (AEI) database. Several factors can be behind this dynamic growth: first, the growing consumer demand for organic products and potential for higher profits

Table 2.1 Land use structure, 1960–2019

Year	*Arable land*	*Garden*	*Orchard*	*Vineyard*	*Grassland (1,000 ha)*	*Forest*	*Reed*	*Fishpond*	*Built-up area*	*Total*
1960	5309.8	107.5	82.3	203.6	1437.9	1306.2	26.1	-	829.7	9303.1
1990	4712.8	341.1	95.1	138.5	1185.6	1695.4	40.4	26.8	1067.5	9303.2
2004	4510.3	96.8	102.6	94.5	1059.6	1823.4	61.5	33.6	1521.1	9303.4
2019	4317.7	38.6	94.4	68.4	790.4	1939.5	34.9	35.2	1984.3	9303.4
2019–1990	−395.1	−302.5	−0.7	−70.1	−395.2	244.1	−5.5	8.4	916.8	
2019/1990 (%)	91.6	11.3	99.2	49.4	66.7	114.4	86.4	131.3	185.9	

Source: Authors' own elaboration based on data from the Hungarian Central Statistical Office.

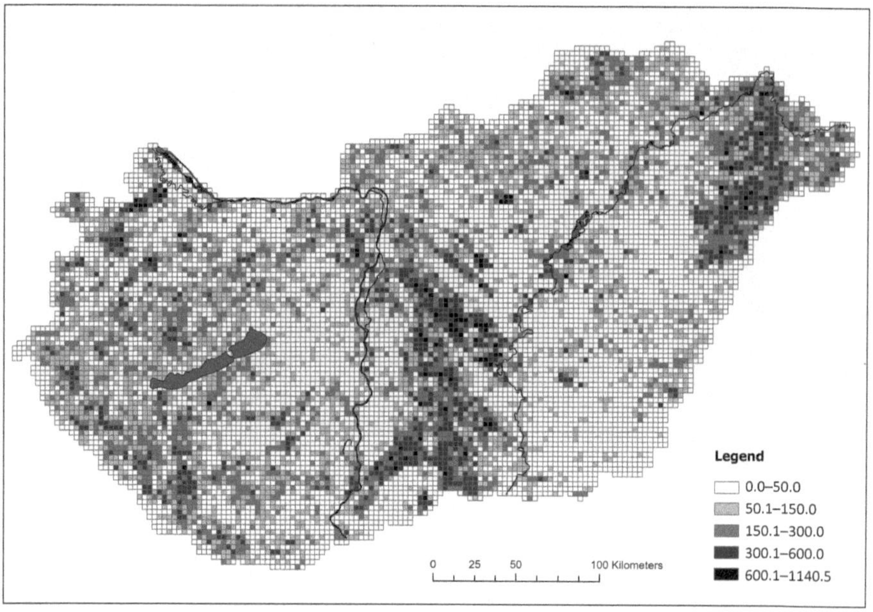

Figure 2.1 Land-cover changes between 1990 and 2018 (hectares/3×3 km grid cell).

Source: Authors' own elaboration based on Corine Land Cover Change, European Environment Agency, Copernicus Land Monitoring Service.

(Takács, 2007), second, soil quality improvements from OF which can encourage producers to switch (Takács et al., 2021), third, regional variety trials show potential for high yields and quality in OF (Földi et al., 2021), and fourth, increased organic soybean production for organic animal husbandry (Hunyadi et al., 2023).

In contrast to the aforementioned processes, it is noteworthy that according to HCSO data, the sale of pesticides increased from 16,000 tonnes to more than 29,000 tonnes between 2001 and 2022. However, it is also important to highlight that the amount of active ingredients sold remained relatively stable between 2011 and 2022, fluctuating around 8.5 million kg. Furthermore, the value of the harmonized pesticide risk indicator, with the 2011–2013 average set at 100, was 70 in 2022 according to Eurostat AEI. This value, however, is still higher than the EU27 average of 62.

The aforementioned positive changes in land use and production practices in Hungarian agriculture are also reflected in the common farmland bird index. Taking the year 2000 as the baseline (100), the index value, after reaching a minimum of 75.11 in 2010, gradually increased to 81.27 by 2020, positioning Hungary in the mid-range among the member countries with available data.

Changing production structure

In the wake of the economic transition, both the output of crop production and livestock breeding suffered a drastic drop, which nearly nullified the growth experienced in previous decades. However, the limits of the socialist agricultural model became apparent well before, and the 1980s were already characterized by stagnation.

After the initial plunge, crop production and livestock breeding followed different trajectories. After reaching its lowest value in 1993, output from crop production started to rise again rapidly, and by the early 2000s, it surpassed the output of the communist era. In contrast, the crisis in livestock breeding was more prolonged, reaching its low point only in 2010. In the early 2010s, both sectors reported growth and reached a plateau by the end of the decade. In recent years, the total output of agricultural production is only slightly behind the peak years of the 1980s (160–170% of the initial 1960 value), but the focus has significantly shifted from livestock breeding to crop production.

Since the 18th century, cereals have made up the lion's share of crop production, accounting for more than a quarter of the total agricultural production value and covering around 49% (2.4 million hectares) of the total agricultural area in 2022. The most prominent crops are wheat and maize (both with a cultivated area of around 1 million hectares), followed by barley (339,000 hectares in 2022). Additional crops of secondary importance are still produced in areas with less favourable natural conditions (e.g. rye in sandy soil areas), with a limited area of cultivation. After the transition, the average yield of the main crops increased significantly (e.g. the yield of maize nearly doubled from 4 tonnes per hectare to 8 tonnes per hectare), narrowing (but not entirely closing) the gap between Hungary and Western countries.

The cultivated area of oilseeds has more than doubled since 1990, covering 17% of the total agricultural area in 2022. Among these, sunflower seeds are the most prominent, especially in the Hungarian Plain, while rapeseed cultivation is preferred in Western Transdanubia, where higher precipitation benefits production. On the other hand, the cultivated area of sugar beet has collapsed to less than a tenth of its original size. This reduction was facilitated by EU accession: the strict quotas on sugar production introduced by the European Union led to the near-complete elimination of sugar production. Of the 12 sugar plants from the state-socialist era, only one remains operational today (in Kaposvár). The reasons behind the reduced cultivation area of potatoes include changes in domestic demand and less favourable environmental conditions – the latter perhaps a warning sign of the ongoing climatic change (Ruby et al., 2018).

The conditions in Hungary are ideally suited for the production of a wide variety of fruits and vegetables. During the state-socialist era, private farming was allowed on small allotment gardens, which also facilitated the development of horticulture. After the economic transition, the volume of produced vegetables only slightly decreased (1991–1993: 1.3 million tonnes; 2020–2022: 1.2 million tonnes). However, significant shifts can be observed in the composition of production. The production volumes of tomatoes, cucumbers, onions, and cabbages have drastically reduced, while the production of green peas, peppers, carrots, and watermelons has remained more or less the same. On the other hand, the production of sweetcorn has more than quadrupled, becoming the most important vegetable (2020–2022:

0.46 million tonnes). Some of these shifts reflect changes in the food production industry. The privatized producers often underwent portfolio rationalization, and instead of covering all segments of domestic demand, they focused on producing fewer products for the global market (such as the Hungarian Bonduelle canning factories focusing on sweetcorn).

Fruit production, however, suffered a more marked decline (1990–1992: 1.27 million tonnes; 2020–2022: 0.58 million tonnes). Apples make up two-thirds of this production, which deeply suffered from the loss of its former socialist markets. Other reasons include limited cooperation between producers, which leads to higher costs, a weak assertion of interests during price negotiations, and a large gap between producer and retail prices (Erdész et al., 2009). Climate change also plays a role: the notable raspberry-producing areas in the valleys of mountainous northern Hungary have almost completely disappeared. Apart from unfavourable weather conditions, the ageing of original farmers and the lack of workforce also contribute to the decline. Since rural micro-regions often depend on the production of certain types of fruits or vegetables, their decline endangers many livelihoods in these areas. Moreover, since these products are often part of the local identity, their loss may contribute to the social erosion of these villages.

The primary aim of grape production in Hungary is winemaking. Following the economic transition, annual output fell in terms of the amount of both grapes and wine. However, a reduction in the volume of wine produced is not necessarily a bad trend. Under communism, the quantity of wine dominated over quality, and that situation could not continue in a competitive Common Market environment. Thus, production shifted to higher quality wine instead of prioritizing quantity.

As it was outlined earlier, the crisis for livestock breeding was more prolonged compared to crop production, and the following recovery was only partial. A key factor in this differentiation lies in the circumstances of privatization. During the communist era, livestock breeding operated within a vertically integrated system. However, during the transition, central facilities (e.g. piggeries, pens, dairies) were often privatized independently of their supporting agricultural areas (e.g. fodder-producing fields). Under these circumstances, the steady rise in fodder prices was especially detrimental to livestock breeding.

In case of cattle, swine, sheep, and poultry, different trends can be observed, which can be also connected to consumption and lifestyle changes. After a moderate reduction, the poultry stock quickly stabilized (50 million head in 1990 comparing with 35 million in 2023), while the number of pigs kept falling (8 million in 1990, 2.6 million in 2023). Prior to the transition, a significant proportion of swine was reared in rural households as a traditional component of family farming, with cooperatives supplying fodder and managing sales (de Haan et al., 1992). However, after the dissolution of cooperatives, these services were also discontinued. Simultaneously, competition from large-scale private companies intensified, and inexpensive imported alternatives became readily available on the shelves of newly established food retail chains. Moreover, tightening animal welfare regulations coupled with the changing lifestyle of rural families also led them to give up the most demanding and inconvenient aspects (e.g. smell, flies) of agricultural activity first, including pig breeding (Vajda & Baksa, 2008; Mroczkowska, 2019; Marczin et al., 2023). The shifting emphasis in livestock production also reflects the changes in consumer habits: the consumption of pork has decreased (1990: 39 kg/person/yr, 2022: 30 kg/person/yr), while the consumption of poultry has markedly increased (1990: 23 kg/person/yr, 2022: 33 kg/person/yr).

Changes in the holding structure

On the precipice of the political and economic transition, 130 state farms and 1,200 cooperatives were responsible for two-thirds of the Hungarian agricultural output, the rest was produced by small family farms on household plots (Kapronczai, 2010; Csáki & Lerman, 1997). The legislative measures following the fall of state socialism completely reshaped this landownership structure. The new political elite emphasized providing compensation for previously confiscated private property. The previous ownership rights of these properties were not restored. Instead, beneficiaries of the new compensation laws (Law XXV of 1991, Law XXIV of 1992, and Law XXXII of 1992) received coupons, which allowed them to bid for land from the former cooperatives and state farms at compensation auctions. Former members and workers of the dissolved entities also received land. On the other hand, new laws regarding cooperatives established a framework for them to function similarly to Western European models.

As a consequence of the political intentions, the transformation of the Hungarian ownership structure was very rapid. In 1996, the majority of utilized area was already cultivated by private holdings. The number of private farmers skyrocketed: during the compensation auctions, 1.5 million new landowners received land. However, most of them received only minuscule parcels, so the holding structure became extremely fragmented. Moreover, many of the new owners did not possess the necessary knowledge or equipment to successfully manage their farms. These drawbacks sowed the seeds for a new wave of land concentration.

After the drastic decline in the 1900s, the number of legal entities and their utilized area is stabilized. In 2020, 8,710 legal entities cultivated 1.9 million hectares of agricultural area. Close to 40% of the legal entities utilized more than 100 hectares; the overwhelming majority of land belonged to legal entities (Figure 2.2). Legal entities without a significant amount of land are often involved in livestock breeding.

This stands in stark contrast with the characteristics of the private farms managed by natural persons. The vast majority (78% in 2020) of natural persons farm on less than ten hectares, and among them, those who utilize less than five hectares are the most numerous. On the other hand, only 3% of the natural persons cultivate more than 100 hectares. However, the distribution of the utilized area by farm size is somewhat more similar to the share of land managed by legal entities. Despite the overwhelming majority of farmers (175,320 in 2020) manage less than ten hectares, their share make up only 12% of the area utilized by natural persons. On the other hand, less than 7,000 farmers manage more than half of the agricultural area cultivated by natural persons.

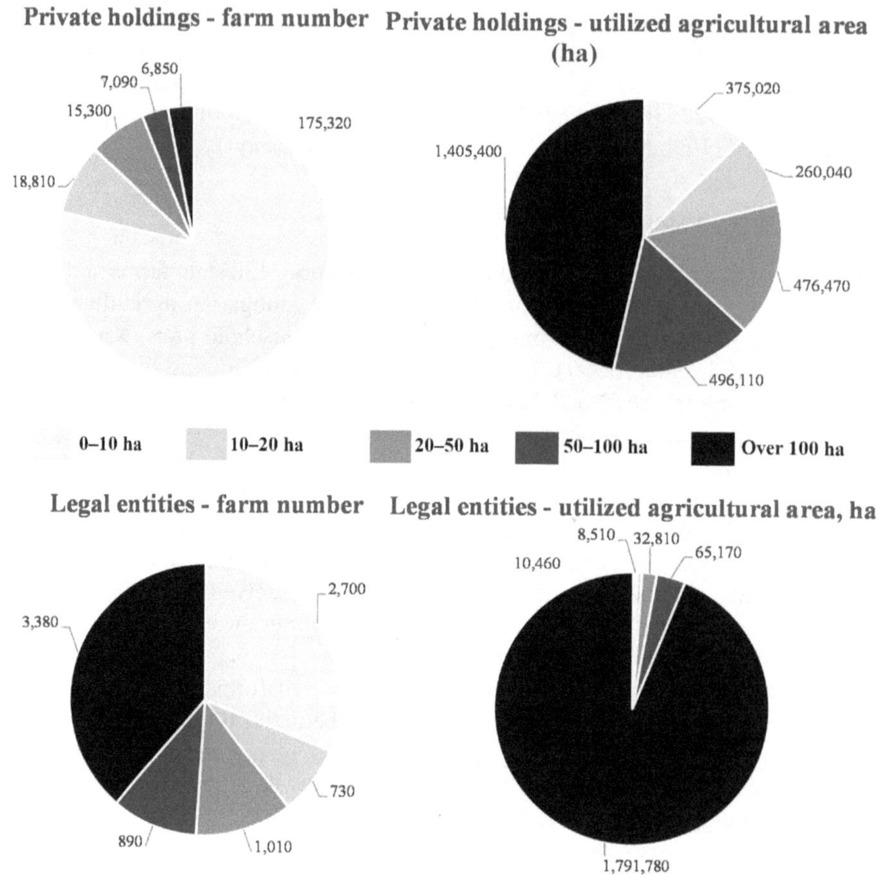

Figure 2.2 The number of farms and the areas of land they utilized, categorized by size and legal status, 2020.

Source: Authors' own elaboration based on Eurostat.

It also needs to be highlighted that the current situation is just a snapshot in the decades-long transformation of the holding structure. Since the 1990s, the number of farmers is constantly decreasing. Following the EU accession, the number of farmers still exceeded 700,000, but by 2020, it had decreased to 223,000. This reduction affected smallholders the most. During the same period, the number of large-scale farmers utilizing more than 100 hectares doubled (2005: 3390; 2020: 6850). Their share from the area utilized by natural persons similarly inflated (2005: 27%; 2020: 47%). The rate of change between 2016 and 2020 (the number of farmers cultivating less than ten hectares were halved) also indicates that this process hasn't reached the point of equilibrium yet.

The land concentration is driven by multiple factors. As it was described, many farmers obtaining land through compensation had neither the knowledge nor the experience to successfully operate their farms. Moreover, during the state socialist era, private household farming existed in symbiosis with the cooperatives (e.g. borrowing equipment, entry to market). The post-socialist conditions required new approaches even from the experienced farmers.

The changing market conditions didn't help either. Small-scale farming is the most competitive in labour-intensive production, like horticulture and viticulture. However, the loss of socialist markets heavily affected fruit production, and the privatization of the food industry also put stress on the sector. Many of the smallholders didn't have the financial means to adapt to the quickly changing conditions.

It has to be added that despite their declared intentions, some measures from the European Union or Hungary did not contribute to the viability of small-scale farming either. EU subsidies provide the most incentives for large-scale crop production. Before the accession, vineyards and orchards could receive targeted financial support for restoration or even for the founding of new plantations, but during the first years in the European Union, these schemes were prohibited, and producers received support for cutting down their plantations. This direction lasted till 2010, only to be followed by a period of a renewed support for founding of vineyards and orchards. These drastic policy changes are unfavourable for smallholders, who often lack the necessary information and capital to quickly adapt to and profit from the altered regulatory framework.

Even after the compensation, a significant amount of land remained in the possession of the Hungarian state. After 2010, the government launched a wave of privatization and long-term lease programme with the dedication to strengthen the small- and middle-scale farming system. However, many of the new farmers (often with ties to the political and economic elite) gained an area well over 100 hectares. József Ángyán, who served as State Secretary at the Ministry of Rural Development between 2010 and 2012, resigned his post, openly challenging the latest government practice of land lease.

Liquidating undivided common landownership may also facilitate the land concentration process. Undivided common landownership is a remnant and unresolved problem of the agricultural transformation of the 1990s. While this type of landownership definitely obstructed proper market functions and agricultural modernization, it is also the key reason behind the large share of land being rented out (Kovách, 2018). In 2012, the Hungarian government declared that the abolition of undivided common landownership was one of its political goals. After a pilot project, the Hungarian Parliament passed a law in 2020 to phase out undivided common land. Again, the declared intentions are to strengthen the position of smallholders (Árvai, 2023). While there is no deadline for the liquidation process, this may also contribute to closing the gap between land use and ownership.

As a result of this land concentration process, the holding structure increasingly resembles the pre-1945 arrangement, characterized by the dominance of large latifundia managed by a narrow, distinct class of landowning elite.

The role of agriculture in the Hungarian economy and rural society

During the accession negotiations with the European Union, agriculture proved to be one of the thorny issues. The two parties had differing objectives, which were resolved through a compromise: Hungary received only a portion of the subsidies from the Single Area Payment Scheme (SAPS) in the first decade and complied with the quota system. However, since 2013, Hungary has received 100% of the subsidies its farmers are eligible for, and the quota system was also phased out, allowing Hungary to be considered one of the greatest beneficiaries of the CAP.

The subsidiaries ensured that after the accession, unlike the decline seen after 1990, the share of agriculture in the Gross Value Added (GVA) did not decrease further. Depending on weather conditions, agriculture contributes between 3.7% and 4.7% of added value each year. In 2022, an unfavourable year due to droughts, the GVA of agriculture was 5.2 billion euros. Also in that year, 262,000 Hungarian beneficiaries received 2.4 billion euros from the CAP, highlighting the critical importance of subsidies for the agricultural sector. The Single Area Payment Scheme (SAPS) and payments for agricultural practices beneficial to the climate and environment ("greening") contribute the largest share to this amount. According to an analysis for the year 2020 (K-Monitor, 2021), 1% of farms received 27% of SAPS and greening subsidies, while the top 10% received 70%. The concentration of these payments is further supported by spatial analysis, utilizing 2022 data from FarmSubsidy.org. Counties with the highest share of smallholders received a low amount of financial support per holding. This remains true even when considering only payments for climate and environmental practices ("greening"), despite these areas being among the most environmentally vulnerable.

On the other hand, nearly 4,500 recipients residing in Budapest (capital of Hungary) received the highest amounts per holding, totalling more than 100 million euros, even though the agricultural area of the city is negligible. These findings further support earlier observations (Kovách, 2012) that a significant portion of agricultural subsidies intended for rural areas is being transferred to city centres, thereby failing to support the sustainability of local rural communities.

For centuries, rural areas have been primarily considered sites of agricultural production. This traditional view runs deep, and agriculture is still often perceived as the main pillar of rural society. Through this lens, agricultural policy and subsidies are seen as the most important aspect of rural development. However, data from the last few decades may challenge the validity of this approach. According to the General Population Census of 2022, 175,000 people were employed in agriculture. While this number is about the same as in the 2011 Census (177,000 people), the overall number of employed people has significantly increased, causing the share of agriculture in total employment to further decrease (from 14.2% in 1990 to 3.7% in 2022). Even in villages, the share of agricultural employees reaches only 7%. On the other hand, more than half of the employees from villages commute to other settlements for work, mostly outside the agricultural sector.

This means the potential role of agriculture in the lives of rural societies is shifting from productivist to post-productivist functions. Instead of being a key source of revenue for the general populace, it can become a source of local identity, an esteemed part of the cultural landscape, and an opportunity for small-scale household production beyond financial rationale. The key question is whether the concurrent changes and looming challenges will allow agriculture to grow into this new role and contribute to the sustainability of rural settlements.

Conclusions

The main goal of this chapter was to provide a general overview of the transformation processes in Hungarian agriculture since the political and economic transition of 1990. In this analysis, we also aimed to explore whether the recent changes indicate a restructuring within the productivist framework or the emergence of a post-productivist agricultural system.

We can conclude that Hungarian agriculture has indeed undergone significant and, in many cases, dramatic changes after 1990. The main drivers of these changes were

- the political desire to dismantle the former state-organized agricultural structure and compensate for previously confiscated private property,
- the loss of protected state socialist markets and integration to Western and global markets,
- the ripple effect of market economy transition in other sectors (especially the food processing industry),
- EU accession (complying with EU regulations, receiving subsidies),
- environmental factors, especially climate change.

The changes reshaped the agricultural profile of the country. The production output of livestock breeding and horticulture declined, while the output of cereal production and certain industrial crops increased. The compensation measures and the dissolution of state farms and cooperatives initially resulted in a very fragmented holding structure. However, subsequent concentration processes reshaped it into a triadic structure. Approximately one-third of the agricultural area is now utilized by 8,710 legal entities, another third by 6,850 large-scale holders (over 100 hectares), and the remaining third by 216,520 smallholders and middle-scale holders. Moreover, the balance is further shifting, with private large-scale holders increasing both in number and in the amount of agricultural area they utilize, while the number of smallholders is rapidly declining. While this concentration process is mostly driven by market forces, certain elements of national and EU policy also facilitate or at least enable it.

In broad terms, these results indicate an agricultural restructuring that remains within the productivist framework. While there are some examples of post-productivist development, these were either temporal (an initial stronger emphasis on the protected areas) mostly side effects of the crisis (a temporal reduction of fertilizer and pesticide use) or didn't become transformative (e.g. ecological farming, agrotourism). Instead, the changes point towards an agricultural landscape dominated by large-scale grain and industrial crop production, which relies on machinery, fertilization, and pesticides to reach high yields and requires little workforce. Most rural residents earn their living outside of agriculture, while a significant share of agricultural profits leaves rural areas.

However, recent years (e.g. the halted increase in production output) indicate that this development path may have plateaued, and several threats challenge the continuation of this course. Urban sprawl has been chipping away at agricultural land at an increasing rate since 1990, but it is more difficult to argue against the maximization of land rent from a productivist perspective. Increasing the desirability of rural areas could help curb urban sprawl, but providing post-material amenities through landscape management lies outside the scope of the productivist framework. The ongoing climate change may be the biggest obstacle in this development path. It has been discussed that rising temperatures have already led to shifting production patterns. Moreover, the short-term outlook – characterized by heatwaves and droughts of unprecedented severity – may turn the cultivation of some key arable crops (e.g. maize) into a risky endeavour.

To address these challenges, adapting elements of Agriculture 4.0 may be more compatible with the current development course than a radical post-productivist turnaround. However, aside from declarations, we found little evidence that precision agriculture and the digitization of agriculture are gaining momentum. A good example is that despite their previous commitments, in 2024 February, the Hungarian Ministry of Agriculture significantly trimmed the financial support for digital transition for the current (2023–2027) CAP period. If national and EU-level agricultural policies effectively address the multifaceted challenges of agriculture through their financial instruments, they could lead to a more resilient agricultural system. However, without fine-tuning, the current policies and financial transfers may even contribute to delaying the necessary adaptation measures.

Acknowledgements

This chapter was supported by the János Bolyai Research Scholarship of the Hungarian Academy of Sciences (BO/00353/21/10 & BO/00583/22/10).

References

Árvai, G. (2023). The liquidation of undivided common land ownership in Hungary. *Journal of Agricultural and Environmental Law*, *18*(35), 7–17. https://doi.org/10.21029/JAEL.2023.35.7

Bowler, I. R. (1986). Intensification, concentration and specialisation in agriculture: The case of the European Community. *Geography*, *71*(1), 14–24. http://www.jstor.org/stable/40571039

Csáki, C., & Lerman, Z. (1997). *Hungarian land reform and farm restructuring*. EC2AU Policy Note (Draft), The World Bank. Retrieved July 30, 2020, from https://www.researchgate.net/publication/304936397_Hungarian_Land_Reform_and_Farm_Restructuring

Dayıoğlu, M. A., & Turker, U. (2021). Digital transformation for sustainable future-agriculture 4.0: A review. *Journal of Agricultural Sciences*, *27*(4), 373–399.

de Haan, C., Van Veen, T. S., & Brooks, K. (1992). *The livestock sector in Eastern Europe: Constraints and opportunities* (World Bank – Discussion Papers). World Bank. Retrieved December 10, 2024, from https://documents1.worldbank.org/curated/en/680301468752077168/text/multi-page.txt

EEA. (2006). *Urban sprawl in Europe – the ignored challenge* (EEA Report No 10/2006). European Environment Agency.

Erdész, F., Jankuné, K., Kozak, A., & Radóczné, K. (2009). *A zöldség- és gyümölcságazat helyzete*. Research Institute of Agricultural Economics.

Földi, M., Bencze, S., Hertelendy, P., Veszter, S., Kovács, T., & Drexler, D. (2021). Farmer involvement in agro-ecological research: Organic on-farm wheat variety trials in Hungary and the Slovakian upland. *Organic Agriculture*, *12*(3), 293–305. Springer Science and Business Media LLC. https://doi.org/10.1007/s13165-020-00335-x

Freshwater, D. (2008). Maintaning farmland: A new focus for agricultural policy. In *Multifunctionality in agriculture: Evaluating the degree of jointness. Policy implications* (pp. 47–65). OECD Publishing.

Goodman, D., & Redclift, M. (2002). *Refashioning nature: Food, ecology and culture*. Routledge.

Hodge, I., Hauck, J., & Bonn, A. (2015). The alignment of agricultural and nature conservation policies in the European Union. *Conservation Biology*, *29*(4), 996–1005.

Hunyadi, B. É., Drexler, D., Berényi Üveges, J., Szigeti, N., Fejes Tóth, P., & Ladányi, M. (2023). Organic soybean production in Hungary. *Acta Agraria Debreceniensis*, (1), 45–52. University of Debrecen/Debreceni Egyetem. https://doi.org/10.34101/actaagrar/1/12514

Ilbery, B., & Bowler, I. (1998). From agricultural productivism to post-productivism. In B. Ilbery (Ed.), *The geography of rural change* (pp. 57–84). Longman.

Kapronczai, I. (Ed.). (2010). A magyar agrárgazdaság az adatok tükrében az EU csatlakozás után. *Agrárgazdasági Információk*, 12. Research Institute of Agricultural Economics.

K-Monitor. (2021). *Kikhez jutnak az agrárpénzek?* Retrieved August 21, 2024, from https://agrar.k-monitor.hu/2021_kikhez-jutnak-az-agrarpenzek/

Kovách, I. (2012). *A vidék az ezredfordulón: A jelen-kori magyar vidéki társadalom szerkezeti és hatalmi változásai*. MTA Tár-sadalomtudományi Kutatóközpont Szociológiai Intézet–Argumentum Kiadó.

Kovách, I. (2018). Földhasználat és földtulajdon-szerkezet. In: T. Kolosi & I. Tóth (Eds.), *Társadalmi Riport 2018* (pp. 248–263). TÁRKI.

Lowe, P., Murdoch, J., Marsden, T., Munton, R., & Flynn, A. (1993). Regulating the new rural spaces: The uneven development of land. *Journal of Rural Studies*, *9*(3), 205–222.

Maffezzoli, F., Ardolino, M., Bacchetti, A., Perona, M., & Renga, F. (2022). Agriculture 4.0: A systematic literature review on the paradigm, technologies and benefits. *Futures*, *142*, 102998.

Marczin, T., Kovács, K., Nagy, A. S., & Vida, V. (2023). A sertéságazat helyzetének bemutatása (jövedelmezőség, hatékonyság). *Gazdálkodás*, *67*(3), 226–246. https://doi.org/10.53079/GAZDALKODAS.67.3.t.pp_226-246

Marsden, T., Whatmore, S., & Munton, R. (1987). Uneven development and the restructuring process in British agriculture: A preliminary exploration. *Journal of Rural Studies*, *3*(4), 297–308.

Mather, A. S., Hill, G., & Nijnik, M. (2006). Post-productivism and rural land use: Cul de sac or challenge for theorization? *Journal of Rural Studies*, *22*(4), 441–455.

Mroczkowska, J. (2019). Pork politics: The scales of home-made food in Eastern Poland. *Appetite*, *140*, 223–230. https://doi.org/10.1016/j.appet.2019.04.022

Rabbinge, R., & van Diepen, C. A. (2000). Changes in agriculture and land use in Europe. *European Journal of Agronomy*, *13*(2–3), 85–100. https://doi.org/10.1016/S1161-0301(00)00067-8

Ruby, R., Senthold, A., Richard, R., Athanasios, P., Gerrit, H., Roberto, Q., Guy, H., & Joost, W. (2018). Climate change impact on global potato production. *European Journal of Agronomy*, *100*, 87–98. https://doi.org/10.1016/j.eja.2017.11.008

Takács, I. (2007). Factors of increasing of organic farming according to demand and supply. *Cereal Research Communications*, *35*(2), 1173–1176. Springer Science and Business Media LLC. https://doi.org/10.1556/crc.35.2007.2.252

Takács, I., Amiri, M., Károly, K., & Mohammed, S. (2021). Assessing soil quality changes after 10 years of agricultural activities in eastern Hungary*. *Irrigation and Drainage*, *70*(5), 1116–1128. Wiley. https://doi.org/10.1002/ird.2623

UNCCD. (2006). *Second national report of the Republic of Hungary on the implementation of the United Nation convention to combat desertification*. Ministry of Environment and Water of the Republic of Hungary.

Vajda, L., & Baksa, A. (2008). Major changes in Hungarian agricultural economy as a result of EU-membership. *Gazdalkodás: Scientific Journal on Agricultural Economics*, *52*(22), 2–17. https://doi.org/10.22004/ag.econ.48086

Van Huylenbroeck, G., Vandermeulen, V., Mettepenningen, E., & Verspecht, A. (2007). Multifunctionality of agriculture: A review of definitions, evidence and instruments. *Living Reviews in Landscape Research*, *1*(3), 1–43.

Wolfert, S., Ge, L., Verdouw, C., & Bogaardt, M. J. (2017). Big data in smart farming – a review. *Agricultural Systems*, *153*, 69–80.

3 Structural transitions of agricultural land use across Northeast Africa from 2001 to 2020

Simon Measho Yhdego

Introduction

General information

The Northeast Africa region shows the highest economic growth with an increasing population in Africa. The region encompasses 12 countries: Egypt, Sudan, South Sudan, Eritrea, Ethiopia, Djibouti, Kenya, Somalia, Rwanda, Burundi, Uganda, and the United Republic of Tanzania. The area covers the Greater Horn of Africa, including Egypt from Northern African countries. The region has an estimated population of 432 million living over 7.2 million square kilometers. Northeast Africa is a tropical area with a warm climate dominated by arid and semi-arid climate systems. The average rainfall is less than 800 mm per year in the semi-arid lands, and it ranges between 800 and 1,300 mm per year in the sub-humid zones, with an average rainfall varying strongly through time (Kalisa et al., 2019; Measho et al., 2023). The land cover types of Northeast Africa based on ESA CCI (European Space Agency Climate Change Initiative) of 2020 extractions include bare land, shrub land, agriculture, forest, and grassland; representing 30.13%, 21.3%, 18.8%, 15.81%, and 8.04%, respectively (Measho et al., 2023).

Most studies of the land use and land cover changes to socioeconomic processes in Northeast Africa give more focus to the impacts on the residents at a local scale and use either hypothetical models or questionnaire-based assessments on households. Many countries in the region heavily depend on agricultural production dominated by smallholder farming for their livelihoods and economic growth. There are some success stories in the introduction of agricultural growth and crop intensification programs such as in Ethiopia and Rwanda. However, there is a need for holistic regional geospatial data development and agricultural transformation information. Specific geospatial literature reviews on land cover dynamics and controlling factors can help identify the research gaps and give further information.

Literature review

Over 25 years before the new millennium, there were intensive land cover changes in sub-Saharan Africa mainly due to agricultural expansion, which increased by

DOI: 10.4324/9781032672779-5

57% (Brink & Eva, 2009). In North and East Africa, the land cover dynamics happened quickly, and the changes were more of agroecosystem variations with essential impact on food security (Maitima et al., 2009). There is a high exposure rate to the effects of contemporary climate change on plant species and croplands in East Africa (Ayana et al., 2016; Measho et al., 2021). There have been reported impacts of climate variability, particularly fluctuations in annual precipitation and an increase in land surface temperature accompanied by recurrent droughts and population pressures on agricultural production and crop dynamics (Abera et al., 2018; Midgley & Bond, 2015; Pellikka et al., 2013). There has been an increase in surface temperature and fluctuations in rainfall patterns with higher frequency and intensity of drought conditions over Africa (Engelbrecht et al., 2013). Niang et al. (2014) in the Intergovernmental Panel on Climate Change (IPCC) 2014 report confirmed that the impacts of climate change are more visible with occurrences of agricultural drought, water scarcity, and increasing threats to human health and biodiversity problems upon which Africa is the most vulnerable region of the world.

Vegetation and agroecosystem change studies are increasing, and their quantification is improving through time. The assessment of such changes and variations in larger areas may increase human understanding of the terrestrial agroecosystem dynamics and climate–biosphere interactions in the environment (Bao et al., 2014; Zhang et al., 2003). The vegetation distribution and crop planting patterns for small areas are possible to assess and monitor with field-based measurements; however, as the spatial coverage increases, the complexity of the vegetation and crop type changes. Hence, the use of multi-source remote sensing datasets and geospatial datasets becomes sound to decrease expense and save time, reducing uncertainties that come with manual data collection and processing (Gong et al., 2015; Wu et al., 2010).

The MODIS-derived images can be used to find key phonological parameters and perform pixel-based analysis on multiple causes affecting the agroecosystem variations at a regional scale, mainly to assess the effect of climate change in the past few decades (Lunetta et al., 2010; Ndayisaba et al., 2016; Xie et al., 2008). In the Sahel Region of Africa, the use of Global Area Coverage using NDVI (Normalized Difference Vegetation Index) maximum value composites was helpful in forecasting and estimating yield (Herrmann et al., 2005; Xie et al., 2008). There was a track record of annual cropland analysis and phonological changes using Landsat imagery from 1984 to 2016 by targeting three small experiment sites in Africa, and the result showed an increasing slope from barren land to irrigated systems in Egypt and smaller and steady cropland dynamics in smallholder farming fields in Ethiopia (Xu et al., 2018).

The Horn of Africa countries, in particular, have been exposed to a series of droughts and aridity, and the region at large is vulnerable to their effects due to dependency on rain-fed agriculture, and the root causes are mainly associated with climate variability and climate change (Gemeda & Sima, 2015; Masih et al., 2014). The integration of NDVI with crop conditions and cropped areas has a high potential for characterizing cropland changes and food production estimations (Wu et al., 2015).

There is an increasing concern about the changing climate and associated effects on the agroecosystems of the drylands of Northeast Africa, where drought is the main environmental problem (Brink & Eva, 2009; Guzha et al., 2018; Measho et al., 2020). To understand the effects of the recent changes, there is a need to utilize multi-source spatial crop-related datasets and assess the dynamics.

Spatial and temporal assessments related to cropland changes and crop yield estimations along with their driving factors at the regional scale are almost absent, and this problem needs specific attention. The changes in agricultural land due to climatic and soil-related factors can have a huge impact on the productivity and phenology of the croplands and agroecosystem shifts. There is a complex interrelationship among the climate variables as the independent variables and the cropland as the dependent variable, and analyzing the patterns, drivers, and their associated impacts has been a challenging research area. The development of a framework that can generalize agroecosystem dynamics using cropland NDVI, combine crop yield estimations, and assess the correlations of cropland changes with climate factors will be essential for agricultural development and planning in the study region and Africa.

Research goals and questions

Considering the background information and the literature reviews, studies related to agricultural transformations, crop yield changes, and climatic factors behind the changes are highly limited. Understanding the agricultural transformation in Northeast Africa is crucial as it can support the efforts to mitigate the impacts of climate change, increase awareness and modernize the agricultural sector, enhance crop production and maintain food security, develop agricultural transformation strategies, and encourage regional geospatial data cooperation. The research goals of this study include the following four themes.

1) Assess the land use and land cover changes and associated transitions;
2) Identify cropland dynamics that are essential to comprehend the patterns and develop strategies for agricultural transformations;
3) Evaluate agricultural land distributions and determine the changes in harvested areas;
4) Characterize NDVI cropland trends and identify the major climate factor that affects the spatial and temporal cropland NDVI patterns.

The specific research questions based on the goals of the study are as follows.

1) Which major land cover types increased or decreased rapidly with gains and losses in areal extents?
2) What are the total areas and percentages of transitions and conversions of croplands?
3) What are the trends of the harvested area and estimated crop productions of the dominant crops?

4) How do the spatial and temporal cropland NDVI trends affect the greening patterns and vegetation dynamics across the region?
5) Which climate factors affect the cropland NDVI distributions most?

Data and methods

The author downloaded the LULC images from the ESA CCI-LC product. The annual LULC images between 2001 and 2020 were masked to the study area and reclassified according to the IPCC legend and key classes (CCI, E. L. C., 2017). The key classes include agriculture, forest, grassland, wetland, settlement, and other categories based on the ESA CCI user guide version 2.0.

To understand agricultural land distribution and changes, areas that grow more perennial crops are considered. The data to evaluate these changes were areas harvested based on the FAOSTAT annual summary report (https://www.fao.org/faostat/en/#data/QCL) (Food and Agriculture Organization of the United Nations, 2021). More focus was directed on the most common and widely spread crop types to reflect the structure of crops across the region. According to Bański (2021), the trends of areas planted with different crop species were modeled in three categories, namely an upward trend, stabilization, and a downward trend. An upward trend happens when the cultivated area increases by more than 10% relative to the base year (2001), stabilization happens as the values are higher or lower than the original value by less than 10%, and a downward trend applies whereby the cultivated area decreased by more than 10% in comparison to the initial value. Moreover, estimations of crop production were generated from the FAOSTAT annual summary data; country-based crop production estimations were converted and compiled for the study region.

The author accessed and downloaded spatial and temporal vegetation-related indices from MODIS (Moderate Resolution Imaging Spectral-radiometer) using Google Earth Engine as a platform. Terra Surface reflectance data, mainly band 1 and band 2 of MOD09Q1 with medium spatial resolution and eight-day temporal resolution from 2001 to 2020, were prepared as monthly NDVI image composites. The author further processed the time-series MODIS NDVI in ArcGIS 10.6. NDVI was calculated using near-infrared and red bands (NIR – Red/NIR + R). The MOD09Q1-based vegetation index from the red and infrared bands is commonly used for estimating vegetation and/or land surface phenology over wider spatial coverages (Leroux et al., 2017; Mechiche-Alami & Abdi, 2020).

The image values with NDVI < 0.1 were masked to remove the non-vegetated surfaces and barren areas (Kalisa et al., 2019; Measho et al., 2021) before the application of the cropland mask. The cropland images were extracted and masked from the ESA CCI land cover product for the years 2001 and 2020 using the ESA CCI product and user guide version 2.0 (CCI, E. L. C., 2017; Mechiche-Alami & Abdi, 2020). The cropland masks that continued to be cropland from 2001 to 2020 were the target croplands for Northeast Africa. A gradual change in the annual NDVI cropland over time for each pixel was estimated using a least square linear regression model, tested at p-value < 0.05 (Piao et al., 2011).

To evaluate the key drivers of the cropland changes and variations across Northeast Africa, the author performed a Pearson correlation analysis for cropland NDVI as a dependent variable against the independent variables. The independent variables include land surface temperature (LST), precipitation, soil moisture, solar radiation (downward), and actual evapotranspiration (ET); these climatic data between 2001 and 2020 were collected and processed using Climate Engine and complemented to the average annual NDVI of the croplands. A detailed investigation was conducted to analyze the percentage contribution of the predictor variables; it explicitly compares only the independent variables and provides a vital feature analysis among climate and soil-related variables using the Random Forest model.

Results and discussion

LULC changes from 2001 to 2020

The LULC dynamics in the last two decades across Northeast Africa showed some critical changes based on ESA land cover products and further extractions (Table 3.1). Forest, grassland, wetland, and 'other' land cover types decreased during 2001–2020. The 'other' land cover category, according to the ESA CCI user guide, includes shrubland, sparse vegetation, bare area, and water (CCI, E. L. C., 2017). Conversely, forest and settlement or built-up areas increased in the same period. The forest experienced the biggest gain (96,360 km^2) followed by the settlement area, while the 'other' category followed by agriculture represents the cropland that experienced the largest loss.

Spatial transitions of the land cover changes converted from the pixel-based spatial variations are summarized in Annex 1. The areas of forest gains and cropland loss are distributed across the southeastern parts of the United Republic of Tanzania, central and western parts of Kenya, southwestern parts of Ethiopia, and central areas of the Nile Delta in Egypt.

Table 3.1 Land cover changes in Northeast Africa during 2001–2020 using ESA land cover products (km^2)

Land cover types	*2001*	*2010*	*2020*	*2001–2010*	*2010–2020*
Agriculture	1,376,380	1,376,820	1,362,130	440	-14,250
Forest	1,048,840	1,081,950	1,145,200	33,110	96,360
Grassland	584,459	587,810	583,109	3,351	- 1,350
Wetland	112,952	113,974	112,494	1,022	- 458
Settlement	4,470.58	6,936.44	11,126.8	2,465.86	6,656.22
Other	4,123,550	4,083,160	4,036,600	- 40,390	- 86,950
Total		**7,250,660**			

Source: Author's extractions from ESA land cover products.

During the first decade between 2001 and 2010, except in the 'other' category, there was no loss in the land cover extent. In the second decade, however, the land cover types, including cropland, grassland, and wetland, experienced a decline that could be associated with different factors. The main reasons that take paths toward climate change and climate variability are the increase in drought patterns, agroecosystem variations, and changes in agricultural practices in the region (Gemeda & Sima, 2015; Winkler et al., 2017).

Compared to other land cover types, the cropland area declined between 2001 and 2020 based on ESA land cover products. About 58,731 km^2 of agricultural area were converted into forest, grassland, wetland, settlement, and 'other' category. The highest conversions were from agriculture to forest, accounting for 18.3% of all the transitions in the region. The changes could be associated with the agroecosystem practices and afforestation programs in the preexisting agricultural lands.

Some areas experienced increasing agricultural areas in the region. About 43,813 km^2 areas were converted into croplands from forest, grassland, wetland, settlement, and 'other' classes. The largest contributor to the cropland increase was grassland, representing 6.7% of all the transitions and changed land covers. The changes in grassland to cropland indicate the expansion of human activities, population pressure, and the need for more agricultural production (Bashaasha et al., 2013; Pricope et al., 2013).

Agricultural land and crop production

There are more than 118 crop varieties in the region, but the most dominant crops are few. The overall harvested area for all crops was 50.93 Mha (million hectares) in 2001, and this raised to 80.09 Mha by 2020. The most common crops with a wider coverage across the region are wheat, sorghum, barley, maize, potatoes, cassava, and sugarcane. The harvested area of selected crops in 2020 covered 39.7% of all crop varieties in the region. The most commonly planted crop in all countries of Northeast Africa is maize. For example, in 2020, the largest harvest areas of maize were in the United Republic of Tanzania (4.2 Mha), Ethiopia (2.36 Mha), and Kenya (2.14 Mha). Wheat, sorghum, and potatoes are dominant crops in the region, common in all countries except in Djibouti and South Sudan.

Barley is not extensively planted and harvested, and it is mainly found in the northern part of the region such as Egypt, Eritrea, Ethiopia, and Kenya with the highest in Ethiopia. The cultivation of cassava is common in the southern part of the region like Burundi, Kenya, Rwanda, Somalia, South Sudan, Uganda, and the United Republic of Tanzania. Uganda followed by the United Republic of Tanzania had the highest harvested areas of cassava in 2020. Sugarcane harvested area reports were available across many countries of the region except from Eritrea, Djibouti, and South Sudan.

Table 3.2 shows the summary of the temporal variations of harvested areas measured in hectares (ha). Except for the barley cereal crop, the harvested areas of all crop types increased between 2001 and 2020. The trend percentage between 2001 and 2020 is shown in the last column. Wheat, maize (corn), potatoes, cassava

Table 3.2 Temporal variations of harvested areas of croplands in Northeast Africa (units in hectare)

Crop types	*2001*	*2005*	*2010*	*2015*	*2020*	*2001–2020 (%)*
Wheat	2,551,558	3,081,281	3,343,264	3,613,774	3,839,403	20.15
Barley	1,046,289	1,344,549	1,385,141	1,225,392	992,750	−2.63
Sorghum	9,089,497	13,279,037	9,837,009	9,943,130	10,555,682	7.46
Maize	6,409,011	8,962,129	9,997,833	11,843,375	11,908,801	30.02
Potatoes	544,328	703,707	702,527	720,212	905,535	24.91
Cassava (fresh)	1,359,770	1,551,110	1,634,527	2,506,509	3,007,924	37.74
Sugarcane	311,589	350,691	404,674	454,065	550,988	27.75

Source: Author compiled from FAOSTAT country-based harvested area estimations.

(fresh), and sugarcane crops experienced an upward trend, while barley and sorghum crops were under stabilization in Northeast Africa. The harvested area of barley decreased by about 53,539 ha in the last two decades, and this demonstrates the abating preference of the crop variety by farmers and differences in intensive farming practices in the region.

There are some notable differences between the area harvest changes and the production of the croplands in Northeast Africa. The data to assess changes in crop yield were produced based on the FAOSTAT annual summary (Food and Agriculture Organization of the United Nations, 2021). Crop production assessment considered only the most common crop types. The evaluations and comparisons in tonnes (t) per hectare (ha) are essential to contemplate the big difference in the spatial extent that exists among the countries.

The overall yield for all crop species was 165.25 million tonnes in 2001, and it increased to 278.93 Mt in 2020. The crop production estimates for the most common crops showed overall increasing trends in Northeast Africa. The highest rising trends were in the yields of potatoes (37.6%) and maize (34.06%). This largely corresponds with the growth of population in the region, such as in Ethiopia. The average potato production increased from 10.68 t/ha to 14.18 t/ha between 2001 and 2020. For example, the highest producers of potatoes per hectare in 2020 were Egypt, Ethiopia, and Sudan, with 24.62 t/ha, 13.28 t/ha, and 13.24 t/ha, respectively.

Similarly, average maize production for the region increased from 2.58 t/ha to 2.83 t/ha between 2001 and 2020 (Table 3.3). Egypt (7.78 t/ha) followed by Ethiopia (4.24 t/ha) and Uganda (3.33 t/ha) had the highest maize production in 2020. The largest record of harvest area for maize was in the United Republic of Tanzania; however, the maize yield record indicated only 1.6 t/ha in 2020. On the other hand, the average yield of barley increased from 1.22 t/ha to 2.51 t/ha, while the harvested area of barley decreased in the last two decades across the region.

The largest regional producers of barley per hectare in 2020 were Egypt, Kenya, and Ethiopia, with 3.78 t/ha, 3.16 t/ha, and 2.52 t/ha, respectively. Even though Ethiopia was the highest producer of barley in the region, the yield per hectare

Table 3.3 Estimated crop production changes in Northeast Africa (units in tons)

Crop produce	*2001*	*2005*	*2010*	*2015*	*2020*	*2001–2020 (%)*	*2020 t/ha*
Wheat	8,567,976	11,271,899	11,081,139	15,411,113	15,852,539	29.83	4.13
Barley	1,279,958	1,680,648	2,077,723	2,285,881	2,490,752	32.11	2.51
Sorghum	8,496,662	9,528,414	9,166,773	10,617,642	11,411,730	14.64	1.08
Maize	16,563,655	18,767,735	23,838,074	29,453,762	33,673,701	34.06	2.83
Potatoes	5,815,708	9,239,880	10,590,171	10,892,692	12,836,308	37.64	14.18
Cassava	11,800,704	13,008,175	9,561,731	14,217,396	15,287,308	12.87	5.08
Sugarcane	30,264,748	35,938,523	35,637,013	39,059,754	38,627,003	12.14	70.10

Source: Author compiled from FAOSTAT country-based crop production estimations.

was not as high as in Egypt and Kenya. The use of intensive farming practices, including irrigation, fertilizer applications, agricultural mechanization, and climate factors such as precipitation variability, could be the reasons for the differences between harvest area and yield variations.

The average production of sugarcane per hectare generally decreased from 97.13 t/ha to 70.1 t/ha between 2001 and 2020. Likewise, the average yield of fresh cassava declined from 8.68 t/ha to 5.08 t/ha in the region. On the contrary, the average production of wheat and sorghum in the region increased from 3.36 t/ha to 4.13 t/ha and 0.93 t/ha to 1.08 t/ha, respectively, with Egypt and Ethiopia being the highest producers per hectare.

Considering the different aspects of agricultural land use transitions, there are some levels of disagreement between the ESA land cover extractions of the cropland decrease and the FAOSTAT report that indicated an increase in estimated harvested area in the last two decades. Most of the land cover conversions were from agriculture to forest; this is mainly due to the rising of agroecosystem plantations across fields of crops generalized as forest cover from mixed pixels in the analysis and due to the practice of intensive and smallholder farming activities that may overestimate the harvested areas under consideration. Moreover, there are notable differences between the harvested area changes and the production of the croplands in the region.

Characterization of NDVI cropland changes

The cropland covers 18.8% of the total study area. The annual mean NDVI and standard deviation per pixel value were 0.344 and 0.117, respectively. There were gradual fluctuations and growth in the mean NDVI for croplands with some significant changes of increasing trends in the last two decades. The distributions of the NDVI in croplands during 2001–2020 revealed high cropland phonological growth in countries, for example, Egypt and Uganda, which had 0.481 and 0.523 average NDVI only in the cropland areas, respectively. These specific results are strongly related to the conclusion of Kalisa et al. (2019); nonetheless, 0.342 average NDVI for the farmlands in all countries of Northeast Africa can be considered weak because the region includes some arid and semi-arid lands.

Nonsignificant increasing pixels were common in the annual NDVI of most croplands across the region in the last two decades with variable browning and greening patterns. Figure 3.1 shows the annual NDVI trends in croplands; the average rate was $1.06 \times 10^{-2}y^{-1}$ at a *p*-value less than 0.05 between 2001 and 2020. Egypt had the highest greening slope that increased at an average rate of

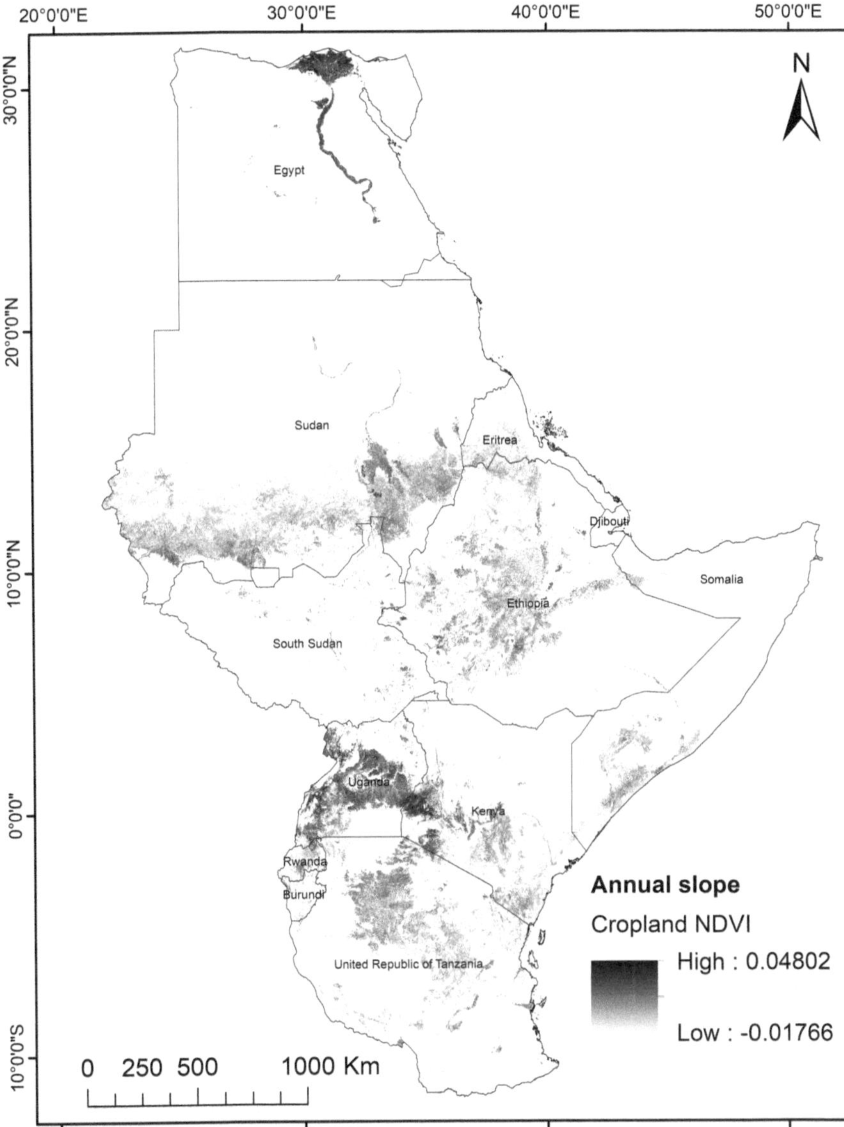

Figure 3.1 Annual cropland NDVI trend in Northeast Africa between 2001 and 2020.

Source: Produced by the author.

$1.58 \times 10^{-2}y^{-1}$, the second and third were Uganda and South Sudan with average rates of $1.45 \times 10^{-2}y^{-1}$ and $1.37 \times 10^{-2}y^{-1}$, respectively.

The largest greening trends were evident along the Nile River delta in Egypt, the southwestern areas of Kenya, and large parts of Uganda. Browning patterns with the lowest NDVI slopes were in Djibouti, Eritrea, and Somalia, where the NDVI in cropland increased at a rate of $4.1 \times 10^{-3}y^{-1}$, $7.7 \times 10^{-3}y^{-1}$, and $7.8 \times 10^{-3}y^{-1}$, respectively. The maximum NDVI variability in cropland was in Ethiopia, and the minimum NDVI variability was in Djibouti during the study period.

The annual NDVI trends in cropland reveal a rising greening pattern in most areas during 2001–2020. The overall increasing NDVI trend or slope in croplands matches with the vegetation dynamics-related studies in the region (Ghebrezgabher et al., 2020; Kalisa et al., 2019; Measho et al., 2021). The largest greening trends in the croplands of Egypt are also in agreement with the results of Xu et al. (2018), which confirmed a rapid and increasing NDVI trend in croplands and changes from barren land to irrigated land in Egypt.

Factors affecting cropland NDVI and its variability

Figure 3.2 demonstrates the correlation plots and pairs for the average cropland NDVI with LST, precipitation, soil moisture, solar radiation, and actual ET during 2001–2020. The mean annual precipitation was 771 mm in the agricultural lands, with a maximum precipitation observed in the southwestern areas of Ethiopia and south-central Uganda. The mean LST value was 34.4 ° C, soil moisture was 32.1 mm,

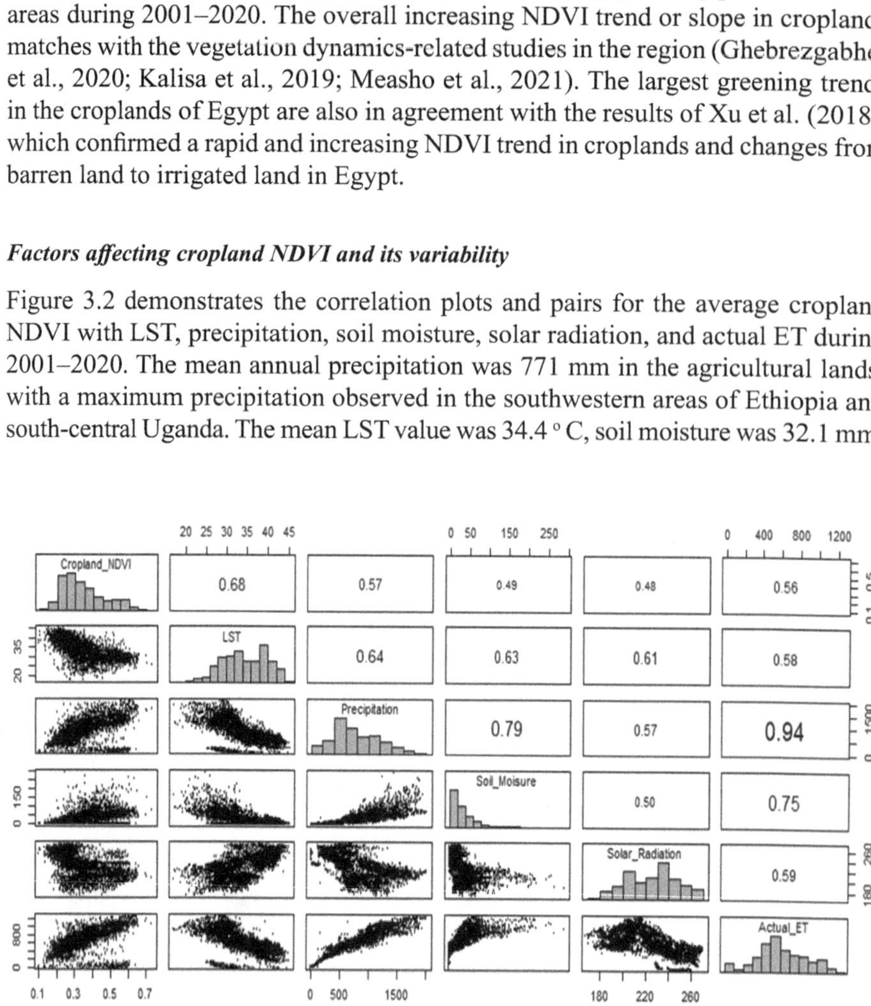

Figure 3.2 Correlation plots among cropland NDVI, LST (C°), precipitation (mm), soil moisture (mm), and solar radiation (w/m²) in Northeast Africa between 2001 and 2020.

Source: Generated by Author.

the downward solar radiation was 225 w/m^2, and actual ET reached 617 mm in the cropland areas of the region.

The LST had the strongest correlation with the cropland NDVI (r = 0.68) followed by precipitation and actual ET in the region (r = 0.57 and r = 0.57, respectively). The three factors (LST, precipitation, and ET) had interrelated contributions to the variability and changes of the agroecosystems in the past two decades. The role of LST was evident in southern Sudan and the northeastern parts of Tanzania. The precipitation was a determinant factor in the greening patterns across Uganda, Rwanda, Eastern Kenya, and central parts of Ethiopia. Moreover, the rising surface temperature in Africa has been playing an important role in cropland dynamics, mainly in the central and northeastern places, where the effects of warmer tropics, such as aridity and drought have been severe (Dai, 2011; Measho et al., 2019; Mpelasoka et al., 2018).

The correlation results (Figure 3.2) correspond with the existing research in the Horn of Africa that carried out specific analyses and results on the farmland NDVI. For example, Ghebrezgabher et al. (2020) indicated that the average farmland NDVI experienced significant changes through time with an r-value of 0.48 during 1998–2013. The same authors found that the mean temperature amplified significantly (r = 0.79) from 1982 to 2013 and reported a strong negative relationship between temperature and farmland NDVI (Ghebrezgabher et al., 2020). The precipitation changes and fluctuations can have a wider influence on the cropland dynamics of the region due to the time-lag effect between the variables. For instance, the ratio of annual NDVI composites to the annual precipitation was higher where the absolute phase-lag effect of croplands was greater than other LULC types in East Africa during 1982–2015 (Zhou et al., 2020). The correlation of soil moisture to the precipitation was strong; however, it was weak with the cropland NDVI (r = 0.49). This showed that the time-lag influence between seasonal rainfall and cropland production is relatively low.

The percentage contribution of the predictor variables and feature importance analysis revealed that LST, solar radiation, precipitation, soil moisture, and ET as independent variables contributed 29.15%, 26.95%, 17.1%, 14.19%, and 12.7%, respectively. The highest importance level was 5.3 for LST with R^2 of 0.85. Generally, all the selected climatic and soil-related variables considerably contributed to the dynamics of the cropland NDVI. However, considering the random forest estimation as a more reliable method, the LST has been the main controlling factor that significantly affected the cropland patterns and agroecosystem changes across Northeast Africa.

Conclusions

The main LULC changes in Northeast Africa based on the satellite image extractions show that forest and settlement areas expanded, while the ‘other’ category and cropland areas experienced losses. The forest experienced the biggest gain followed by the settlement area, while the ‘other’ category followed by agriculture experienced the largest loss. Most agricultural areas are converted into forests, grasslands, wetland, settlements, and the 'other' category. The highest conversions were from agriculture to forest, and they represent 18.3% of all the transitions. On the other hand, the largest contributor to the cropland increase was the grassland area, which accounts for 6.7% of all the transitions in the region. The harvested area for all crops increased from 50.93 Mha to 80.09 Mha, with a relative change of +57.26% between 2001 and 2020. Similarly, the overall crop production for all crops increased from 165.25 Mt in 2001 to 278.93 Mt in 2020, with a relative change of +68.79. The estimates for the most common crops indicated overall rising trends between 2001 and 2020.

The characterization of croplands using NDVI showed an average of 0.344 cropland NDVI for the whole region, with the highest cropland NDVI in Uganda, followed by Egypt. There were nonsignificant increasing NDVI trends for most croplands in Northeast Africa, and the average rate was 1.06×10^{-2} per year at p-value < 0.05 in the last two decades. The highest greening trends were along the delta of the Nile River in Egypt, central areas of Uganda, and south-western parts of Kenya; on the other hand, browning trends were common in Djibouti, Eritrea, and Somalia. The main driving factor for the cropland NDVI and agroecosystem changes was LST; however, human activities may also contribute to the agroecosystem changes.

In general, there is spatial and temporal variability in the agricultural land use changes and crop production, and the main controlling factor is mainly climatic, which is the rising of LST as one of the reflections of climate variability. The challenges of agricultural transformations are climate change as reflected by the driving factor, low crop production in comparison with the population growth, and lack of regional institutions for agricultural development and food security. The main strategies for agricultural transformation in the region may include the adoption of geospatial technology, development of agricultural infrastructure, policy reformation, engagement of the private sector, and capacity building. In the future, crop-specific models and crop yield predictions that can deal with the changing climate in the region are highly required, and the development of regional geospatial datasets of high spatial resolution can be more helpful for the detailed assessment of agricultural land use transitions and patterns.

References

Abera, T. A., Heiskanen, J., Pellikka, P., & Maeda, E. E. (2018). Rainfall–vegetation interaction regulates temperature anomalies during extreme dry events in the Horn of Africa. *Global and Planetary Change*, *167*, 35–45. https://doi.org/10.1016/j.gloplacha.2018.05.002

Ayana, E. K., Ceccato, P., Fisher, J. R. B., & DeFries, R. (2016). Examining the relationship between environmental factors and conflict in pastoralist areas of East Africa. *Science of the Total Environment*, *557–558*, 601–611. https://doi.org/10.1016/j.scitotenv.2016.03.102

Bański, J., & Kamińska, W. (2021). Trends for agricultural land-use in the CEECs following the collapse of the Eastern Bloc. *Land Use Policy*, 105794. https://doi.org/10.1016/j.landusepol.2021.105794

Bao, G., Qin, Z., Bao, Y., Zhou, Y., Li, W., & Sanjjav, A. (2014). NDVI-based long-term vegetation dynamics and its response to climatic change in the Mongolian plateau. *Remote Sensing*, *6*, 8337–8358. https://doi.org/10.3390/rs6098337

Bashaasha, B., Thomas, T., Waithaka, M., & Kyotalimye, M. (2013). *East African agriculture and climate change: A comprehensive analysis – Uganda*. International Food Policy Research Institute, Washington, DC.

Brink, A. B., & Eva, H. D. (2009). Monitoring 25 years of land cover change dynamics in Africa: A sample based remote sensing approach. *Applied Geography*, *29*, 501–512. https://doi.org/10.1016/j.apgeog.2008.10.004

CCI, E. L. C. (2017). *Product user guide version 2.0*. UCL-Geomatics: London, UK.

Dai, A. (2011). Drought under global warming: A review. *Wiley Interdisciplinary Reviews: Climate Change*, *2*, 45–65. https://doi.org/10.1002/wcc.81

Engelbrecht, C. J., Engelbrecht, F. A., & Dyson, L. L. (2013). High-resolution model-projected changes in mid-tropospheric closed-lows and extreme rainfall events over Southern Africa. *International Journal of Climatology*, *33*, 173–187. https://doi.org/10.1002/joc.3420

Food and Agriculture Organization of the United Nations. (2021). *FAOSTAT* [WWW Document]. Retrieved November 20, 2024, from https://www.fao.org/faostat/en/#data/QCL

Gemeda, D. O., & Sima, A. D. (2015). The impacts of climate change on African continent and the way forward. *7*, 256–262. https://doi.org/10.5897/JENE2015.

Ghebrezgabher, M. G., Yang, T., Yang, X., & Eyassu Sereke, T. (2020). Assessment of NDVI variations in responses to climate change in the Horn of Africa. *Egyptian Journal of Remote Sensing and Space Sciences*. https://doi.org/10.1016/j.ejrs.2020.08.003

Gong, Z., Kawamura, K., Ishikawa, N., Goto, M., Wulan, T., Alateng, D., Yin, T., & Ito, Y. (2015). MODIS normalized difference vegetation index (NDVI) and vegetation phenology dynamics in the Inner Mongolia grassland. *Solid Earth*, *6*, 1185–1194. https://doi.org/10.5194/se-6-1185-2015

Guzha, A. C., Rufino, M. C., Okoth, S., Jacobs, S., & Nóbrega, R. L. B. (2018). Impacts of land use and land cover change on surface runoff, discharge and low flows: Evidence from East Africa. *Journal of Hydrology: Regional Studies*, *15*, 49–67. https://doi.org/10.1016/j.ejrh.2017.11.005

Herrmann, S. M., Anyamba, A., & Tucker, C. J. (2005). Recent trends in vegetation dynamics in the African Sahel and their relationship to climate. *Global Environmental Change*, *15*, 394–404. https://doi.org/10.1016/j.gloenvcha.2005.08.004

Kalisa, W., Igbawua, T., Henchiri, M., Ali, S., Zhang, S., Bai, Y., & Zhang, J. (2019). Assessment of climate impact on vegetation dynamics over East Africa from 1982 to 2015. *Scientific Reports*, *9*, 1–20. https://doi.org/10.1038/s41598-019-53150-0

Leroux, L., Bégué, A., Lo Seen, D., Jolivot, A., & Kayitakire, F. (2017). Driving forces of recent vegetation changes in the Sahel: Lessons learned from regional and local level analyses. *Remote Sensing of Environment*, *191*, 38–54. https://doi.org/10.1016/j.rse.2017.01.014

Lunetta, R. S., Shao, Y., Ediriwickrema, J., & Lyon, J. G. (2010). Monitoring agricultural cropping patterns across the Laurentian Great Lakes Basin using MODIS-NDVI data. *International Journal of Applied Earth Observation and Geoinformation*, *12*, 81–88. https://doi.org/10.1016/j.jag.2009.11.005

Maitima, J. M., Mugatha, S. M., Reid, R. S., Gachimbi, L. N., Majule, A., Lyaruu, H., Pomery, D., Mathai, S., & Mugisha, S. (2009). The linkages between land use change, land degradation and biodiversity across East Africa. *African Journal of Environmental Science and Technology*, *3*, 310–325. https://doi.org/10.5897/AJEST08.173

Masih, I., Maskey, S., & Trambauer, P. (2014). A review of droughts on the African continent : A geospatial and long-term perspective. 3635–3649. https://doi.org/10.5194/hess-18-3635-2014

Measho, S., Chen, B., Pellikka, P., Guo, L., Zhang, H., Cai, D., Sun, S., Kayiranga, A., Sun, X., & Ge, M. (2021). Assessment of vegetation dynamics and ecosystem resilience in the context of climate change and drought in the horn of Africa. *Remote Sensing*, *13*. https://doi.org/10.3390/rs13091668

Measho, S., Chen, B., Pellikka, P., Trisurat, Y., Guo, L., Sun, S., & Zhang, H. (2020). Land use/land cover changes and associated impacts on water yield availability and variations in the Mereb-Gash river basin in the Horn of Africa. *Journal of Geophysical Research: Biogeosciences*, *125*, 1–16. https://doi.org/10.1029/2020JG005632

Measho, S., Chen, B., Trisurat, Y., Pellikka, P., Guo, L., Arunyawat, S., Tuankrua, V., Ogbazghi, W., & Yemane, T. (2019). Spatio-temporal analysis of vegetation dynamics as a response to climate variability and drought patterns in the semiarid region, Eritrea. *Remote Sensing*, *11*, 724. https://doi.org/10.3390/rs11060724

Measho, S., Li, F., Chen, G., & Hirwa, H. (2023). Characterizing cropland patterns across North-East Africa using time series vegetation indices. *Journal of Geophysical Research: Biogeosciences*, *128*, 1–18. https://doi.org/10.1029/2022JG007075

Mechiche-Alami, A., & Abdi, A. M. (2020). Agricultural productivity in relation to climate and cropland management in West Africa. *Scientific Reports*, *10*, 3393. https://doi.org/10.1038/s41598-020-59943-y

Midgley, G. F., & Bond, W. J. (2015). Future of African terrestrial biodiversity and ecosystems under anthropogenic climate change. *Nature Climate Change*, *5*, 823–829. https://doi.org/10.1038/nclimate2753

Mpelasoka, F., Awange, J. L., & Zerihun, A. (2018). Influence of coupled ocean-atmosphere phenomena on the Greater Horn of Africa droughts and their implications. *Science of Total Environment*, *610–611*, 691–702. https://doi.org/10.1016/j.scitotenv.2017.08.109

Ndayisaba, F., Guo, H., Bao, A., Guo, H., Karamage, F., & Kayiranga, A. (2016). Understanding the spatial temporal vegetation dynamics in Rwanda. *Remote Sensing*, *8*, 1–17. https://doi.org/10.3390/rs8020129

Niang, I., Ruppel, O. C., Abdrabo, M. A., Essel, A., Lennard, C., Padgham, J., & Urquhart, P. (2014). *Climate change 2014–impacts, adaptation and vulnerability: Regional aspects* (pp. 1199–1265). Cambridge Univ. Press.

Pellikka, P. K. E., Clark, B. J. F., Gosa, A. G., Himberg, N., Hurskainen, P., Maeda, E., Mwang'ombe, J., Omoro, L. M. A., & Siljander, M. (2013). Agricultural expansion and its consequences in the Taita Hills, Kenya. *Developments in Earth Surface Processes*, *16*, 165–179. https://doi.org/10.1016/B978-0-444-59559-1.00013-X

Piao, S., Wang, X., Ciais, P., Zhu, B., Wang, T., & Liu, J. (2011). Changes in satellite-derived vegetation growth trend in temperate and boreal Eurasia from 1982 to 2006. *Global Change Biology*, *17*, 3228–3239. https://doi.org/10.1111/j.1365-2486.2011.02419.x

Pricope, N. G., Husak, G., Lopez-Carr, D., Funk, C., & Michaelsen, J. (2013). The climate-population nexus in the East African Horn: Emerging degradation trends in rangeland and pastoral livelihood zones. *Global Environmental Change*, *23*, 1525–1541. https://doi.org/10.1016/j.gloenvcha.2013.10.002

Winkler, K., Gessner, U., & Hochschild, V. (2017). Identifying droughts affecting agriculture in Africa based on remote sensing time series between 2000–2016: Rainfall anomalies and vegetation condition in the context of ENSO. *Remote Sensing*, *9*. https://doi.org/10.3390/rs9080831

Wu, B., Gommes, R., Zhang, M., Zeng, H., Yan, N., Zou, W., Zheng, Y., Zhang, N., Chang, S., Xing, Q., & van Heijden, A. (2015). Global crop monitoring: A satellite-based hierarchical approach. *Remote Sensing*, *7*, 3907–3933. https://doi.org/10.3390/rs70403907

Wu, W., Yang, P., Tang, H., Zhou, Q., Chen, Z., & Shibasaki, R. (2010). Characterizing spatial patterns of phenology in cropland of China based on remotely sensed data. *Agricultural Sciences in China*, *9*, 101–112. https://doi.org/10.1016/S1671-2927(09)60073-0

Xie, Y., Sha, Z., & Yu, M. (2008). Remote sensing imagery in vegetation mapping: A review. *Journal of Plant Ecology*, *1*, 9–23. https://doi.org/10.1093/jpe/rtm005

Xu, Y., Yu, L., Zhao, F. R., Cai, X., Zhao, J., Lu, H., & Gong, P. (2018). Tracking annual cropland changes from 1984 to 2016 using time-series Landsat images with a change-detection and post-classification approach: Experiments from three sites in Africa. *Remote Sensing of Environment*, *218*, 13–31. https://doi.org/10.1016/j.rse.2018.09.008

Zhang, X., Friedl, M. A., Schaaf, C. B., Strahler, A. H., Hodges, J. C. F., Gao, F., Reed, B. C., & Huete, A. (2003). Monitoring vegetation phenology using MODIS. *Remote Sensing Environment*, *84*, 471–475. https://doi.org/10.1016/S0034-4257(02)00135-9

Zhou, J., Jia, L., Menenti, M., van Hoek, M., Lu, J., Zheng, C., Wu, H., & Yuan, X. (2020). Characterizing vegetation response to rainfall at multiple temporal scales in the Sahel-Sudano-Guinean region using transfer function analysis. *Remote Sensing of Environment*, *252*, 112108. https://doi.org/10.1016/j.rse.2020.112108

Appendix

Annex 1. Major land cover transitions and changes in Northeast Africa from 2001 to 2020.

Class values	*Class_name*	*Class_from*	*Class_to*	*Pixel counts*	*Change (%)*
0	Agriculture->Forest	Agriculture	Forest	480,671	18.29
1	Agriculture->Grassland	Agriculture	Grassland	35,954	1.37
2	Agriculture->Wetland	Agriculture	Wetland	1,327	0.05
3	Agriculture->Settlement	Agriculture	Settlement	39,448	1.50
4	Agriculture->Other	Agriculture	Other	76,724	2.92
5	Forest->Agriculture	Forest	Agriculture	128,877	4.90
6	Forest->Grassland	Forest	Grassland	12,010	0.46
7	Forest->Wetland	Forest	Wetland	29,055	1.11
8	Forest->Settlement	Forest	Settlement	5,500	0.21
9	Forest->Other	Forest	Other	118,810	4.52
10	Grassland->Agriculture	Grassland	Agriculture	175,700	6.69
11	Grassland->Forest	Grassland	Forest	113,182	4.31
12	Grassland->Wetland	Grassland	Wetland	2	0.00
13	Grassland->Settlement	Grassland	Settlement	10,079	0.38
14	Grassland->Other	Grassland	Other	107,717	4.10
15	Wetland->Agriculture	Wetland	Agriculture	697	0.03
16	Wetland->Forest	Wetland	Forest	38,316	1.46
17	Wetland->Grassland	Wetland	Grassland	13	0.00
18	Wetland->Settlement	Wetland	Settlement	836	0.03
19	Wetland->Other	Wetland	Other	2,979	0.11
25	Other->Agriculture	Other	Agriculture	167,772	6.38
26	Other->Forest	Other	Forest	719,206	27.37
27	Other->Grassland	Other	Grassland	341,832	13.01
28	Other->Wetland	Other	Wetland	7,557	0.29
29	Other->Settlement	Other	Settlement	13,819	0.53
30	No Change	Same	Same	75,194,851	

N.B. The percentage change considered only the classes that changed relative to the total changed area, excluding no-change areas.

4 Land fragmentation and agricultural productivity in Tanzania

Francisco M. P. Mugizi

Introduction

In sub-Saharan Africa (SSA), most people derive their livelihood from agriculture. As a result, land is one of the most essential natural resources for livelihood, especially for rural households in the region. However, even though SSA has abundant land compared to other regions, evidence shows that arable land per person is declining (Jayne & Muyanga, 2012; Otsuka & Place, 2015). Moreover, although the agricultural sector is considered the backbone of most of the economies in the region, its performance is not impressive (Otsuka & Larson, 2016; Senbet & Simbanegavi, 2017). Consequently, most SSA countries have remained net food importers (Mugizi & Matsumoto, 2020). Moreover, both the region's food insecurity, at 27%, and undernourishment, at 23%, are still high (FAO et al., 2023). Moreover, rural poverty remains a serious problem in most parts of the region. Policymakers and scholars have continued to stress the need to increase agricultural productivity. Indeed, increasing agricultural productivity is considered extremely important, at least for two main reasons. First, boosting agricultural productivity is important in improving food security. Second, increasing agricultural productivity is crucial in the fight against poverty.

Like most other SSA countries, agriculture employs about two-thirds of Tanzania's population. However, its contribution to the Gross Domestic Product is only 27% (World Bank, 2017, 2020). Moreover, food insecurity, undernourishment, and malnutrition remain serious problems in Tanzania (Mkupete et al., 2023; FAO et al., 2023). All these suggest that agricultural productivity in the country is still low. Several reasons may be attributed to low agricultural productivity in the country. Land fragmentation which is prevalent in SSA and has been claimed to be one of the obstacles to agricultural development in most SSA countries (Alemu et al., 2017; Ali et al., 2019; Veljanoska, 2018; Cholo et al., 2019; Ntihinyurwa et al., 2019; Knippenberg et al., 2020) may also hinder agricultural productivity in Tanzania.

This chapter examines the correlation between land fragmentation and agricultural productivity in Tanzania. It uses a more recent, large, and unique nationally representative data set from the National Sample Census of Agriculture (NSCA) collected in 2020 from all regions of Tanzania. The data has various

DOI: 10.4324/9781032672779-6

agricultural-related information collected from smallholders across the country. Amongst others, it has detailed information on harvests both from annual and permanent crops. Unlike previous studies, this chapter's unique aspect and contribution is that it examines whether land fragmentation has a differential effect on the productivity of permanent and annual crops. The unique data used makes this separation possible and provides the most reliable data on yields of permanent and annual crops. Unlike most surveys that combine annual and permanent crops, the NSCA data is well-suited for this analysis because it has different files for the two categories of crops. The main results show a robust negative relationship between land fragmentation and agricultural productivity. More precisely, land fragmentation is negatively correlated with the productivity of both permanent and seasonal crops.

Conceptually, land fragmentation, a practice by which the household cultivates more than one dispersed parcel of land, can have two opposing effects on agricultural productivity. It can reduce agricultural productivity due to the following reasons. First, land fragmentation is considered a constraint and a source of inefficiencies. It is blamed for its inefficiency in resource allocation and production costs, such as extra travel time needed to travel from one parcel to another. It is argued that distance from homestead to dispersed parcels wastes labour time by travelling between homesteads and parcels, hence making it difficult and costly to transport inputs from homestead to parcels. This, in turn, increases the production costs in terms of travel time, loss of working hours, and monitoring (Hung et al., 2007; Tan et al., 2008; Kawasaki, 2010; Deininger et al., 2016; Ali et al., 2019; Zheng et al., 2023). It also hinders the transport of agricultural implements and mechanisation (Demetriou et al., 2013; Alemu et al., 2017; Mwesigye & Barungi, 2021). Second, fragmentation may make protecting the dispersed parcels difficult, leading to land invasion and conflicts (Demetriou et al., 2013). Land conflicts tend to be associated with lower yields (Mugizi & Matsumoto, 2021a). Third, land fragmentation leads to inadequate monitoring, making supervision and crop tending difficult (Alemu et al., 2017). Indeed, empirical studies have documented that fragmentation increases production costs and has a negative impact on efficiency (see, e.g., Manjunatha et al., 2013; Deininger et al., 2016 in India; Tan et al., 2008 in Jiangxi Province of China, Niroula & Thapa, 2007 on Nepal).

On the other hand, land fragmentation is believed to have potential benefits (Shaw, 1963; Blarel et al., 1992; Veljanoska, 2018). First, dispersed parcels are likely to have varying soils, ecological characteristics, and biophysical conditions. This may allow households to grow a greater variety of crop types. Land fragmentation may also allow farmers to produce crops with different ripening times and production seasons. This is beneficial because it may allow farmers to concentrate their labour on different plots at different times (Alemu et al., 2017). Moreover, having dispersed parcels in different locations can reduce risks like droughts, floods, pests, and fire. In a nutshell, under this school of thought, the potential benefit of land fragmentation is risk reduction (Goland, 1993; Nguyen et al., 1996; Kawasaki, 2010; Alemu et al., 2017; Veljanoska, 2018).

The two theoretical arguments highlighted here suggest that fragmentation may be costly and beneficial. Consequently, its net effect on agricultural productivity is ambiguous a prior. Several studies, however, argue that the negative effects of land fragmentation on crop productivity outweigh its potential benefits (see, e.g., Kawasaki, 2010). Notwithstanding, empirical studies on land fragmentation and agricultural productivity document mixed results. Elsewhere, a bunch of studies in regions other than SSA document a negative association between land fragmentation and productivity (see, e.g., Nguyen et al., 1996; Wan & Cheng, 2001; Tan et al., 2010 – China; Jha et al., 2005; Deininger et al., 2016 – India; Hung et al., 2007 – Vietnam; Rahman & Rahman, 2009 – Bangladesh).

In SSA, Ali et al. (2019) document the negative effects of fragmentation as measured by travel cost on crop yield, the intensity of labour use, and technical efficiency in Rwanda. In Uganda, land fragmentation was found to be detrimental for households not exposed to rainfall anomalies; however, it decreased crop yield loss when households experienced rainfall irregularities (Veljanoska, 2018). Using cross-sectional data collected from 240 respondents during the 2015/2016 production seasons in north-western Ethiopia, Alemu et al. (2017) found a negative relationship between land productivity and land fragmentation measured by both the average distance of parcels from homestead to the cultivated land and the number of parcels. Similar results are documented by Oyebanjo et al. (2023) in Nigeria. In Ghana, Blarel et al. (1992) found that distance to parcel positively impacted yields, but the Sympson index was insignificant. A positive association between yield and land fragmentation is also documented in Ethiopia by Paul and Gĩthĩnji (2017).[1] On the other hand, Blarel et al. (1992) found no significant relationship between fragmentation and crop yield in Rwanda. In Tanzania, the only study known to us is by Rao (2019), which documents a positive relationship between fragmentation and technical efficiency.

Even though studies have examined the effect of land fragmentation on agricultural productivity, none of the existing studies separately look at permanent and seasonal crops. The proposed study intends to fill this void in the literature. As explained in the subsequent paragraph, fragmentation could impact permanent and seasonal crops differently. Paul and Gĩthĩnji (2017) argue that fragmentation could have different impacts depending on whether the household mainly grows cash crops or subsistence crops. Therefore, exploring whether land fragmentation has differential impacts on permanent and seasonal crops is an area of policy relevance. Another contribution of this study is that most existing studies in SSA have examined land fragmentation and productivity in relatively land-constrained countries. Tanzania is relatively land-abundant compared to countries like Rwanda, Ethiopia, Kenya, Uganda, and many other SSA countries. Hence, a study on relatively land-abundant African settings may provide more insights. Moreover, even though studies have examined the impact of land fragmentation on some SSA countries, we know of no study in Tanzania that links fragmentation and agricultural productivity. The only study on fragmentation known to us is Rao (2019). However, Rao's (2019) study examines the effect of land fragmentation on production risk and technical efficiency. Lastly, unlike the existing studies in SSA, which use small sample sizes and look at some villages or regions within the country, the present study uses a large nationally representative data set. Therefore, its results may be more informative and may shed more light.

Tanzania is an exciting and ideal case study. With an area of about 942,600 km^2 and a land surface of 888,200 km^2, of which 487,100 km^2 are arable land (The United Republic of Tanzania [URT], 1997), Tanzania used to be one of the countries with abundant land per person. However, due to population growth, arable land per person has continued to shrink in many regions. In addition, over 80% of the country's population that live in rural areas depend exclusively on land for their livelihoods. Due to population growth and the resultant subdivision of land across generations, farm sizes per smallholder farmer are declining, and most of these farms are now fragmented. Therefore, a study on land fragmentation and agricultural productivity is of policy relevance.

The chapter proceeds as follows. The following section briefly explains why land fragmentation may impact permanent and annual crops differently. The third section describes the data. The fourth section lays out the estimation strategy. The fifth section presents and discusses the results. The last section concludes and gives policy recommendations.

Why could land fragmentation affect permanent and seasonal crops differently?

The first reason centres on monitoring costs as to why land fragmentation could have different impacts on the productivity of permanent and annual crops centres on monitoring costs. Unlike seasonal crops, which take only a few months before harvesting and have fewer monitoring costs, permanent crops take several years before harvesting. Therefore, monitoring permanent crops is more costly. Since land fragmentation tends to increase monitoring costs, it may be more detrimental to permanent crops than seasonal crops. Relatedly, the more fragmented the land is, the higher the likelihood of crop theft. This may disincentivise farmers to produce valuable permanent crops, which are, in most cases, cash crops. This is likely to reduce crop yield.

Third, compared to annual crops, permanent crops require significant investments whose returns come after a long time (Mwesigye & Barungi, 2021). Because fragmentation hinders or makes it challenging to transport investments such as manure, farmers will be discouraged from producing permanent crops on fragmented parcels; if they do, they are likely to get low yields because of the higher likelihood of underinvesting in those parcels. In contrast, land fragmentation increases crop diversification, especially for annual crops, which do not need significant investment (Mwesigye & Barungi, 2021). For this reason, land fragmentation is likely to increase the production of annual crops.

A study by Mwesigye and Barungi (2021) on land fragmentation and crop choice finds that land fragmentation reduces the probability of growing permanent crops and trees and is associated with fewer hectares allocated for the same crops in Uganda. In addition, they find that a one-minute increase in the distance to a parcel from the homestead reduces the likelihood of growing permanent commercial crops by 0.2%. Conversely, the distance to the parcel increased the likelihood of growing annual crops. Overall, their study suggests that land fragmentation hurts the production of permanent crops more than it does for annual crops.

Data and descriptive statistics

Data

The study uses unique and large nationally representative data from the National Sample Census of Agriculture (NSCA) of 2019/2020 collected by the National Bureau of Statistics (NBS). Its sample design was based on the National Master Sample developed by NBS and the Office of the Chief Government Statistician to serve as a national framework for conducting household surveys in Tanzania. The 2019/2020 NSCA adopted a two-stage design with census enumeration areas (EAs) as Primary Sampling Units (PSUs) and households as second-stage units. The stage one sampling frame comprises a selection of urban and rural EAs from the 2012 Population and Housing Census frame. The EAs were sorted by region and district before employing a Probability Proportionate to Size (PPS). The second stage was the selection of agricultural farming households from the selected EAs for data collection.

A total of 2,820 PSUs were selected from the 2012 Population and Housing Census frame. Of the 2,820 PSUs, 260 were from urban areas, and 2,560 were from rural areas. The number of households differed from one PSU to another; it ranged from 5 to 30 households. The data set has 30,650 households from 31 regions of Tanzania. For each region, a sample of PSUs was selected from the total number of PSUs in the region, classified into rural and urban PSUs, with more weight given to the rural PSUs because of their intensity in agriculture practices. The probability of selecting a household depended on the total number of households in the PSU. The sample was higher for rural EAs than urban EAs.

The 2019/2020 NSCA data contains an array of information, including household demographics and landownership. It also contains information on permanent and annual crops, farm input use, and poverty indicators at the household level, such as asset ownership and access to infrastructure and other services. Moreover, it has community-level information, such as whether the community has farmer field schools, farmers' associations, and irrigation schemes.

Although the data set has a total of 30,650 households, some of them did not grow any crops; these are dropped from the analysis. In addition, some households grew annual crops but reported no permanent crops. Therefore, the analytical sample for the former is larger than the latter's. Moreover, there is missing information on village identification for a few households. Since we control for some relevant village information and cluster the standard errors at the village level (Abadie et al., 2023), households whose villages could not be identified are excluded. This cleaning process leaves us with 25,070 households.

The outcome variables of interest are measures of agricultural productivity. For each crop type – permanent and seasonal crops – the study uses three alternative productivity measures: the value of harvests (kg/acre) for both permanent and seasonal crops, permanent crop income (Tanzanian Shillings [TZS]/acre), seasonal crop income (TZS/acre), as well as permanent and seasonal crop income per capita. The value of crops is measured as a product of crop yield (kg) and the crop price per kg. We divide this product by farm size (area cultivated) to get the value of crops produced per acre. Crop income is measured by taking the value of the respective crops produced less the cost of paid inputs such as land preparation costs, planting costs, irrigation costs, weeding costs, harvesting costs, seed costs, fertiliser costs, herbicides costs, fungicides costs, insecticides costs, and other costs. Using this variable, we calculate crop income per acre and crop income per capita, variables of interest.

Our explanatory variable of interest is land fragmentation. We measure land fragmentation by using two variables – distance from parcel to homestead and Simpson index. Regarding the former, since the analysis is at the household level, we use the average distance from all the parcels cultivated by the household to the homestead. Concerning the Simpson Index, we follow extant literature (see, e.g., Blarel et al., 1992; Hung et al., 2007; Tan et al., 2008; Kawasaki, 2010; Deininger et al., 2016; Knippenberg et al., 2020; Wang et al., 2020; Mwesigye & Barungi, 2021) and construct the index as $SI = 1 - \frac{\sum_{i=1}^{n} A_i^2}{A^2}$. SI is the Simpson Index, i denotes individual plot, A_i is the area of the ith plot, $A = \sum_{i=1}^{n} A_i$ is the total sum of individual plot areas, and $A = \sum_{i} A_i$ is the total landholding in acres. A value of zero ($SI = 0$) indicates complete land consolidation (one parcel only). In contrast, the value of one ($SI = 1$) is approached by holdings of numerous parcels of equal size.

Descriptive statistics

The following is the descriptive statistics[2] of smallholder farmers cultivating perennial and annual crops. 22% and 21% of households cultivating perennial and annual crops, respectively, are female-headed. On average, the household head's age is 50 years. Regarding educational attainment, household heads' average years of schooling is about 5.8 years. On average, each household has five members. Concerning human capital, on average, each household has about three working-age members (between 15 years and 65 years) whose years of schooling are, on average, six years. Regarding landholding, a key asset for smallholders, each household has 5.7 acres of land on average. Only 34% of the households who cultivated annual crops consider themselves to have sufficient land for agriculture; for those who cultivated perennial crops, 43% have sufficient land for agriculture. On average, only 27% of our sample households used organic fertiliser, while only 12% practised irrigation systems. Regarding agricultural productivity, on average, the value of harvests of all seasonal crops per acre (g/acre) was TZS 563,307 (USD241); for perennial crops, the value was TZS 4,278,995 (USD1,832).[3] On the other hand, crop income per acre was TZS 484,531 (USD207) and TZS 4,171,586 (USD1,786) for annual crops and perennial crops, respectively. Crop income per capita was TZS 284,682 (USD122) and TZS 1,075,391 (USD460) for annual and permanent crops, respectively. One noteworthy observation is a huge variation in productivity between annual and permanent crops.

Estimation strategy

Land fragmentation and agricultural productivity

In examining the impact of land fragmentation on agricultural productivity, a major concern is that land fragmentation is likely to be endogenous. The endogeneity may arise from omitted variable problems and reverse causality. Regarding the former, some unobserved individual characteristics, such as management ability and entrepreneurial spirit, may influence farmer's choice of fragmentation as well as agricultural productivity (Veljanoska, 2018). Although we control for several relevant household-level characteristics, in the absence of panel data, we cannot control such unobserved individual time-invariant characteristics. Regarding reverse causality, farmers may choose their level of land fragmentation to cope with production risk (Veljanoska, 2018). Relatedly, fragmentation is in no way exogenous. Indeed, farmers can choose the number, size, and location of parcels subject to the constraints imposed by technology and factor markets. Therefore, due to the potential endogeneity of land fragmentation, our estimates offer insights into correlation and should not be interpreted as a causal relationship. To this end, we estimate the following model to relate land fragmentation and agricultural productivity.

$$lnY_{ij} = \alpha + \beta F_{ij} + \delta H_{ij} + \varphi V_j + \varepsilon_{ij} \tag{4.1}$$

where i and j stand for household and village, respectively. For each crop type – permanent and seasonal crops – the outcome variable, Y_{ij} takes three variants of measures of agricultural productivity: the value of crops produced per acre (TZS/acre) – measured as a product of crop yield (kg) and the crop price per kg, crop income per acre, and crop income per capita. F_{ij} is a vector of land fragmentation indicators at the household level: the Simpson fragmentation index and the average distance from the homestead to all parcels cultivated by the household.

The distance variable used is not continuous; it is proxied by six dummies – less than 300 m, between 300 m and 500 m, between 500 m and 1 km, between 1 km and 2 km, between 2 km and 3 km, over 3 km, we use less than 300 m as a base category. H_{ij} is a vector of household-level characteristics, including female headship, household head's age, household head's years of schooling, household members of working age (labour), and average years of schooling of adults. In H_{ij}, we also include other relevant controls such as asset index and landholding – the two proxies for household wealth. Other controls are the amount of inorganic fertiliser used per acre, the amount of organic fertiliser used per acre, whether improved seeds were used, and whether the household accessed agricultural extension advice for crop production during the 2019/2020 agricultural year. V_j is a set of village characteristics, including whether there were farmer field schools during the 2019/2020 agricultural year, whether there were farmers' associations, the presence of an irrigation scheme, and the distance to district headquarters from the village centre. ε_{ij} is an error term. Since the error term may be heteroskedastic and correlated within the village, we use robust standard errors clustered at the village level.

Pathways analysis

As highlighted earlier, land fragmentation may make it difficult and costly to transport heavy inputs such as manure and other inputs from homesteads to parcels. Thus, farmers are likely to be discouraged from using such investment materials on fragmented land. We, therefore, also examine how land fragmentation relates to agricultural intensification. To do so, we estimate the following model.

$$AI_{ij} = \alpha + \beta F_{ij} + \delta H_{ij} + \varphi V_j + \varepsilon_{ij} \tag{4.2}$$

where i and j stand for household and village, respectively. The outcome variable, AI_{ij} denotes three variants of intensification: the log of the quantity of chemical fertiliser used per acre of land cultivated, the log of the quantity of manure used per acre of land cultivated, and a dummy variable equal to 1 if improved seeds were used and 0 otherwise. F_{ij} is a vector of land fragmentation indicators at the household level – Simpson fragmentation index and the average distance from homestead to all parcels cultivated by the household as defined earlier. H_{ij} is a vector of household-level characteristics as defined in Equation 4.1. V_j is a set of village characteristics defined earlier in Equation 4.1, and ε_{ij} is an error term.

It is worth noting that many farmers in Tanzania do not use chemical fertilisers and manure. We, therefore, include these farms as zeros. Because of this, these variables' distributions have a huge peak at 0 and are very right-skewed. For these reasons, we transform them into logarithms. Because ln(0) is undefined, and our dependent variables have many zeros, to define the logarithms of these variables when the variable has zeros, we use ln(min$(x - e)$, where e is a very small positive number.[4] We also use the inverse hyperbolic sign (IHS) Bellemare and Wichman (2020) proposed. Following Bellemare and Wichman (2020), we take the variables' IHS and regress the variables' IHS version. The intention is to examine the zero and non-zero inputs together. As a robustness check, since chemical fertiliser and manure are rarely used, we also consider the extensive choice to use or not to use (the "binary" choices). To do so, we create and use the binary variables, that is 1 if chemical fertiliser was used and 0 otherwise; and 1 if manure was used and 0 otherwise (Mugizi & Matsumoto, 2021b).

Estimation results

Land fragmentation and agricultural productivity

In the first section, we argued that agricultural extension might be detrimental and, at the same time, beneficial to agricultural productivity, and its net effect may be ambiguous. This section presents the results of the relationship between land fragmentation and agricultural productivity. We use three alternative measures of agricultural productivity: value of crop harvests (TZS/acre), crop income (TZS/acre), and crop income per capita. As explained earlier, we look at two crop categories separately – permanent and seasonal and use two measures of land fragmentation – the Simpson index and dummies for the average distance from the homestead to all the parcels cultivated by the household.

Table 4.1 starts by assessing whether and how fragmentation affects permanent crops (columns 1–3). In column 1, the relationship between land fragmentation and the value of crop harvests is negative. However, the Simpson index is not statistically significant. The higher the average distance from the homestead to the parcels, the less the harvest value of permanent crops the household gets. More specifically, compared to the reference category (average distance below 300 m), households whose parcels' average distance is between 300 m and 500 m, between 500 m and 1 km, between 1 km and 2 km, between 2 km and 3 km, and over 3 km receive less harvest by about 18, 21, 29, 28 and 22 percentage points, respectively. In column 2, we measure agricultural productivity using crop income (TZS/acre). Like in column 1, the Simpson index is negative but not statistically correlated with crop income. Similarly, compared to households whose parcels' average distance from the homestead is below 300 m, households whose parcels' average distance is between 300 m and 500 m, between 500 m and 1 km, between 1 km and 2 km, between 2 km and 3 km, and over 3 km receive less crop income by about 23, 33, 37, 31, and 28 percentage points, respectively. Lastly, column 3 measures agricultural productivity using crop income per capita. The results show a strong negative correlation between the Simpson index and crop income per capita. Likewise, compared to households whose parcels' average distance from the homestead is below 300 m, households whose parcels' average distance is between 300 m and 500 m, between 500 m and 1 km, between 1 km and 2 km, between 2 km and 3 km, and over 3 km receive less crop income by about 17, 16, 16, 23, and 19 percentage points, respectively.

Next, in columns 4–6, we examine the relationship between land fragmentation and the productivity of seasonal crops. The results are essentially and qualitatively similar to those presented in columns 1–3. In column 4, the more fragmented the land is as measured by the Simpson index, the less the harvest value of seasonal crops (TZS/acre) the household receives. Similarly, compared to the households whose parcels' average distance from the homestead is below 300 m, households whose parcels' average distance is between 500 m and 1 km, between 1 km and 2 km, between 2 km and 3 km, and over 3 km receive less harvest by 7, 24, 24, and 34 percentage points, respectively.

Table 4.1 Land fragmentation and agricultural productivity

	Permanent crops			*Seasonal crops*		
	Log of value of harvests (TZS/acre)	*Log of crop income (TZS/acre)*	*Log of crop income per capita*	*Log of value of harvests (TZS/acre)*	*Log of crop income (TZS/acre)*	*Log of crop income per capita*
	(1)	(2)	(3)	(4)	(5)	(6)
Simpson index	−0.091	−0.111	−0.219***	−0.377***	−0.512***	−0.319***
	(−1.394)	(−1.437)	(−3.502)	(−6.833)	(−7.293)	(−4.253)
Average distance from homestead to parcels§						
Between 300 m and 500 m	−0.181***	−0.229***	−0.168***	−0.001	−0.030	0.006
	(−3.602)	(−3.457)	(−2.673)	(−0.024)	(−0.786)	(0.135)
Between 500 m and 1 km	−0.207***	−0.326***	−0.156**	−0.069**	−0.161***	−0.139***
	(−3.258)	(−4.288)	(−2.106)	(−2.039)	(−3.550)	(−2.628)
Between 1 km and 2 km	−0.293***	−0.367***	−0.161*	−0.192***	−0.328***	−0.380***
	(−4.522)	(−4.542)	(−1.835)	(−4.415)	(−5.148)	(−5.623)
Between 2 km and 3 km	−0.275***	−0.310***	−0.233**	−0.198***	−0.402***	−0.453***
	(−3.868)	(−3.363)	(−2.527)	(−3.254)	(−5.110)	(−5.188)
Over 3 km	−0.222***	−0.275***	−0.188***	−0.269***	−0.510***	−0.560***
	(−3.856)	(−4.031)	(−2.806)	(−5.949)	(−8.533)	(−7.886)
1 female-headed household	−0.005	−0.015	0.102**	−0.063**	−0.128***	−0.071
	(−0.105)	(−0.283)	(2.120)	(−2.117)	(−3.028)	(−1.574)
Household head's years of schooling	0.008	0.004	0.010	0.008*	0.007	0.004
	(1.236)	(0.513)	(1.312)	(1.757)	(1.053)	(0.505)
Household head's age	−0.000	0.000	0.003*	−0.002**	−0.003**	−0.001
	(−0.328)	(0.002)	(1.717)	(−2.368)	(−2.343)	(−0.833)
Number of adult working population	−0.013	−0.020	−0.290***	−0.007	−0.003	−0.186***
	(−1.223)	(−1.441)	(−24.102)	(−0.975)	(−0.336)	(−16.131)
Average years of schooling of adults	0.002	0.000	0.015	0.013**	0.008	0.016*
	(0.251)	(0.013)	(1.503)	(2.410)	(1.032)	(1.674)
Ln inorganic fertiliser per acre	−0.024	−0.092***	−0.045**	0.081***	0.035***	−0.016

(*Continued*)

Table 4.1 (Continued)

	Permanent crops			*Seasonal crops*		
	Log of value of harvests (TZS/ acre)	*Log of crop income (TZS/ acre)*	*Log of crop income per capita*	*Log of value of harvests (TZS/ acre)*	*Log of crop income (TZS/ acre)*	*Log of crop income per capita*
	(−1.137)	(−3.193)	(−2.467)	(10.577)	(3.363)	(−1.099)
Ln organic fertiliser per acre	0.057***	0.046***	0.053***	0.030***	0.034***	0.025***
	(5.739)	(3.388)	(4.252)	(6.684)	(5.367)	(3.791)
1 if used improved seeds	0.222***	0.022	−0.030	0.130***	−0.153***	−0.178***
	(3.362)	(0.231)	(−0.402)	(3.859)	(−3.400)	(−3.191)
Ln landholding in acres	0.075***	0.059***	0.052***	−0.015**	0.003	−0.010
	(4.937)	(3.110)	(3.249)	(−2.145)	(0.374)	(−0.892)
Ln farm size (planted area in acres)	−0.334***	−0.348***	0.700***	0.016	−0.035**	−0.019
	(−25.535)	(−23.215)	(42.209)	(1.564)	(−2.458)	(−1.109)
Asset index	0.041***	0.044***	0.068***	0.354***	0.392***	1.108***
	(3.777)	(3.363)	(5.969)	(12.175)	(11.525)	(33.375)
1 if received any extension advice for crop production during 2019/2020	0.080	0.040	−0.003	0.035	0.056	0.094
	(1.203)	(0.477)	(−0.044)	(0.818)	(0.928)	(1.351)
1 if there were farmer field schools during 2019/2020, agr. yr	−0.032	−0.010	−0.052	0.104*	0.131*	0.016
	(−0.407)	(−0.105)	(−0.560)	(1.824)	(1.754)	(0.165)
1 if there were farmers' associations	−0.111	−0.088	−0.056	−0.099*	−0.028	0.260***
	(−1.144)	(−0.771)	(−0.563)	(−1.712)	(−0.366)	(3.142)
1 if there was an irrigation scheme	−0.060	0.003	−0.088	−0.008	0.057	0.080
	(−0.276)	(0.014)	(−0.320)	(−0.074)	(0.400)	(0.339)
Constant	13.993***	13.819***	12.583***	12.041***	11.779***	10.550***
	(67.736)	(57.973)	(60.963)	(82.035)	(58.782)	(79.042)
Observations	6,163	6,163	6,087	25,042	25,042	25,042
R-squared	0.283	0.227	0.569	0.108	0.099	0.156
Regional dummies	*Yes*	*Yes*	*Yes*	*Yes*	*Yes*	*Yes*

Robust t-statistics in parentheses. ***, **, * Significant at 1%, 5%, and 10%, respectively. Robust standard errors are clustered at the community level. §Reference category is less than 300 m.

In column 5, the Simpson index is negative and statistically correlated with crop income. In addition, compared to households whose parcels' average distance from the homestead is below 300 m, households whose parcels' average distance is between 500 m and 1 km, between 1 km and 2 km, between 2 km and 3 km, and over 3 km receive less crop income by 17, 39, 45, and 60 percentage points, respectively. Similar results are revealed in column 6, where agricultural productivity is measured using crop income per capita. The Simpson index is statistically negatively correlated with agricultural productivity as measured by crop income per capita. Likewise, our second measure of fragmentation reveals the same relationship. More precisely, compared to households whose parcels' average distance from the homestead is below 300 m, households whose parcels' average distance is between 500 m and 1 km, between 1 km and 2 km, between 2 km and 3 km, and over 3 km receive less crop income by 14 percentage points, 38 percentage points, 44 percentage points, and 57 percentage points, respectively.

Taken together, the results presented in columns 1–3 and columns 4–6 show that fragmentation is negatively related to agricultural productivity. These findings are in line with previous findings in the literature (see, e.g., Hung et al., 2007; Rahman & Rahman, 2009; Tan et al., 2010; Deininger et al., 2016; Ali et al., 2019; Veljanoska, 2018; Alemu et al., 2017; Oyebanjo et al., 2023) but are in contrast with a few studies that find a positive relationship (see, e.g., Paul & Gĩthĩnji, 2017).

Pathways analysis

Table 4.2 presents the correlation between land fragmentation and agricultural intensification (Equation 4.2). The outcome variables are three measures of agricultural intensification: application of chemical fertilisers, application of organic fertilisers, and whether improved seeds were used. We first present the relationship between fragmentation and intensification on permanent crops (columns 1–7). The Simpon index is negatively related to the log of chemical fertiliser used (kg/acre), but the relationship is not statistically significant (column 1). In column 2, the Simpon index is negatively and statistically significantly related to the log of organic fertiliser used (kg/acre). Similarly, in line with the conventional wisdom, compared to households whose parcels' average distance from the homestead is below 300 m, households whose parcels' average distance is between 2 km and 3 km and over 3 km use fewer organic fertilisers by about 20 and 16 percentage points, respectively (column 2). In column 3, there is no significant relationship between land fragmentation and the use of improved seeds.

In columns 4 and 5, we report the results when the outcome variables – chemical and organic fertilisers – are the IHS-version of the variables. The results are largely similar to those presented in columns 1 and 2. Finally, columns 6 and 7 present the results when the two outcome variables are binary – the extensive choice to use or not to use fertilisers. The results in columns 6 and 7 are consistent with those in columns 1 and 2. The negative relationship between land fragmentation and the use of organic fertilisers is consistent with the conventional view that land fragmentation may hinder or make transporting heavy inputs such as manure difficult.

Table 4.2 Correlation between land fragmentation and agricultural intensification

	Intensification on permanent crops						
	(1)	*(2)*	*(3)*	*(4)*	*(5)*	*(6)*	*(7)*
	Ln Chemical (kg/acre)	*Ln Manure (kg/acre)*	*1 if used improved seeds*	*IHS of Chemical (kg/acre)*	*HS of Manure (kg/acre)*	*Chemical fertilizer dummy*	*Manure dummy*
Simpson index	−0.102	−0.218***	0.008	−0.082	−0.289***	−0.014	−0.040**
	(−1.106)	(−2.657)	(0.612)	(−1.130)	(−2.669)	(−1.042)	(−2.000)
Average distance from homestead to parcels§							
Between 300 m and 500 m	0.034	0.057	−0.006	0.027	0.077	0.006	0.003
	(0.979)	(0.796)	(−0.524)	(0.952)	(0.822)	(1.034)	(0.168)
Between 500 m and 1 km	0.029	−0.129	−0.002	0.019	−0.174	0.007	−0.049**
	(0.762)	(−1.473)	(−0.149)	(0.644)	(−1.505)	(1.176)	(−2.565)
Between 1 km and 2 km	0.027	−0.019	−0.016	0.019	−0.029	0.005	−0.054***
	(0.533)	(−0.220)	(−1.176)	(0.500)	(−0.251)	(0.650)	(−2.988)
Between 2 km and 3 km	−0.059	−0.197**	−0.013	−0.048	−0.249**	−0.007	−0.085***
	(−0.848)	(−2.342)	(−0.845)	(−0.847)	(−2.222)	(−0.665)	(−3.977)
Over 3 km	0.086	−0.155**	0.003	0.069	−0.216**	0.012	−0.092***
	(1.498)	(−2.277)	(0.166)	(1.537)	(−2.423)	(1.291)	(−6.795)
1 female-headed household	0.021	−0.129**	−0.018**	0.016	−0.160**	0.004	−0.007

(*Continued*)

Table 4.2 (Continued)

	Intensification on permanent crops						
	(1)	*(2)*	*(3)*	*(4)*	*(5)*	*(6)*	*(7)*
	Ln Chemical (kg/acre)	*Ln Manure (kg/acre)*	*1 if used improved seeds*	*IHS of Chemical (kg/acre)*	*HS of Manure (kg/acre)*	*Chemical fertilizer dummy*	*Manure dummy*
	(0.726)	(−2.515)	(−2.012)	(0.726)	(−2.321)	(0.768)	(−0.542)
Household head's years of schooling	−0.002	−0.022**	0.001	−0.002	−0.030**	−0.000	0.003
	(−0.519)	(−2.059)	(0.563)	(−0.528)	(−2.089)	(−0.531)	(1.102)
Household head's age	−0.001	0.000	−0.000	−0.001*	0.000	−0.000	−0.000
	(−1.598)	(0.032)	(−1.594)	(−1.712)	(0.002)	(−1.099)	(−0.820)
Number of adult working population	−0.015**	−0.001	−0.004*	−0.011**	−0.002	−0.003***	0.009**
	(−2.453)	(−0.080)	(−1.667)	(−2.260)	(−0.085)	(−2.933)	(2.494)
Average years of schooling of adults	0.013**	0.050***	0.002	0.010**	0.065***	0.002**	0.009***
	(2.249)	(4.362)	(0.803)	(2.187)	(4.328)	(2.262)	(2.664)
Ln landholding in acres	−0.002	0.015	0.008***	−0.001	0.017	−0.000	0.011***
	(−0.315)	(0.798)	(2.745)	(−0.240)	(0.674)	(−0.446)	(2.678)
Ln farm size (planted area in acres)	0.043***	−0.014	0.010***	0.033***	−0.006	0.007***	0.007*
	(3.315)	(−0.708)	(3.071)	(3.203)	(−0.247)	(3.735)	(1.806)
Asset index	0.002	0.078***	0.017***	0.002	0.105***	−0.000	0.027***
	(0.288)	(3.455)	(4.833)	(0.383)	(3.485)	(−0.065)	(5.758)

(Continued)

Table 4.2 (Continued)

	Intensification on permanent crops						
	(1)	*(2)*	*(3)*	*(4)*	*(5)*	*(6)*	*(7)*
	Ln Chemical (kg/acre)	*Ln Manure (kg/acre)*	*1 if used improved seeds*	*IHS of Chemical (kg/acre)*	*HS of Manure (kg/acre)*	*Chemical fertilizer dummy*	*Manure dummy*
1 if received any extension advice for crop prod. during 2019/2020	−0.033	0.355***	0.058***	−0.027	0.474***	−0.004	0.134***
	(−0.612)	(2.925)	(3.108)	(−0.644)	(3.027)	(−0.485)	(6.327)
1 if there were farmer field schools during 2019/2020, agr. yr	0.007	−0.008	0.009	0.006	−0.004	0.001	−0.008
	(0.091)	(−0.091)	(0.651)	(0.100)	(−0.032)	(0.046)	(−0.378)
1 if there were farmers' associations	0.026	−0.072	−0.024	0.017	−0.077	0.006	−0.018
	(0.629)	(−0.850)	(−1.597)	(0.560)	(−0.674)	(0.848)	(−0.849)
1 if there was an irrigation scheme	−0.095	−0.338**	−0.007	−0.076	−0.407*	−0.013	0.038
	(−1.309)	(−2.083)	(−0.265)	(−1.348)	(−1.897)	(−1.175)	(1.361)
Constant	−1.620***	2.229***	0.117**	0.413*	1.163**	0.062**	0.545***
	(−5.938)	(5.597)	(2.139)	(1.792)	(2.231)	(2.049)	(6.254)
Observations	6,163	6,163	6,163	6,163	6,163	6,163	6,163
R-squared	0.111	0.185	0.061	0.109	0.188	0.109	0.269
Regional dummies	*Yes*	*Yes*	*Yes*	*Yes*	*Yes*	*Yes*	*Yes*

Robust t-statistics are in parentheses. ***, **,* Significant at 1%, 5%, and 10%, respectively. Robust standard errors are clustered at the community level. §Reference category is less than 300 m.

Table 4.2 (Continued)

	Intensification on seasonal crops						
	(8)	*(9)*	*(10)*	*(11)*	*(12)*	*(13)*	*(14)*
	Ln chemical (kg/acre)	*Ln manure (kg/acre)*	*1 if used improved seeds*	*IHS of chemical (kg/acre)*	*HS of manure (kg/acre)*	*Chemical fertilizer dummy*	*Manure dummy*
Simpson index	−0.499*** (−5.073)	−0.497*** (−5.263)	−0.049*** (−3.663)	−0.380*** (−5.279)	−0.419*** (−5.283)	−0.066*** (−4.452)	−0.066*** (−4.718)
Average distance from homestead to parcels§							
Between 300 m and 500 m	0.053 (1.055)	−0.310*** (−4.142)	−0.004 (−0.421)	0.042 (1.121)	−0.257*** (−4.066)	0.006 (0.832)	−0.041*** (−4.762)
Between 500 m and 1 km	−0.069 (−1.176)	−0.625*** (−8.595)	−0.028*** (−2.674)	−0.051 (−1.178)	−0.521*** (−8.556)	−0.010 (−1.162)	−0.090*** (−9.284)
Between 1 km and 2 km	−0.009 (−0.126)	−0.681*** (−8.741)	−0.032*** (−2.692)	−0.006 (−0.127)	−0.572*** (−8.703)	−0.001 (−0.111)	−0.086*** (−8.347)
Between 2 km and 3 km	−0.048 (−0.618)	−0.845*** (−9.232)	−0.048*** (−3.901)	−0.041 (−0.713)	−0.710*** (−9.268)	−0.004 (−0.338)	−0.103*** (−8.461)
Over 3 km	−0.203*** (−2.713)	−0.976*** (−11.525)	−0.081*** (−6.938)	−0.163*** (−2.957)	−0.820*** (−11.466)	−0.022** (−1.973)	−0.122*** (−11.392)
1 female-headed household	−0.058 (−1.453)	−0.244*** (−4.414)	−0.031*** (−4.031)	−0.045 (−1.537)	−0.207*** (−4.486)	−0.007 (−1.167)	−0.030*** (−3.960)

(*Continued*)

Table 4.2 (Continued)

	Intensification on seasonal crops						
	(8)	*(9)*	*(10)*	*(11)*	*(12)*	*(13)*	*(14)*
	Ln chemical (kg/acre)	*Ln manure (kg/acre)*	*1 if used improved seeds*	*IHS of chemical (kg/acre)*	*HS of manure (kg/acre)*	*Chemical fertilizer dummy*	*Manure dummy*
Household head's years of schooling	−0.009	−0.002	0.001	−0.007	−0.002	−0.001	0.000
	(−1.486)	(−0.228)	(0.590)	(−1.545)	(−0.239)	(−1.299)	(0.282)
Household head's age	−0.003**	0.003**	−0.000	−0.002**	0.002**	−0.000**	0.001**
	(−2.321)	(2.142)	(−1.454)	(−2.431)	(2.059)	(−2.021)	(2.509)
Number of adult working population	−0.035***	0.126***	0.002	−0.025***	0.106***	−0.006***	0.016***
	(−3.248)	(7.490)	(0.911)	(−3.165)	(7.482)	(−3.455)	(7.269)
Average years of schooling of adults	0.088***	0.032***	0.013***	0.064***	0.027***	0.013***	0.006***
	(8.273)	(2.642)	(7.222)	(8.250)	(2.653)	(8.216)	(3.691)
Ln landholding in acres	−0.005	0.075***	−0.001	−0.003	0.063***	−0.001	0.016***
	(−0.416)	(5.839)	(−0.317)	(−0.320)	(5.775)	(−0.707)	(9.375)
Ln farm size (planted area in acres)	0.298***	0.092	0.055***	0.225***	0.074	0.041***	0.004
	(7.015)	(1.607)	(8.044)	(7.115)	(1.554)	(6.653)	(0.507)
Asset index	0.167***	0.100***	0.036***	0.124***	0.083***	0.024***	0.027***
	(9.649)	(5.371)	(15.170)	(9.587)	(5.352)	(9.607)	(9.816)
1 if received any extension advice for crop prod. during 2019/2020	0.784***	0.648***	0.175***	0.595***	0.542***	0.106***	0.135***
	(7.093)	(6.689)	(12.729)	(6.976)	(6.664)	(7.276)	(9.910)

(*Continued*)

Table 4.2 (Continued)

	Intensification on seasonal crops						
	(8)	*(9)*	*(10)*	*(11)*	*(12)*	*(13)*	*(14)*
	Ln chemical (kg/acre)	*Ln manure (kg/acre)*	*1 if used improved seeds*	*IHS of chemical (kg/acre)*	*HS of manure (kg/acre)*	*Chemical fertilizer dummy*	*Manure dummy*
1 if there were farmer field schools during 2019/2020,agr.yr	−0.279*	−0.213*	−0.042**	−0.198*	−0.180*	−0.045**	−0.018
	(−1.933)	(−1.656)	(−1.971)	(−1.849)	(−1.673)	(−2.170)	(−0.833)
1 if there were farmers' associations	0.171	−0.176	−0.008	0.125	−0.149	0.025	−0.004
	(1.066)	(−1.439)	(−0.489)	(1.055)	(−1.451)	(1.091)	(−0.193)
1 if there was an irrigation scheme	0.292*	−0.177	−0.028	0.221*	−0.151	0.039*	−0.082
	(1.683)	(−0.561)	(−1.000)	(1.670)	(−0.574)	(1.657)	(−1.513)
Constant	−2.278***	−0.428	0.148***	0.161	1.384***	0.027	0.220***
	(−15.501)	(−1.359)	(4.035)	(1.491)	(5.265)	(1.203)	(5.833)
Observations	25,042	25,042	25,042	25,042	25,042	25,042	25,042
R-squared	0.357	0.107	0.199	0.357	0.107	0.341	0.123
Regional dummies	*Yes*	*Yes*	*Yes*	*Yes*	*Yes*	*Yes*	*Yes*

Robust t-statistics are in parentheses. ***, **,* Significant at 1%, 5%, and 10%, respectively. Robust standard errors are clustered at the community level. §Reference category is less than 300 m.

Next, in columns 8–14, we present the estimation results of the relationship between land fragmentation and agricultural intensification on seasonal crops. In column 8, the Simpon index negatively correlates with the log of chemical fertiliser (kg/acre). In addition, compared to households whose parcels' average distance from the homestead is below 300 m, households whose parcels' average distance is over 3 km use fewer chemical fertilisers by about 20 percentage points. In column 9, the Simpson index is negatively and statistically significantly correlated with using organic fertiliser. Similarly, in line with the conventional wisdom, compared to households whose parcels' average distance from the homestead is below 300 m, households whose parcels' average distance is between 300 m and 500 m, between 500 m and 1 km, between 1 km and 2 km, between 2 km and 3 km, and over 3 km use fewer organic fertilisers by about 31, 63, 68, 85, and 97 percentage points, respectively (column 9).

In column 10, a statistically negative relationship exists between land fragmentation and the use of improved seeds on seasonal crops. In columns 11 and 12, we report the results when the outcome variables – chemical fertiliser and organic fertiliser – are the variables' IHS version. The results are largely similar to those presented in columns 1 and 2. Finally, columns 13 and 14 present the results when the outcome variable is binary – the extensive choice to use or not to use organic fertilisers. The results in columns 13 and 14 are consistent with those in columns 8 and 9. The negative relationship between land fragmentation and agricultural intensification observed in Table 4.2 is consistent with the conventional view that land fragmentation may make it difficult and costly to transport heavy inputs such as manure and other inputs from homesteads to parcels.

Conclusion and policy implications

Although the relationship between land fragmentation and agricultural productivity has been examined, none of the existing studies have looked at permanent and seasonal crops separately. Land fragmentation could impact permanent and seasonal crops differently. This chapter examines the relationship between land fragmentation and agricultural productivity by separately looking at its effect on permanent and seasonal crops in Tanzania. We used a unique, large, nationally representative data set from Tanzania, a relatively land-abundant country.

Two main findings emerge. First, we find a negative relationship between land fragmentation and the agricultural productivity of permanent crops. This relationship is robust across different measures of agricultural productivity and across different measures of land fragmentation, that is the Simpson index and average distance from the homestead to all the parcels cultivated by the household. Second, we find a significant negative correlation between land fragmentation and the productivity of seasonal crops. Taken together, the findings of this study suggest that fragmentation is negatively related to agricultural productivity. These findings are in line with the conventional view that fragmentation reduces yield. The findings are also in line with many previous findings in the literature (see, e.g., Hung et al., 2007; Rahman & Rahman, 2009; Tan et al., 2010; Deininger et al., 2016;

Ali et al., 2019; Veljanoska, 2018; Alemu et al., 2017; Oyebanjo et al., 2023) but are in contrast with a few studies that find a positive relationship (see, e.g., Paul & Gĩthĩnji, 2017). Examining the potential pathways, we find a negative and statistically significant relationship between land fragmentation and the use of organic fertilisers on both permanent and seasonal crops. Indeed, this relationship aligns with the conventional view that land fragmentation may make it difficult and costly to transport heavy inputs such as manure from homesteads to parcels. Also, we find a negative relationship between land fragmentation and the use of chemical fertiliser and improved seeds on seasonal crops but not permanent crops.

Two key implications can be drawn from these findings. First, as long as population growth continues, land fragmentation is likely to continue due to the resultant subdivision of land across generations. Therefore, greater support should be given to smallholders to boost the yields of their fragmented lands. This support could include the promotion of the adoption of modern farming technologies. Second, policymakers should promote non-farm income activities to improve the welfare of smallholders. One way to achieve this is by developing supportive rural infrastructure, such as rural electrification.

Notes

1 Other studies, including Knippenberg et al. (2020) and Cholo et al. (2019), have found that land fragmentation improves food security in Ethiopia.
2 The descriptive statistics table is not reported to economise space.
3 Exchange rate 1 USD = 2,335 TZS.
4 In our case, $e = 0.0001$.

References

Abadie, A., Athey, S., Imbens, G. W., & Wooldridge, J. M. (2023). When should you adjust standard errors for clustering? *The Quarterly Journal of Economics*, *138*(1), 1–35.

Alemu, G. T., Ayele, Z. B., & Berhanu, A. A. (2017). Effects of land fragmentation on productivity in Northwestern Ethiopia. *Advances in Agriculture*, *2017*, 4509605.

Ali, D. A., Deininger, K., & Ronchi, L. (2019). Costs and benefits of land fragmentation: Evidence from Rwanda. *The World Bank Economic Review*, *33*(3), 750–771.

Bellemare, M. F., & Wichman, C. J. (2020). Elasticities and the inverse hyperbolic sine transformation. *Oxford Bulletin of Economics and Statistics*, *82*(1), 50–61.

Blarel, B., Hazell, P., Place, F., & Quiggin, J. (1992). The economics of farm fragmentation: Evidence from Ghana and Rwanda. *The World Bank Economic Review*, *S*(2), 233–254.

Cholo, C. T., Fleskens, L., Sietz, D., & Peerlings, J. (2019). Land fragmentation, climate change adaptation, and food security in the Gamo Highlands of Ethiopia. *Agricultural Economics*, *50*, 39–49.

Deininger, K., Monchuk, D., Nagarajan, H. K., & Singh, S. K. (2016). Does land fragmentation increase the cost of cultivation? Evidence from India. *The Journal of Development Studies*, *53*(1), 82–98.

Demetriou, D., Stillwell, J., & See, L. (2013). A new methodology for measuring land fragmentation. *Computers, Environment and Urban Systems*, *39*, 71–80.

FAO, IFAD, UNICEF, WFP, & WHO. (2023). *The state of food security and nutrition in the world 2023. Urbanization, agrifood systems transformation and healthy diets across the rural–urban continuum*. FAO. https://doi.org/10.4060/cc3017en

Goland, C. (1993). Field scattering as agricultural risk management: A case study from Cuyo Cuyo, Department of Puno, Peru. *Mountain Research and Development*, *13*(4), 317–338.

Hung, P. V., Gordon, T., MacAulay, T. G., & Marsh, S. P. (2007). The economics of land fragmentation in the North of Vietnam. *Australian Journal of Agricultural and Resource Economics*, *51*(2), 195–211.

Jayne, T. S., & Muyanga, M. (2012). Land constraints in Kenya's densely populated rural areas: Implications for food policy and institutional reform. *Food Security*, *4*, 399–421.

Jha, R., Nagarajan, H. K., & Prasanna, S. (2005). *Land fragmentation and its implications for productivity: Evidence from Southern India* (ASARC Working Paper 2005/01).

Kawasaki, K. (2010). The costs and benefits of land fragmentation of rice farms in Japan. *The Australian Journal of Agricultural and Resource Economics*, *54*, 509–526.

Knippenberg, E., Jolliffe, D., & Hoddinott, J. (2020). Land fragmentation and food insecurity in Ethiopia. *American Journal of Agricultural Economics*, *102*(5), 1557–1577.

Manjunatha, A. V., Anik, A. R., Speelman, S., & Nuppenau, E. A. (2013). Impact of land fragmentation, farm size, land ownership and crop diversity on profit and efficiency of irrigated farms in India. *Land Use Policy*, *31*, 397–405.

Mkupete, J., Donath, L., & Mugizi, F. M. P. (2023). Household resilience to food and nutrition insecurity during COVID-19 in Tanzania. *GeoJournal*, *88*, 1721–1735.

Mugizi, F. M. P., & Matsumoto, T. (2020). Population pressure and soil quality in Sub-Saharan Africa: Panel evidence from Kenya. *Land Use Policy*, *94*.

Mugizi, F. M. P., & Matsumoto, T. (2021a). From conflict to conflicts: War-induced displacement, land conflicts, and agricultural productivity in post-war Northern Uganda. *Land Use Policy*, *101*.

Mugizi, F. M. P., & Matsumoto, T. (2021b). A curse or a blessing? Population pressure and soil quality in Sub-Saharan Africa: Evidence from Rural Uganda. *Ecological Economics*, *179*.

Mwesigye, F., & Barungi, M. (2021). *Land tenure insecurity, fragmentation and crop choice: Evidence from Uganda* (AERC Research Paper 419). African Economic Research Consortium.

Nguyen, T., Cheng, E., & Findlay, C. (1996). Land fragmentation and farm productivity in China in the 1990s. *China Economic Review*, *7*(2), 169–180.

Niroula, G. S., & Thapa, G. B. (2007). Impacts of land fragmentation on input use, crop yield and production efficiency in the mountains of Nepal. *Land Degradation and Development*, *18*, 237–248.

Ntihinyurwa, P. D., de Vries, W. T., Chigbu, U. E., & Dukwiyimpuhwe, P. A. (2019). The positive impacts of farmland fragmentation in Rwanda. *Land Use Policy*, *81*, 565–581.

Otsuka, K., & Larson, D. F. (2016). *In pursuit of an African green revolution: Views from rice and maize farmers' fields*. Springer.

Otsuka, K., & Place, F. (2015). Land tenure and agricultural intensification in Sub-Saharan Africa. In *The Oxford handbook of Africa and economics: Vol. 2: Policies and practices*. Oxford University Press.

Oyebanjo, O., Ologbon, O. A. C., & Dada, O. M. (2023). Effect of land fragmentation and socioeconomic factors on food crop productivity in Ogun State, Nigeria. *EuroEconomica*, *1*, 42.

Paul, M., & wa Gĩthĩnji, M. (2017). Small farms, smaller plots: Land size, fragmentation, and productivity in Ethiopia. *The Journal of Peasant Studies*, *45*(4), 757–775.

Rahman, S., & Rahman, M. (2009). Impact of land fragmentation and resource ownership on productivity and efficiency: The case of rice producers in Bangladesh. *Land Use Policy*, *26*, 95–103.

Rao, X. (2019). Land fragmentation with double dividends – the case of Tanzanian agriculture. *European Review of Agricultural Economics*, *46*(4), 609–635.

Senbet, L. W., & Simbanegavi, W. (2017). Agriculture and structural transformation in Africa: An overview. *Journal of African Economies*, *26*(AERC Suppl. 1), i3–i10.

Shaw, J. (1963). The problem of land fragmentation in the mediterranean area: A case study. *Geographical Review*, *53*(1), 40–51.

Tan, S., Heerink, N., Kruseman, G., & Qu, F. (2008). Do fragmented landholdings have higher production costs? Evidence from rice farmers in Northeastern Jiangxi province, P.R. China. *China Economic Review*, *19*, 347–358.

Tan, S., Heerink, N., Kuyvenhoven, A., & Qu, F. (2010). Impact of land fragmentation on rice producers' technical efficiency in South-East China. *NJAS – Wageningen Journal of Life Sciences*, *57*, 117–123.

The United Republic of Tanzania (URT). (1997). *National land policy*. Ministry of Lands and Human Settlements Development Dar es Salaam, Tanzania.

Veljanoska, S. (2018). Can land fragmentation reduce the exposure of rural households to weather variability? *Ecological Economics*, *154*, 42–51.

Wan, G. H., & Cheng, E. (2001). Effects of land fragmentation and returns to scale in the Chinese farming sector. *Applied Economics*, *33*(2), 183–194.

Wang, Y., Li, X., Lu, D., & Yan, J. (2020). Evaluating the impact of land fragmentation on the cost of agricultural operation in the southwest mountainous areas of China. *Land Use Policy*, *99*.

World Bank. (2017). *Zanzibar poverty assessment*. World Bank. http://hdl.handle.net/10986/28851

World Bank. (2020). *Tanzania Mainland poverty assessment*. World Bank. https://doi.org/10.1596/33543

Zheng, L., Su, L., & Jin, S. (2023). Reducing land fragmentation to curb cropland abandonment: Evidence from rural China. *Canadian Journal of Agricultural Economics*, 1–19.

5 Structural transitions of agricultural land use in China after 1978

Shiqi Wang, Zhi Cao, and Ruan Yizhu

Introduction

The transition of agricultural land use is one of the core topics of global land change research, which is directly related to food security, ecological balance, and sustainable development. Agricultural land changes in different countries and regions show diversity and complexity. For example, the overall trend of agricultural land use in Europe is decreasing, mainly due to the abandonment of part of the land as a result of agricultural intensification and the implementation of environmental protection policies (Winkler et al., 2021). Africa is experiencing a rapid expansion of agricultural land due to the arid climate as well as a continuously growing population (Knauer et al., 2017).

China, with its huge population and limited agricultural land resources concentrated in the southeast of the Hu Huanyong Line (Liu et al., 2017), is of special significance to study its agricultural land change characteristics. Many scholars have conducted in-depth studies on agricultural land change in China from different perspectives, including spatial and temporal dynamics, driving mechanisms, policy implications, and salient features of change such as non-grain and abandonment. These studies have provided valuable theoretical and empirical foundations for understanding China's agricultural land change.

To comprehensively demonstrate that China's agricultural land is characterized by fluctuating quantitative changes, declining quality, functional shifts, and obvious regional differences under the influence of multiple factors, this chapter analyzes the policy background, spatial and temporal characteristics, and driving factors through a combination of quantitative and qualitative methods.

Study area and methods

China is a vast country with a complex and diverse geographical environment. Based on natural and human factors, China can be divided into six administrative subregions, namely the northeastern, northern, eastern, southcentral, southwestern, and northwestern regions (Figure 5.1). Agricultural land's changes in each region show different characteristics and trends.

DOI: 10.4324/9781032672779-7

Figure 5.1 Provincial-level map of China.

Note: This map is based on standard maps downloaded from the Department of Natural Resources Standard Map Service, with no modifications to the base map. Review No.: GS (2019) 1673.

This chapter demonstrates the changes in China's agricultural land from a number of perspectives. It combines data from national statistical information to quantitatively assess the spatial and temporal fluctuations of agricultural land. Also, it summarizes a large amount of related literature, comparatively analyzes typical regional cases, and qualitatively describes driving mechanisms.

Stages of China's agricultural and rural policies

Establishment of the household contract responsibility system (1978–1984)

Following the Reforms and Opening-up, the fundamental management system in rural areas gradually transitioned from the people's commune system to a basic management framework based on family contractual management, integrated with a two-tier management system. The rise of the household contract responsibility system in Anhui greatly released the production potential of the countryside, and it was strongly promoted by the Central Committee of the Communist Party of China (CPC). The system made the farm household a unit of economic accounting, largely increasing the freedom of operation of the farm household and boosting agricultural production, thus leading to the later establishment of township governments (Wei & Liu, 2019). It was not until 1983 that *the Circular of the Central Committee of the Communist Party of China and the State Council on the*

Separation of the Administration from the Society and the Establishment of the Township Government formally declared the dissolution of the people's commune and established the model of township and village governance (Lv, 2019). As for the urban–rural relationship, the implementation of the dual urban–rural household registration management system greatly restricts the movement of population and hinders China's urbanization.

Transition from planned to market economy (1985–2004)

During this period, the CPC Central Committee adopted a series of reforms to gradually move from a planned economy to a market economy, and in 1985 the state began to reform the system for the distribution of agricultural products. The previous system of unified marketing and purchasing was gradually replaced by the marketization of agriculture, and the prices of agricultural products began to be set in accordance with market supply and demand, improving the irrational price relations between agricultural and industrial products. Farmers' freedom of operation was greatly enhanced, and the liberalization of the agricultural market further contributed to the improvement of social productivity in the countryside. At the same time, the development of China's economy during this period was also characterized by the rapid development of township and village enterprises. This rural industrialization of peasants on collective land greatly influenced the pattern of industrialization in China at that time, forming a new economic structure of rural agriculture, rural industry, and urban industry (Han, 2009). In addition, in terms of urban–rural relations, the state has helped to promote the development of urbanization by relaxing the urban–rural population movement and the management of the household registration system, among other measures.

Construction of a new socialist countryside (2005–2012)

As China enters the 21st century, the solution of the three rural issues is urgent. In response, under the policy of "give more, take less, and be more proactive", the state has promoted the construction of a new socialist countryside in many ways in an attempt to develop a new type of urban–rural relationship. First, the government has been reducing the burden on farmers by reforming agricultural taxes and fees. Second, it has adopted a series of policies to strengthen agriculture and favor farmers, vigorously promoting the development of agriculture and rural areas and increasing farmers' incomes and wealth. For example, agricultural technology inputs and subsidies have been increased, and farmers' rights to education and medical care have been guaranteed. In addition, with regard to urban–rural relations, since 2002, when the state clarified the status of small towns in the construction of new rural areas and the development of urbanization, a large number of small towns have emerged and developed rapidly. This phenomenon has both weakened the boundaries between urban and rural areas and laid the foundation for rapid urbanization (Han, 2009).

Comprehensive deepening of rural reform (since 2013)

After the Third Plenary Session of the 18th CPC Central Committee, China's rural reform has entered a comprehensive and deepening phase. For example, in response to the problem of rural poverty, the CPC Central Committee first put forward the strategy of precise poverty alleviation, and then gradually established a working mechanism of precise identification, precise assistance, precise management, precise assessment, precise withdrawal, and precise supervision. This provided institutional safeguards for winning the battle against poverty and building a moderately prosperous society in all respects. After socialism with Chinese characteristics entered a new era, the 19th CPC National Congress put forward the strategy of rural revitalization in line with the trend, aiming to crack the prominent problems of unbalanced urban and rural development and insufficient rural development (Liu, 2018). On this basis, the 20th CPC National Congress further emphasized the importance of comprehensively promoting rural revitalization and put forward new development requirements and policy measures. The report to the 20th CPC National Congress points out that it is necessary to put agricultural and rural development first and steadily promote the revitalization of businesses, elite, culture, ecosystems, and organizations in the countryside.

Structural transformations in agricultural land use and production

Changes in agricultural land

1) **Quantitative changes.** First of all, from the data in Figure 5.2, there are certain fluctuations in the change of the amount of agricultural land in China from 1978 to 2023, which can be roughly divided into three phases: (a) The stage of decrease from 1978 to 1995. During this stage, the area of agricultural land continued to decrease, which was mainly due to two reasons. One is the serious loss of agricultural land resources due to the sudden land grant in 1986 in

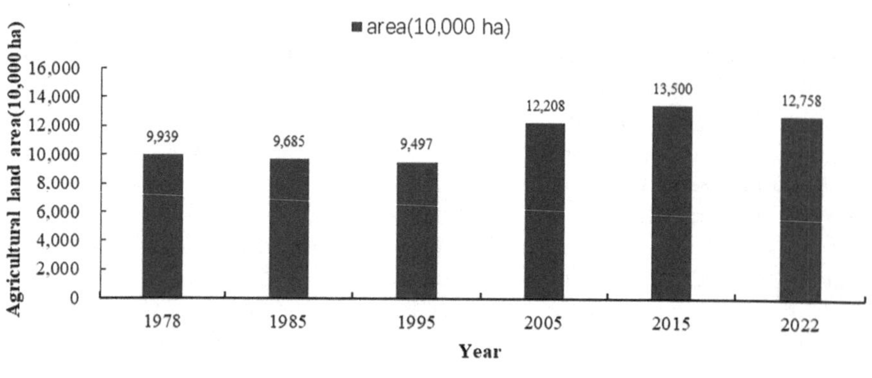

Figure 5.2 Changes in China's agricultural land in years 1978–2022.

Data source: China Statistical Yearbook and *China Land Resources Bulletin.*

various places; the other is the development of urbanization in China and the relatively loose policy environment that led to the occupation of agricultural land by the real estate industry (Liu et al., 2003). (b) The stage of increase from 1995 to 2015. In this stage, there was a steady increase in the area of agricultural land. This was mainly due to the fact that the government noticed the trend of large-scale reduction of agricultural land and implemented a relatively strict agricultural land protection system, so that the number of it has realized an effective rebound. Meanwhile, the cultivation of grassland, forestland, and unutilized land has been the most important source of new agricultural land, and the occupation of unutilized land has been increasing (Zhao et al., 2014). (c) The stage of decrease from 2015 to 2022. In this stage, the amount of agricultural land had a relatively large reduction. In addition to the occupation of agricultural land by urban construction, this was largely related to the nonagricultural, non-grain, and marginalization of agricultural land (Ye et al., 2023).

2) **Spatial changes.** The changes in agricultural land since the Reform and Opening-up have also shown large regional differences. (a) Between the late 1980s and 2000, there was a net increase in the area of agricultural land in the northern region and a net decrease in the area of agricultural land in the southern region. And this trend has become more pronounced over time (Zhang et al., 2003). The new agricultural land was mainly concentrated in the agricultural and forestry intertwined areas of northeast, north, and northwest China (Liu et al., 2014). (b) After entering the 21st century, the spatial change of China's agricultural land has gradually strengthened the characteristics of "decrease in the south and increase in the north, decrease in the east and increase in the west" (Yuan et al., 2023). For example, the area of agricultural land in north and northwest China increased, while the area of agricultural land in east and south China decreased during the period from 2009 to 2017 (Qu et al., 2022).

Structural changes in agricultural land use

Characteristics of urban agricultural land use

Since the Reform and Opening-up, the agricultural land in cities and their surroundings has undergone great changes in terms of planting structure, industrial layout, and territorial functions under economic, social, and ecological driving factors. Numerous cities have completed the transformation from peri-urban agriculture characterized by emphasis on production to urban agriculture with multifunctionality as the main feature in the process of development, which has gradually embodied the characteristics of modernization, industrialization, and specialization. Typical areas include metropolises such as Beijing, Shanghai, and coastal cities in provincial capitals. Take Shanghai as an example; before the Reform and Opening-up, Shanghai's agricultural production mainly focused on planting, and its production content was relatively simple, mainly producing bulk products such as grain, cotton, and pork, and the layout of agricultural production was relatively simple (Tang, 2012). After the Reform and Opening-up, the content of agricultural production was gradually enriched in the stage of peri-urban agriculture. Until the early 1990s, the main function of agriculture was not only to meet the self-sufficiency needs of suburban farmers but also to provide the city with major by-products. The production layout became more complex and diverse, and the regionalized layout of agricultural production was gradually formed (Deng, 2009). In the stage of urban agriculture, due to the drastic reduction of agricultural land resources in the suburbs and the division of non-agricultural land, the spatial distribution of agricultural land has evolved into the characteristics of fragmented small blocks in the suburbs and compact clusters and strips in the middle and far suburbs (Deng, 2009). At the same time, the function of agricultural derivative services has been expanding. Agricultural tourism is one of the main development directions of Shanghai agriculture. Leisure agriculture, as a new industry, new consumption, and new business form vigorously promoted by Shanghai, is striving to transform from quantitative, rough, and low-cost to intensive, fine, and high-grade (Bian, 2017).

Characteristics of agriculture land use in traditional agricultural areas

Traditional agricultural areas have many years of farming history, and most of the areas have flat terrain and fertile soil, constituting the characteristics of a large area of local agricultural land and a high degree of scale. Typical agricultural areas in China include the Three River Plain and Songnen Plain in Heilongjiang and the Huang-Huai-Hai area. In recent years, the area of agricultural land in traditional agricultural areas has been gradually declining, and the uniformity of the spatial distribution of various land types has been increasing. Among them, conversion between forestland and construction land and agricultural land is frequent, with the most drastic degree of conversion and the widest distribution. Grassland, mudflat, and other unutilized land are also highly susceptible to conversion with agricultural land because of their small proportion of area and fragmentation (Li et al., 2016). In addition, with the acceleration of industrialization and urbanization, there is a phenomenon of hollowing villages resulting from the vacancy and idleness of rural housing in the core of villages in the vast rural areas of China (Long et al., 2009). In traditional agricultural areas, the development of hollow villages will lead to the nonagricultural occupation of high-quality agricultural land, thus affecting national food security. Moreover, after entering the 21st century, the speed and scale of rural land transfer have been increasing in the context of national policies. However, in general, most traditional agricultural areas are still characterized by low overall transfer efficiency and a clear trend toward non-grain after land transfer (Zhang, 2014).

Characteristics of agriculture land use in mountainous areas

As a mountainous country, China is characterized by significant mountainous areas, with mountainous counties distributed to varying degrees in both the east and west central regions. These mountainous counties are mainly located in the cities of Guizhou, Ningxia, Chongqing, Yunnan, and Sichuan (Ying & Fang, 2017). Since Reform and Opening-up, with the advancement of industrialization and urbanization, there has been a massive loss of young rural workers in mountainous areas. In addition, due to the influence of topography, local agricultural land is highly fragmented, and large-scale agricultural mechanization is limited, resulting in farmers engaging in agriculture with higher input costs and generally low efficiency of agricultural land use. Mountainous areas gradually reflect the trend of marginalization of agricultural land, the extreme manifestation of which is a large number of agricultural lands abandoned (Li & Li, 2018). The phenomenon of agricultural land abandonment appeared in China as early as in the 1980s, and the main agricultural land type in mountainous areas – slope agricultural land, as a type of agricultural land with poorer farming conditions, was the first to experience abandonment (Wang et al., 2024). According to the 2015–2016 National Sample Survey of Mountainous Areas, the rate of agricultural land abandonment in mountainous counties in China was 14.3% in 2014–2015, and among more than 200 surveyed mountainous villages, agricultural land abandonment occurred in 80.0% of the surveyed villages (Li et al., 2017). Nowadays, under the environment of rapid economic development and the precipitation of agricultural laborers, various agricultural land abandonment phenomena are intensifying and presenting more and more complex characteristics (Duan et al., 2018). This will lead to a decreasing amount of agricultural land and thus be detrimental to food security. At the same time, the functional role of agricultural land in mountainous areas has also been transformed in the process of land use change. In the past, the agricultural population in mountainous areas mainly focused on agricultural production, and the functions of agricultural land were mainly basic livelihood security and food security. Subsequently, under the influence of urbanization and other factors, the cultivation of agricultural land in mountainous areas increasingly pursued high yields. The direction of production also changed from grain and oil crops to vegetables, tea, medicinal herbs, and other cash crops. The function of agricultural land gradually shifted to employment security, social stability, economic development, ecological security, and other directions (Su et al., 2019).

Structural changes in agricultural products

As shown in Figure 5.3, after decades of development, China's agricultural structure has tended to diversify, and the output of various types of agricultural products has also shown an increase. Compared with 1978, China's grain output is generally on an upward trend, and its cultivation area has shifted from the middle and lower reaches of the Yangtze River to the Northeast Plain and the Huang-Huai-Hai Plain (Lv & Wang, 2019). Vegetable and fruit output has increased significantly. And the area under vegetable cultivation is still increasing, with the cultivation area shifting from the middle and lower reaches of the Yangtze River Plain to the Yunnan-Guizhou Plateau area (Zhu et al., 2023). Meat, dairy, and egg production in general rose steadily, increasingly meeting the population's demand for food of animal origin.

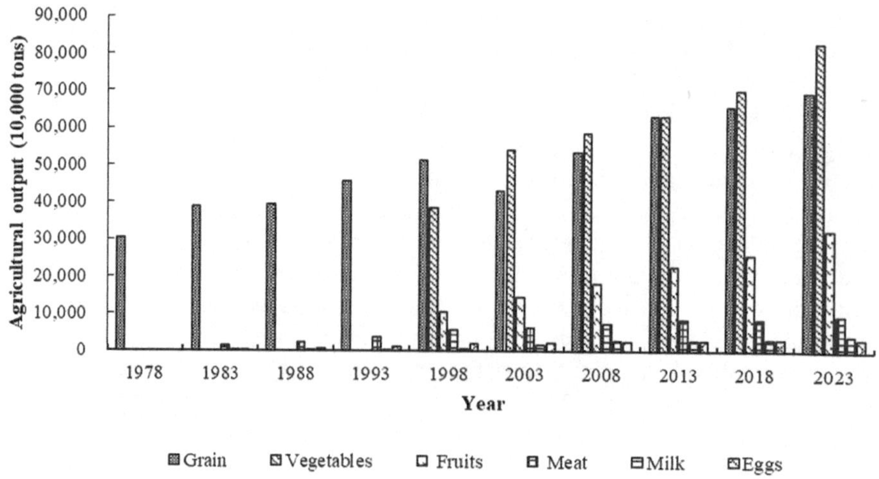

Figure 5.3 Annual changes in China's agricultural products in years 1978–2023.

Data source: *China Agricultural Statistics Yearbook*.

Note: Partial production data missing for 1978–1993.

Mechanisms of structural transformations of agricultural land use

Agricultural land use change is a dynamic process influenced by a variety of economic, social, and climatic factors and so on. This part is based on a qualitative approach to introduce the driving mechanisms of agricultural land use change from the following aspects.

Urbanization driving

After the Reform and Opening-up, China ushered in a period of rapid development, the socioeconomic level continued to rise, and the urbanization rate of the country grew from 17.9% in 1978 to 64.7% in 2021 (Huang, 2023). Accompanied by the rapid increase in the level of urbanization and the substantial increase in the proportion of urban population, the quantity and quality of China's agricultural land resources and the functional use of such aspects have also changed. The main changes included a large reduction in the quantity of agricultural land, the continuous decline in the quality of agricultural land, and the diversification of functions of agricultural land. Regarding the changes in the quantity of agricultural land, on the one hand, the ever-increasing urban population has led to a concomitant increase in the demand for land for construction, which has forced urban land to continue to expand outwards. As a result, a large amount of agricultural land has been replaced by construction land, and the rate of transformation between the two is increasing. On the other hand, the development of urbanization attracts a large number of agricultural people to migrate to neighboring towns and big cities, and the outflow of agricultural people exacerbates the phenomenon of abandonment of agricultural land to a certain extent, thus wasting many agricultural land resources. Regarding changes in the quality of agricultural land, from the viewpoint of the spatial layout of urbanization development, the more urbanized areas are highly overlapped with high-quality agricultural land areas, for example, the Yangtze River Delta region and the Huang-Huai-Hai region. The occupation of high-quality agricultural land by construction land is prominent in these areas (Liu et al., 2008). In addition, the decrease in the amount of agricultural land deepens the pressure upon its availability and quality. To ensure sufficient food supply, many agricultural lands use chemical fertilizers and pesticides more frequently. This will have many unfavorable effects on the quality of agricultural land, such as nutrient loss and soil caking. Regarding the change of agricultural land function, with the emergence of new urbanization nowadays, people are no longer satisfied with the basic needs of life but are gradually pursuing diversified services and high-quality living environment. Different scales of regions are increasingly concerned about the high efficiency and multifunctionality of the utilization of agricultural land, and the potential functions and derivatives of agricultural land are constantly explored on the basis of its intensive management. Nowadays, the types of utilization of agricultural land are becoming increasingly diversified and complex. The agricultural land has changed from the basic production function in the past to a land type with multiple functions, such as production, ecology, and leisure.

Policy driving

In the context of the planned economy system in the past, changes in China's agricultural land resources were largely influenced by national policies and government behavior (Peng et al., 2011). In the early years of Reform and Opening-up, China centered on economic construction, and various economic development policies and regional development strategies prompted increased investment and construction development. This led to a large amount of agricultural land resources being replaced by construction land, and the contradiction between agricultural land protection and economic development became increasingly acute. After noticing the drastic changes in agricultural land, the national government began to realize the necessity of agricultural land protection and introduced a series of policies to slow down the loss of agricultural land and ensure food security. From the perspective of the whole development process of agricultural land protection policies, the content of China's policies has been continuously improved. Changes in agricultural land resources have also been characterized by stages in the context of national policies, developing from sharp changes in the early period to fluctuating changes in the later period and then to relative stability nowadays. At first, as early as the end of the 20th century, the state proposed to strictly control the occupation of agricultural land by building houses in the countryside, and then published a series of policies and regulations on land use planning, protection of basic farmland and dynamic balance of agricultural land etc. The means of protecting the quantity of agricultural land have been continuously enriched. Subsequently, on the basis of the protection of the quantity of agricultural land, China began to pay attention to the quality of agricultural land. As a series of problems such as occupying superiority and compensating for inferiority appeared in the process of agricultural land compensation in the early stage (Sun et al., 2014), the government added the quality of agricultural land to the examination of agricultural land occupation and compensation balance, farmland protection, and land planning as well. Later on, through the annual survey of agricultural land quality and agricultural land use and maintenance combination and other measures to increase protection, the quality of agricultural land has been more comprehensive protection (Wang, 2021). Nowadays, China's development has been adjusted from a single economic benefit goal to a comprehensive goal of economic, social, and ecological benefits (Wang & Liu, 2023). While strengthening the management and monitoring of the quantity and quality of agricultural land, the ecological protection of agricultural land is also urgent. Since the 18th CPC National Congress, the government has constructed a three-in-one protection pattern of agricultural land quantity, quality and ecology, and deepened the ecological protection of agricultural land through the delineation of protection red line, compensation for ecological protection, and use control of natural ecological space (Wang et al., 2020). Thus, China's system of agricultural land protection, with use control as the core and permanent basic farmland protection, high-standard farmland construction, and agricultural land occupation and replenishment balance as the main components, is becoming complete (Liu & Chen, 2020).

Market driving

Since the Reform and Opening-up, people's living standards have been improving, and food demand has tended to be diversified and nutritious. Relevant studies have shown that the per capita consumption of cereal grains in China accounted for 24.2% of the total food consumption in 2017, down from 2000. Vegetable consumption and the combined consumption of other foods such as fruits, meat, and milk accounted for 46.7% and 29.1% of the total food consumption, respectively, up from 2000 (Xie, 2020). Based on the changes in the consumption end, the agricultural behavior in the production end has also produced corresponding changes, and one of the main features is the non-grain of agricultural land. Here, non-grain mainly refers to the narrow sense – the non-grain of the cultivation structure of agricultural land, that is planting cash crops such as vegetables, fruits, flowers, and so on (Zhang & Jiang, 2016). The decrease in demand for traditional food rations and the increase in demand for sideline food in the dietary consumption market have led to low comparative benefits of planting food. Thus, this contributed to the continuous decrease in the area planted with bulk food crops such as wheat and rice, and the increase in the area planted with non-grain and cash crops such as vegetables, fruits, and oilseeds. On a national scale, the proportion of China's grain-sown area has decreased from 80.3% to 69.7% from 1978 to 2020 (Huang et al., 2022). On a provincial scale, large areas of non-food are mainly in Xinjiang, Heilongjiang, Henan, Sichuan, Guangxi, and the two lakes, with all nonagricultural areas larger than 3.0×10^6 hm^2 in 2018 (Chen et al., 2021).

Climate driving

Under the general environment of global climate change, agricultural land resources, influenced by temperature and precipitation, have produced profound changes in terms of quantitative changes, spatial patterns, and planting structures. First, in terms of quantitative changes, on the one hand, climate warming has largely contributed to the increase in the amount of agricultural land in some regions of China, and these new agricultural lands are mainly distributed in the northeast and northwest regions. According to statistics, in 2023, the average surface temperature in China increased by 0.8 °C compared with the average value of previous years, and the warming rate in the northern region was higher than that in the southern region, and the warming rate in the western region was higher than that in the eastern region (Li, 2024). The local light and heat conditions in the northwest and northeast China are more suitable for crop growth under the conditions of warmer temperatures and increased cumulative temperatures, thus driving the conversion of large areas of grassland and unutilized land to agricultural land (Liu et al., 2009). Meanwhile, increased precipitation also favors the reclamation of agricultural land resources in arid and semi-arid regions. Research shows that since the 1990s, the climate in most areas of Tibet has developed in the direction of warmth and humidity (Yang et al., 2013). And according to the statistics, the agricultural land in Tibet has also increased from 3,252.0 km^2 in 1980 to 3,703.0 km^2 in 2010 (Yang et al., 2015), the reason for which is inextricably linked to the local climate change. On the other hand, the negative impacts on agricultural production brought about by the emergence of extreme climates may also lead to a decrease in the amount of agricultural land. The frequent occurrence of extreme weather has brought negative outcomes to agricultural production, such as increased climatic disasters, destruction of farmland facilities, and decreased food production, which have seriously exacerbated the vulnerability of food production and the instability of agricultural land utilization. This often forces the rural population to abandon farming, thus leading to an invisible reduction in agricultural land resources. Second, in terms of spatial pattern, the center of gravity of China's agricultural land showed the spatial characteristics of northward shift, and this center migrated 112.0 km northward in general from 2010 to 2020 (Zhu et al., 2024), and the reasons for this are inseparable from the changes in the accumulated temperature and precipitation. In addition, in terms of cultivation structure, agricultural cultivation in different regions shows different directions of transformation due to changes in temperature and precipitation. For example, to adapt to the drought phenomenon brought about by climate change, agricultural cultivation in the southern region has gradually shifted from being dominated by rice to rotating rice with drought-resistant crops such as wheat and corn. In the northern region, to fully utilize the increased heat resources, the planting ratio of food crops has generally changed from being dominated by wheat and maize to being dominated by maize and rice (Li & Xu, 2017).

Conclusion and discussion

This chapter reveals the trend of changes in China's agricultural land through an in-depth analysis of its structural transformation and driving mechanisms. It is found that China's agricultural land has gone through different stages, from the establishment of the household contract responsibility system to the transformation of the market economy, to the construction of new rural areas and the comprehensive deepening of rural reform. The policy orientation and implementation effects of these stages have had a profound impact on agricultural land. At the same time, factors such as the urbanization process, market demand, and climate change have also significantly contributed to the structural adjustment of agricultural land.

These changes have not only affected the layout and efficiency of agricultural production, but also posed challenges to national food security, rural social stability, and ecological environmental protection. In the future, China's agricultural land management and policymaking will need to pay more attention to the sustainable use of land and ecological protection to ensure a harmonious symbiosis between agricultural production and the ecological environment and to promote the overall economic and social progress of rural areas.

References

Bian, Y.-Q. (2017). *Study on the spatial layout of leisure agriculture in Shanghai*. Shanghai Jiao Tong University.

Chen, F., Liu, J.-N., Chang, Y.-Y., Zhang, Q., Yu, H.-C., & Zhang, S.-L. (2021). Spatial pattern differentiation of non-grain agricultural land and its driving factors in China. *China Land Science*, *35*(09), 33–43.

Deng, C.-X. (2009). *Research on sustainable development of urban agriculture in Shanghai*. East China Normal University.

Duan, Y.-M., Zhou, H., Liu, X.-H., & Wang, H.-L. (2018). Research progress and outlook on agricultural land abandonment in China. *Jiangsu Agricultural Science*, *46*(13), 13–17.

Han, J. (2009). Evolution of the relationship between rural and urban areas in the past sixty years: Looking back and prospect. *Reform*, (11), 5–14.

Huang, J.-H. (2023). China's urbanization in seventy years and modernization of state governance. *Research on Urbanization in China*, (01), 106–128.

Huang, Z.-H., Li, Y.-Y., & Mao, X.-H. (2022). The current situation and countermeasures of "non-agriculturization" and "non-grain" of agricultural land in China. *Jiang Huai Forum*, (04), 13–21.

Knauer, K., Gessner, U., Fensholt, R., Forkuor, G., & Kuenzer, C. (2017). Monitoring agricultural expansion in Burkina Faso over 14 years with 30m resolution time series: The role of population growth and implications for the environment. *Remote Sensing*, *9*(2), 132.

Li, C.-H., Xin, G.-X., Yang, C.-X., & Cheng, X.-Y. (2016). Land use and land cover change (LUCC) and its environmental effects of traditional farm area. *Journal of Southwest University (Natural Science Edition)*, *38*(05), 139–145.

Li, J. (2024). How will climate change affect China in the future? *Ecological Economy*, *40*(09), 9–12.

Li, K., & Xu, Y.-L. (2017). Research on agricultural cultivation structure adjustment in China for climate change adaptation. *Journal of Agricultural Science and Technology*, *19*(01), 8–17.

Li, S.-F., & Li, X.-B. (2018). Economic characteristics and the mechanism of farmland marginalization in mountainous areas of China. *Acta Geographica Sinica*, *73*(05), 803–817.

Li, S.-F., Li, X.-B., Xin, L.-J., Tan, M.-H., Wang, X., Wang, R.-J., Jiang, M., & Wang, Y.-H. (2017). Extent and distribution of agricultural land abandonment in Chinese mountainous areas. *Resources Science*, *39*(05), 1801–1811.

Liu, J.-L., Liu, Y.-S., & Li, Y.-R. (2017). Classification evaluation and spatial-temporal analysis of "production-living-ecological" spaces in China. *Journal of Geographical Sciences*, *72*(07), 1290–1304.

Liu, J.-Y., Kuang, W.-H., Zhang, Z.-X., Xu, X.-L., Qin, Y.-W., Ning, J., Zhou, W.-C., Zhang, S.-W., Li, R.-D., Yan, C.-Z., Wu, S.-X., Shi, X.-Z., Jiang, N., Yu, D.-S., Pan, X.-Z., & Chi, W.-F. (2014). Spatiotemporal characteristics, patterns and causes of land use changes in China since the late 1980s. *Acta Geographica Sinica*, *69*(01), 3–14.

Liu, J.-Y., Zhang, Z.-X., Xu, X.-L., Kuang, W.-H., Zhou, W.-C., Zhang, S.-W., Li, R.-D., Yan, C.-Z., Yu, D.-S., Wu, S.-X., & Jiang, N. (2009). Spatial patterns and driving forces of land use change in China in the early 21st century. *Acta Geographica Sinica*, *64*(12), 1411–1420.

Liu, J.-Y., Zhang, Z.-X., Zhuang, D.-F., Wang, Y.-M., Zhou, W.-C., Zhang, S.-W., Li, R.-D., Jiang, N., & Wu, S.-W. (2003). A study on the spatial-temporal dynamic changes of land-use and driving forces analyses of China in the 1990s. *Geographical Research*, *22*(01), 1–12.

Liu, T.-J., & Chen, M.-Q. (2020). The status-quo and improvement path for the implementation of agricultural land protection policy in China. *China Land Science*, *34*(09), 32–37+47.

Liu, Y.-S. (2018). Research on the urban-rural integration and rural revitalization in the new era in China. *Acta Geographica Sinica*, *73*(04), 637–650.

Liu, Y.-S., Wang, L.-J., & Long, H.-L. (2008). Spatio-temporal analysis of land-use conversion in the eastern coastal China during 1996–2005. *Journal of Geographical Sciences*, *18*, 274–282.

Long, H.-L., Li, Y.-R., & Liu, Y.-S. (2009). Analysis of evolutive characteristics and their driving mechanism of hollowing villages in China. *Acta Geographica Sinica*, *64*(10), 1203–1213.

Lv, D.-W. (2019). 70 Years of rural governance: A perspective of the modernization of state governance. *Journal of Nanjing Agricultural University (Social Sciences Edition)*, *19*(04), 11–19+156.

Lv, J., & Wang, Y.-M. (2019). The state of international food economy and China's food security. *Journal of the CCPS (CAG)*, *23*(04), 131–136.

Peng, L., Ding, E.-J., & Xie, D.-T. (2011). An empirical research of the relationship between changes in agricultural land quantity and agricultural land protection policies in China. *Journal of Southwest University (Natural Science Edition)*, *33*(11), 103–110.

Qu, Q.-Y., Yang, H., Zhao, Y.-H., & Han, L. (2022). Temporal and spatial changes and its characteristics of agricultural land and grain yield in China from 2009 to 2017. *Hubei Agricultural Sciences*, *61*(08), 29–34+76.

Su, K.-C., Yang, Q.-Y., Zhang, B.-L., & Zhang, Z.-X. (2019). The coupling mechanism between rural land use transition and small-scale peasant economy change in mountainous areas. *Geographical Research*, *38*(02), 399–413.

Sun, R., Sun, P., Wu, J.-X., & Zhang, J.-Q. (2014). Effectiveness and limitations of agricultural land requisition-compensation balance policy in China. *China Population, Resources and Environment*, *24*(03), 41–46.

Tang, J.-H. (2012). *Study on the evolution and evaluation of agricultural industry structure in Shanghai suburbs in the process of urbanization*. East China Normal University.

Wang, W.-X. (2021). *The evolution process and performance evaluation of farmland protection policy in China since the reform and opening-up*. China University of Geosciences.

Wang, W.-X., Cao, Y.-G., Su, R.-Q., Qiu, M., Song, L., & Zhou, W. (2020). Agricultural land protection policy in China: Background, effect and future trend. *Chinese Journal of Agricultural Resources and Regional Planning*, *41*(10), 40–51.

Wang, Y.-H., Yang, A.-X., Xin, L.-J., Dong, S.-J., & Li, X.-B. (2024). Spatial and temporal differentiation and influencing factors of fallowing in different types of farmland in China: Based on survey data from 346 mountainous counties. *Geographical Research, 43*(08), 1977–1993.

Wang, Y.-J., & Liu, D.-F. (2023). Multi-scenario simulation of land use change and its effects under requisition-compensation balance of agricultural land. *Geomatics and Information Science of Wuhan University, 48*(9), 1499–1511.

Wei, H.-K., & Liu, C.-Q. (2019). The outline, experiences and prospect of China's rural reform. *Chinese Rural Economy, 2*, 2–18.

Winkler, K., Fuchs, R., Rounsevell, M., & Herold, M. (2021). Global land use changes are four times greater than previously estimated. *Nature Communications, 12*(1), 2501–2501.

Xie, J.-L. (2020). Residents' food consumption upgrade and China's agricultural transformation and development. *Rural Economy*, (06), 66–73.

Yang, C.-Y., Shen, W.-S., & Lin, N.-F. (2013). Temp-spatial changes of temperature and precipitation over the Tibet Plateau during the last 50 years Journal of Arid Land. *Resources and Environment, 27*(12), 167–172.

Yang, C.-Y., Shen, W.-S., & Wang, T. (2015). Spatial-temporal characteristics of agricultural land in Tibet in recent 30 years. *Transactions of the Chinese Society of Agricultural Engineering, 31*(01), 264–271.

Ye, S.-J., Song, C.-Q., Cheng, C.-X., Gao, P.-C., Shen, S., & Mu, W.-S. (2023). Five issues and countermeasures of China cropland resource use. *Bulletin of Chinese Academy of Sciences, 38*(12), 1962–1976.

Ying, B., & Fang, Y.-P. (2017). The classification of China mountain areas and its spatial distribution characteristics. *Journal of Guizhou Normal University (Natural Sciences), 35*(05), 7–14.

Yuan, J.-T., Chen, W.-X., & Zeng, J. (2023). Spatio-temporal differentiation of agricultural land use change and its impact on agricultural land NPP in China. *Journal of Natural Resources, 38*(12), 3135–3149.

Zhang, G.-P., Liu, J.-Y., & Zhang, Z.-X. (2003). Spatial-temporal changes of agricultural land in China for the past 10 years based on remote sensing. *Acta Geographica Sinica, 58*(03), 323–332.

Zhang, L.-W. (2014). *Research on the exit problem of agricultural land in traditional agricultural areas*. Henan Agricultural University.

Zhang, O.-X., & Jiang, C.-Y. (2016). Analysis on differences of "non-grain" of different types farmers in transfer in farmland. *Finance and Trade Studies, 27*(04), 24–31+67.

Zhao, X.-L., Zhang, Z.-X., Wang, X., Zuo, L.-J., Liu, B., Yi, L., Xu, J.-Y., & Wen, Q.-K. (2014). Analysis of Chinese agricultural land's spatial-temporal changes and causes in recent 30 years. *Transactions of the Chinese Society of Agricultural Engineering, 30*(03), 1–11.

Zhu, P.-T., Ma, Y.-H., Chen, P.-X., Li, L., & Guo, R.-X. (2024). Migration characteristics of agricultural land gravity center and climate suitability evaluation in China. *Journal of Beijing Normal University (Natural Science), 60*(03), 397–404.

Zhu, X.-H., Zhang, Y., & Zhu, Y.-Y. (2023). Regional regulation and interregional coordination of agricultural land protection in China from the perspective of "Greater Food" approach. *Acta Geographica Sinica, 78*(09), 2147–2162.

6 Growth of tea plantations and its impact on agricultural land use in Assam (India)

Chandra Kala Magar, Dhanjit Deka, and Bimal Kumar Kar

Introduction

Tea (*Camellia sinensis*), a tropical cash crop, is grown in more than 32 countries in the world. It is a very stimulating drink known for its refreshing, revitalizing, and medicinal values. Although tea cultivation began in China about 4,000 years ago, it was in the 8th century AD that the Japanese discovered it and later others (Evans, 1992). Tea is popular as one of the most liked and cheap non-alcoholic beverages in the world (Parida et al., 2024). In India, tea cultivation was first started in Assam by the British in the middle of the 19th century. India is the second largest producer and consumer of tea globally, accounting for around 20.80% of world production (ITC, 2022). Among all the states of India, Assam is the country's major producer of tea (50.79% of total production and over 56.12% of total area) (Tea Board of India, 2023). It is the largest tea-growing area in the world, with over 123,177 tea estates producing 698.38 million kg of tea (Tea Board of India, 2023).

Tea plays a major role in the economic development of Assam, as it constitutes about 13.24% of the world's tea production (Tea Board of India, 2023). It employs a huge labour population (more than one million) and engages a significant number of indigenous people directly or indirectly. In addition, more than one million people in the state derive their livelihood from ancillary activities associated with the industry (Chakraborty, 2013). In the current situation, tea cultivation has become an alternative livelihood for rural farmers who traditionally depended upon paddy cultivation (Borah & Bora, 2023). Further, the development of transport networks, such as roads, railways, waterways and other communication systems in the state, has also been associated to a great extent with the promotion of tea plantations (Biggs et al., 2018). Hence, the position of tea is highly significant towards the overall socio-economic development of the country in general and the state of Assam in particular. As the single largest sector, the tea industry adds significantly to the nation's economy in terms of GDP and foreign exchange earnings (Chetry & Kar, 2022). Moreover, tea cultivation in Assam significantly affects land use in general and agricultural land use in particular. Tea cultivation in Assam acts as a double-edged sword, given its economic benefits on the one hand and substantial challenges to land use and ecological sustainability on the other (Baruah, 2008).

DOI: 10.4324/9781032672779-8

Hence, maintaining a balance between the production system and ecological sustainability has become a great challenge for the future of the state's tea industry.

Several studies have been conducted globally on the growth of the tea industry, its role in uplifting a region's economy and its various impacts (Sun et al., 2024; Yu et al., 2024; Huang et al., 2024; Sachitra, 2016; Thushara, 2015; Ranweera et al., 2020; Sandika, 2018; Koros et al., 2023; Gesimba et al., 2005). These studies revealed that the growth of tea industries in different countries has faced various challenges with respect to their economic sustainability and ecological impacts. Historically, many Indian studies also investigated the growth and development of the tea industry in India in varying spatial contexts (Das & Zirmire, 2018; Kumar et al., 2008; Dutta, 2024; Hazarika, 2013). George (1980), in a study, focused on the evolution of the marketing system of tea in India with particular emphasis on the nature of domestic demand and export performance. Baroowah (2006), based on his study of the past and present, looked at the future of Indian tea as very bright and viable. However, Guha's (1967) historical study of tea plantations from a political-economy perspective viewed the role of the British for their own benefit of land-grabbing policy adopted through the Wasteland Rules of 1838 in the name of development of tea plantations in the state of Assam. Goswami and Goswami's (2008) work is primarily devoted to the potentiality of small-scale tea cultivation, considering suitable soil, climate, medicinal properties, and economic benefits in Assam. Saikia (2008) discussed the more significant issue of inclusive development of the tea community population inhabiting the enclave-type plantation economy in Assam and associated problems. Sharma et al. (2012) analysed the historical context of forest cover change within the political economy and demographic milieu during the upper Brahmaputra valley's pre-colonial, colonial, and post-colonial periods. His study revealed that the recent expansion of small tea growers has pushed the marginal cultivators towards forest encroachment to mitigate the impending risk in agriculture and meet their livelihood. Sarma (2013) conducted an in-depth analysis of tea plantations and their production and processing in relation to the emerging social, economic, and ecological problems. The recent satellite imagery-based study reveals the expansion of tea cultivated area by 1,280.37 km^2 during the last 32 years, from 1990 (1,637.05 km^2) to 2022 (2,917.52 km^2) in Assam (Parida et al., 2024).

With this background, the present chapter attempts to analyse the trend and spatial patterns of growth of tea plantations in Assam, particularly during the post-independence period, and to examine its impact on changing agricultural land use in the state. The role of growing small-scale tea cultivation in different parts of the state, mainly during the last three decades, towards conversion of traditional rice areas has also been considered. The study is largely based on secondary data obtained from the Tea Board of India and the Directorate of Economics and Statistics, Government of Assam.

Geographical Context of the Study Area

The study area (the state of Assam) is located in the north-eastern part of India being engulfed on three sides by hills and mountains. Covering an area of 78,438 km^2, the state of Assam lies between 24.3°N–28°N latitudes and 89.5°E–96.1°E longitudes (Taher, 2022) (Figure 6.1). It is surrounded by Arunachal Pradesh and Bhutan to the north; Nagaland and Manipur to the east; Tripura, Meghalaya, and Mizoram to the south; Bangladesh to the west and south-west; and West Bengal on the west. Assam is characterized by a sub-tropical monsoon climate with high humidity and heavy rainfall ranging from 1,500 mm to 3,800 mm and an annual temperature of about 8 °C to 38 °C. This makes Assam very rich in diverse flora and fauna. Even though Assam is the second commercial tea production region after southern China, tea cultivation is found to be concentrated in the undivided districts of Nagaon (comprising present districts of Nagaon and Morigaon), Kamrup (comprising present districts of Kamrup, Kamrup Metro, Baksa, Barpeta, and Nalbari), Sibsagar (comprising present districts of Golaghat, Majuli, Jorhat, and Sibsagar), Darrang (comprising the present districts of Sonitpur, Udalguri, and Darrang), Lakhimpur and Dibrugarh (comprising present districts of Lakhimpur, Dhemaji, Charaideo, Tinisukia, and Dibrugarh), Goalpara (comprising the present districts of Goalpara, South Salmara-Mankachar, Bongaigaon, Dhubri, Chirang, and Kokrajhar) in the Brahmaputra valley; Cachar (comprising the present districts of Karimganj, Hailakandi, and Cachar) in the Barak valley; and KarbiAnglong and N. C. Hills (comprising the present districts of KarbiAnglong, West Karbi-Anglong, and Dima Hasao) in the hill region. Thus, physiographically, Assam is

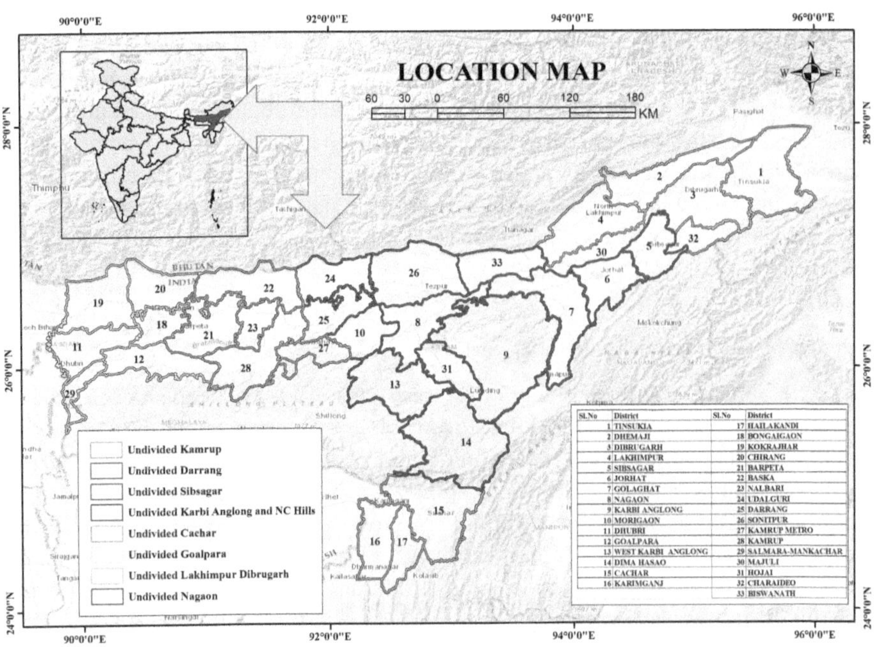

Figure 6.1 Location of the study area (The state of Assam, India).

Source: Prepared by author

characterized largely by river plains and partly by hilly and mountainous terrain. Moreover, Assam is well-known as the largest contiguous tea-growing region in the whole world, which contributes more than 50% of the total tea produced in the country (Biggs et al., 2018; Deka & Goswami, 2021).

Materials and Methods

The study is primarily based on secondary data drawn from various research papers, books, reports, journal articles, and publications from the Tea Board of India, the Census of India, and the Directorate of Economics and Statistics, Government of Assam. The collected data have been processed, analysed, and presented using simple and meaningful statistical and cartographic techniques. In view of changes in the district boundaries over time, the data have been analysed and presented based mostly on the undivided district boundaries of the state of Assam.

Results and Discussion

Growth of Tea Plantations

The tea industry grew tremendously in Assam during the last three decades of the 19th century. The total land area in possession of the sector doubled and accounted for some one-seventh of the entire settled area in Assam plains. The land under tea in the province increased from a little over 21,986 hectares in 1872 to 1,32,705 hectares in 1901, with a consequent increase in the output of tea from about 12 million lbs to 134 million lbs during the same period. This tremendous growth was in response to the rising British demand for Indian tea, in both absolute and relative terms. In the latter period, particularly after independence, there was a massive increase in the number of tea gardens along with the area under tea cultivation in the state due to the inclusion of small tea growers partly through utilization of fallow and wastelands and conversion of less-productive paddy fields (Tea Board of India).

Trend of Growth of Area Under Tea

The development of the tea industry of Assam has not always been smooth since its inception, having faced major difficulties such as labour shortage, labour recruitment, price fluctuations, dissatisfaction of labourers, changes in ownership, ageing of the tea bushes, and so on. Further, the growing global competition from other tea-producing countries, such as Kenya, Sri Lanka, China, Vietnam, and others, has resulted in declining prices in domestic and international markets. However, the industry has successfully overcome such difficulties and still occupies an important part of Assam's economy.

Assam has witnessed a rising trend in the growth of tea-growing area in the state and the country as a whole during the period 1951–2023. It is worth mentioning that the number of tea gardens in the state has increased from 785 covering 1,55,674 hectares of land (49.13% of the country's land under tea plantations) in 1951 to 848 covering 2,33,284 hectares of land (55.48% of the country's total tea area) in 1991. In 2001, the number of tea gardens increased sharply to 40,795 as against the country's total of 1,16,659 (34.96% of the country); and it became 51,605 as against the country's total of 1,43,217 in 2006 (36.03% of the country). Further, it reached 1,23,177 as against the country's total of 2,31,093 in 2023 (53.30% of the country). Consequently, the area under tea plantation in the state witnessed an increase from 180,000 hectares to 343,000 hectares during the last five decades (1970–2020) (Table 6.1). Such a phenomenon, which has been primarily due to emergence of a large number of small-scale tea gardens (each one with area below 10.12 hectare), contributed to a marked increase in the area under tea plantations and tea production in both Assam and India.

Spatial Variation in Growth of Area Under Tea Plantations

Concerning the spatial variation in area under tea cultivation in the state of Assam, undivided Dibrugarh district constitutes the highest proportion of the state's total (31.92%), followed by undivided Sivasagar district (28.65%), undivided Darrang district (19.18%), and undivided Cachar district (10.64%), and these four undivided districts together constitute 90.39% of total tea area of the state and contribute 92.86% of the state's total production of tea as in 2020. As regards inter-district variation in the growth of area under tea, all the districts, except Cachar, registered more than a 12% increase during 1981–1990. On the other hand, during the recent decade (2010–2020), the growth of area under tea had been very high in Darrang (32.26%) and Sivasagar (27.34%) as compared to Dibrugarh and Cachar (only around 15%). It is thus indicative that the gradually expanding area under tea has become more significant in the central and western parts of the Brahmaputra valley.

Table 6.1 Trend of tea gardens and area under tea plantations in Assam, 1970–2020

Year	*No. of tea gardens*	*Tea area (hectares)*	*% of area to India's total tea area*
1970	750	180,065	50.85
1975	756	188,793	51.96
1980	777	200,569	52.63
1985	Not available	216,117	54.03
1990	848	230,363	55.34
1995	1196	226,280	52.98
2000	39,151	266,512	52.84
2005	49,102	300,502	57.63
2010	78,856	281,043	50.78
2015	83,566	304,400	47.17
2020	1,22,440	343,256	53.89

Source: Statistical Handbook, Assam, 1973–2023, Directorate of Economics and Statistics, Government of Assam, Guwahati.

The rapid expansion of tea areas in the state in recent times has also been largely associated with the emergence of many small tea growers (STGs) involved in tea cultivation in smallholdings. It may be mentioned in this context that the statistics of STGs have been included in all such figures since 1998. Accordingly, the districts with a large number of STGs, such as Dibrugarh and Sivasagar, registered high percentage increase in area, that is 34% and 13%, respectively, during 1991–2000.

Emergence of Small-Scale Tea Cultivation and Its Changing Spatial Dimensions

The structural change in agriculture due to the advent of small tea cultivation in the tea industry of Assam is a very recent phenomenon. The state has been experiencing a revolution in tea cultivation for the past 20 years. The first commercial small tea garden was reported to be established in Golaghat district in 1978. After its historical formation, the small tea growers gradually became an organized group. According to the Tea Board of India, small tea growers are persons or groups with a plantation area of up to 10.12 hectares. The All Assam Small Tea Growers Association (AASTGA) was registered in 1987. AASTGA has brought about a green revolution in rural Assam by cultivating tea in unutilized and underutilized uplands. According to the members of AASTGA, it has contributed to about 29% of the total tea produced in Assam, which is approximately 14% of the total tea production in India. Almost 0.9 million people are engaged directly and indirectly in the state's small tea garden system. Due to its positive impact today, it has started employing a large number of educated unemployed youth, generating additional earnings for farmers, including women, contributing large quantities to the state's total production and helping the state's economy, thereby bringing overall socio-economic progress to the state. The contribution of the large tea estates and Tea Board of India and the establishment of Bought Leaf Factories (BLF) are no less significant in the growth of small-scale tea plantations in the state having sizes varying from less than 1 hectare to 10.12 hectares.

At present, small tea gardens have become an essential constituent of tea production, as they produce almost 29% of Assam's total tea, mainly of CTC variety (Economic Times Bureau, 2014). Their contribution has been projected to increase to 50% of the state's total tea production in another five to seven years. Nearly 158,000 STGs are directly engaged in tea cultivation in the state. In the past two decades, the number of STGs in the state has grown to an astounding number of 65,000 from around 14 districts, where more than 0.9 million families are associated with the small tea gardens, including ex-tea garden workers or excess workers from the nearby big tea estates, specifically women pluckers are actively engaged. Around 0.25 million hectares of land has been recorded under small tea gardens in the state. Further, considering the global demand in the tea market, small tea gardens have slowly started converting themselves to organic farming, which augments a bright future for the tea industry in Assam.

It may be mentioned here that as most of the small tea growers cultivate tea on leased-in land, they are not allowed to register under the Tea Board of India, and in view of this, many small tea plantations in the state remain unrecorded. Most of the small tea gardens in the state are confined to upper and central Assam, the core areas of tea plantations. The major plantations of small-scale tea are concentrated mostly in the Upper-Assam districts of Charaideo, Tinsukia, Dibrugarh, Golaghat, Sivasagar, and Jorhat; and Middle-Assam districts of Biswanath and Udalguri; and the minor concentrations in the Middle-Assam districts of Sonitpur and Nagaon (Table 6.2 and Figure 6.2). However, due to increased economic sustainability and consequent

Table 6.2 District-wise small tea growers of Assam registered with Tea Board of India, 2022–2023

District	*No. of small tea growers*	*Area in hectare*	*% of tea area to district's total area*
Kokrajhar	471	1,275.48	0.49
Dhubri	81	52.66	0.03
Goalpara	221	249.94	0.12
Barpeta	3	6	0.03
Morigaon	2	6.45	0.01
Nagaon	2,234	3,370.17	1.33
Sonitpur	3,881	4,294	1.24
Lakhimpur	3,066	2,295.58	0.79
Dhemaji	1,855	1,847.32	0.73
Tinsukia	2,4650	23,143	6.02
Dibrugarh	2,8555	21,475.2	6.35
Sivasagar	5,037	4,488.73	2.83
Jorhat	8,629	6,708.97	3.53
Golaghat	1,3578	1,3316.8	4.06
KarbiAnglong	2,672	3,740.77	0.51
Dima Hasao	4	10.47	0.02
Cachar	204	575.56	0.15
Karimganj	30	105.05	0.06
Hailakandi	9	28.68	0.02
Bongaigaon	594	492.8	0.43
Chirang	30	30.05	0.02
Kamrup	5	17.53	0.01
Kamrup Metro	0	0	0
Nalbari	0	0	0
Baksa	359	489.5	0.19
Darrang	34	40.92	0.02
Udalguri	7,580	8,612.4	4.37
Charaideo	1,0997	10,047.4	9.59
Biswanath	1,0767	10,914.8	6.11
Hojai	0	0	0
Majuli	0	0	0
South Salmara	0	0	0
West K. Anglong	0	0	0

Source: Tea Board of India, Guwahati, 2023.

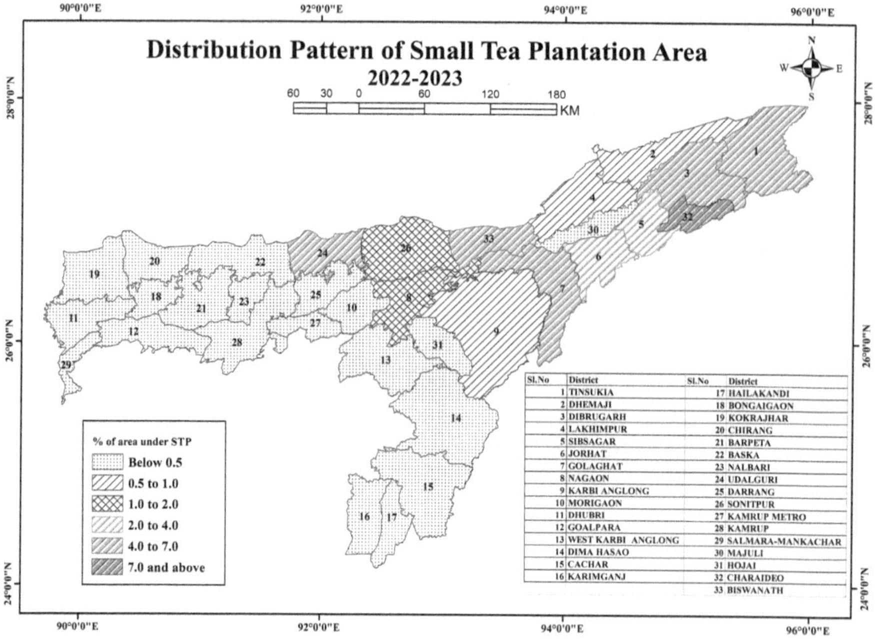

Figure 6.2 District-wise distribution pattern of STP (Small Tea Plantation) area.

Source: Prepared by author

popularity, the STP has been further spread to the Upper-Assam districts of Lakhimpur and

Dhemaji; Lower-Assam districts of Bongaigaon, Kamrup, Baksa and Kokrajhar; Barak-valley district of Cachar; and hill-district of Karbi Anglong (Table 6.2). Among the districts in Assam, Dibrugarh accounts for 30% of such tea gardens, followed by Tinsukia with 22%, Jorhat with 13%, Golaghat with 12%, and Sivasagar with 11%. In comparison, the other districts contribute to around 11% of the total small tea gardens in the state (Economic Survey of Assam, 2009–10).

The current statistics reveal that there are 1,25,548 small tea growers in Assam, and the total area under small tea holding cultivation stands at 1,17,635.76 hectares (Tea Board India, 2023). With a staggering 28,555 tiny tea farmers, Dibrugarh district records the largest area under small tea cultivation (21,475.16 hectares). Tinsukia district stands next with 24,650 small tea cultivators, followed by Golaghat (13,578), Charaideo (10,997), Biswanath (10,767), Jorhat (8,629), and Udalguri districts (7,580) (Table 6.2). The remaining districts also have small tea growers, but the number is quite insignificant. These statistics though offer authentic data about STGs in the form of registered growers, many other STGs have registered themselves as agriculturists instead of tea growers, ultimately hiding the actual picture of small tea cultivation in the state.

Growth of Tea Area and Changes in Agricultural Land Use

There has been a marked change in the agricultural land use pattern during the last few decades in Assam as a result of the conversion of agricultural land, including areas under rice cultivation, into smallholding tea plantations in different parts of the state. Such a trend is largely associated with higher revenue earnings in tea as compared to rice and other traditional crops. This trend has become such that many people in rural areas have even started converting kitchen gardens in the homestead area into small tea gardens by involving all the family members and seasonally engaging wage labourers on a temporary basis. The importance of small tea gardens in the state has increased due to the increased participation of the unemployed, educated youth of the rural and semi-urban areas. Moreover, the government is also encouraging the growth of small-scale tea cultivation by extending subsidies, loans, and other incentives to attract more and more educated unemployed youth. Many new business opportunities and services, such as the supply of garden implements, transportation of green leaves, and retail outlets of agro-chemicals and manures, leaf agents, packaging, and so on, have been created through the expansion of smallholding tea cultivation, which has further contributed to the socio-economic development of the state.

Tea cultivation in Assam has significantly transformed the state's overall land use and land cover. Initially, most parts of the state were largely covered by forests and a variety of traditional crops including rice for subsistence. However, the establishment of large tea plantations, mainly in the upper Brahmaputra Valley and Barak Valley during the 19th century, caused large-scale deforestation and the conversion of diverse ecosystems into monoculture tea estates. Land use change as a consequence of deforestation and tea plantations in Assam reflects a complex interplay of agricultural development, industrialization, and environmental degradation. Initially, tea plantations were practised in the highlands with slopes and good drainage, but nowadays, smallholding tea cultivation is mainly done in rice fields and scrub and barren lands, where there is no water stagnation during the summer monsoon season. It may be mentioned here that initially the conversion of paddy field into smallholding tea plantation started largely among the farmers with large landholding, but later it happened among many small farmers as well. A significant portion of the small tea farmers in rural Assam, who own relatively large tracts of land, rent out their paddy fields on a share-cropping (*adhi*) basis to others, including wage earners without land (Sharma & Barua, 2017). This has happened because these growers are finding it challenging to manage rice farming concurrently as they start plantations in the highlands – both homesteads and the landholdings away from their habitation. In the foothills of Udalguri, Sonitpur, and Biswanath districts of Assam, several farmers belonging to the Nepali community, who are traditionally engaged in cattle rearing, are facing the challenges posed by the indiscriminate encroachment of tea plantations into their grazing land (Sharma & Barua, 2017). Rapid deforestation through the expansion of tea estates has dramatically changed the LULC, even in the eastern Himalayan region, including Assam (Prokop, 2018).

The impact of the expansion of tea plantations, including its smallholding venture, can be observed in the marked decline of the net sown area and area under rice cultivation in the state of Assam during the last five decades (1970–2020). As regards change in area under tea plantation in the state of Assam as a whole is concerned, it has always been higher than the rates of change in net sown area and area under rice cultivation during the entire period. In fact, the change became negative in the case of both net sown area and area under rice during 2010–2020 (Table 6.3). This phenomenon is even more significant in the districts of Darrang, Lakhimpur, Sivasagar, and Nagaon.

As the scope of tea plantations grows from the traditional core areas towards the west of the state, it generates more enthusiasm among the youth for further improvement of tea cultivation, including the production of organic tea.

Conclusion

The foregoing discussion reveals that tea cultivation has been largely carried out in Assam almost since the beginning of British rule due to its prevailing favourable geographical conditions. Apart from having large tea estates, it is dotted with numerous small-scale tea gardens providing livelihood to many farmers in general and educated unemployed youth in particular. Due to limited benefits generated from the rice-dominant traditional agricultural sector, many farmers of rural Assam have been shifting their focus to smallholding tea cultivation as a sustainable and profitable source of livelihood. This transformation was initially more observed in the traditionally rice-cultivated areas in relatively highland devoid of adequate stagnant water. In this way, the small-scale tea plantation has opened a wide range of opportunities to the rural farming communities in the state. It is further observed

Table 6.3 Temporal change in area under tea plantations in relation to the temporal change in net sown area and area under rice in selected districts of Assam, 1970–2020

District (Undivided)	*% change in net sown area*		*% change in area under rice*		*% change in area under tea plantations*	
	1970–2010	*2010–2020*	*1970–2010*	*2010–2020*	*1970–2010*	*2010–2020*
1) Goalpara	24.07	– 9.45	9.93	– 14.92	114.29	24.44
2) Kamrup	18.88	– 18.40	22.27	– 22.69	53.85	– 12.50
3) Darrang	19.66	– 3.01	33.56	0.92	38.94	58.19
4) Lakhimpur	45.95	– 1.25	87.93	– 25.98	247.62	17.80
5) Dibrugarh	39.54	15.03	18.43	– 47.65	123.27	– 10.10
6) Sivasagar	17.46	9.63	28.34	– 16.84	87.57	4.10
7) Nagaon	27.22	– 14.93	10.27	1.99	30.16	37.80
8) Cachar	22.00	– 1.16	28.67	– 17.10	5.19	13.27
Assam	**27.17**	**– 3.94**	**31.45**	**– 16.76**	**79.10**	**7.79**

Source: Computed based on data from Statistical Handbook, Assam, 1973, 2012, and 2023, Directorate of Economics and Statistics, Govt. of Assam, Guwahati.

that the tea cultivation, whether it is large- or small-scale, has a profound impact on agricultural land use change in the state, driving marked transformation in farming ecosystems and local economies in rural Assam. Although this phenomenon brings about many economic advantages, it has some negative impacts including decline in crop diversity, which is likely to threaten food security. However, emerging sustainable practices offer a pathway to mitigate these likely adverse effects. By adopting environment-friendly farming practices, such as organic farming and agro-forestry along with continuation of traditionally practised rice and seasonal vegetable farming in the lowlands, the negative impacts of tea cultivation can be addressed to a great extent by way of enriching biodiversity and preserving ecosystem health. In any case, achieving a balance between economic gains and environmental stewardship is essential for ensuring the long-term viability of tea cultivation along with a traditionally rich agricultural landscape in the state.

References

Baroowah, G. P. (2006). *Tea-legend, life and livelihood of India, in Red River*. LBS Publications.

Baruah, P. (2008). *The tea industry of Assam: Origin and development* (pp. 1–318). EBH Publishers. ISBN: 81-903834-6-2.

Biggs, E. M., Gupta, N., Saikia, S. D., & Duncan, J. M. (2018). Tea production characteristics of tea growers (plantations and smallholdings) and livelihood dimensions of tea workers in Assam, India. *Data in Brief*, *17*, 1379–1387.

Borah, D., & Bora, A. K. (2023). Quantifying agricultural characteristics using several indicators and extensive field surveys in an ecologically fragile watershed of India. *Australian Journal of Crop Science*, *17*(11), 867–875.

Chakraborty, S. (2013). Tea garden workers are now entrepreneurs. *Employment News*, *37*(52).

Chetry, R., & Kar, B. K. (2022). Tea industry of Assam. In A. K. Bhagabati, A. K. Bora, & B. K. Kar (Eds.), *Geography of Assam* (pp. 309–330). Rajesh Publications.

Das, P., & Zirmire, J. (2018). Tea industry in India: Current trends and future prospective. *Journal of Pharmacognosy and Phytochemistry*, *7*(1), 407–409.

Deka, N., & Goswami, K. (2021). Economic sustainability of organic cultivation of Assam tea produced by small-scale growers. *Sustainable Production and Consumption*, *26*, 111–125.

Directorate of Economics and Statistics, Government of Assam. (2023). *Economic survey of Assam, 2009–10*. Directorate of Economics and Statistics, Government of Assam.

Dutta, S. K. (2024). A study on the production and marketing scenario of tea industry in India. In S. Choudhury, J. Sharma, & M. Sabnam (Eds.), *Indian economy in the globalised era from the lenses of millennials & Gen Z* (pp. 38–47). Kavya Publications.

Economic Times Bureau. (2014, June 20). *Sarkar, Debashish, 12th Sept. 2024 at 12:10 pm*. www.economictimes.indiatimes.com

Evans, J. C. (1992). Tea in China: The history of China's national drink. In *Contributions to the study of world history, number 33*. Greenwood Press.

George, T. (1980, October–December). The marketing of Indian tea. *Indian Manager*, *XI*(4).

Gesimba, R. M., Langat, M. C., Liu, G., & Wolukau, J. N. (2005). The tea industry in Kenya: The challenges and positive developments. *Journal of Applied Sciences*, *5*(2), 334–336.

Goswami, A. K., & Goswami, M. K. (2008). *ChahKheti*. GranthaSanskriti.

Guha, A. (1967, December). Colonisation of Assam: Second phase 1840–1859. *Indian Economic and Social History Review (IESHR)*, *4*, 1–2.

Hazarika, K. (2013). Tea industry is the backbone of Indian economy: It's present and future. *ZENITH International Journal of Business Economics & Management Research*, *3*(5), 58–69.

Huang, P., Broto, V. C., & Westman, L. (2024). Harnessing social innovation for a just transition: A case study of tea industrialization in China's era of ecological civilization. *Environmental Innovation and Societal Transitions*, *53*, 100917.

ITC. (2022). *Annual bulletin of statistics*. International Tea Committee.

Koros, D. K., Tshelang, C. K., & Masangano, M. (2023). The current trend in Kenya's tea industry. *Agricultural Socio-Economics Journal*, *23*(2), 231–238.

Kumar, P., Badal, P. S., Singh, N. P., & Singh, R. P. (2008). Tea industry in India: Problems and prospects. *Indian Journal of Agricultural Economics*, *63*(1).

Parida, B. R., Mahato, T., & Ghosh, S. (2024). Monitoring tea plantations during 1990–2022 using multi-temporal satellite data in Assam (India). *Tropical Ecology*, *65*(3), 387–398.

Prokop, P. (2018). Tea plantations as a driving force of long-term land use and population changes in the Eastern Himalayan piedmont. *Land Use Policy*, *77*, 51–62.

Ranweera, S. H., Siriwardena, D., Silva, R., Bandara, N., & Arachchige, U. (2020). Environmental pollution by tea processing. *Journal of Research Technology & Engineering*, *1*(4), 95–104.

Sachitra, K. M. V. (2016). Critical success factors for innovation an empirical analysis on tea industry in Sri Lanka. *Global Journal of Business and Management*, *3*(3), 066-073.

Saikia, B. (2008, April–June). Tea garden community and Adivasi assertion in Assam. *Literature, Society, Polity: Trends and Perspectives*, *5*(I).

Sandika, A. L. (2018). Tea cultivation in Sri Lanka as a legacy of the British colonization: Impacts and trends. *Journal of Social Sciences and Humanities Review*, *3*(1).

Sarma, G. (2013). A historical background of tea in Assam. *Pratidhwani: The Echo*, An online *Journal of Humanities and Social Science*, Dept. of Bengali, Karimganj College, Assam, *1*(4), 123−131. Retrieved June 18, 2013, from www.thecho.in

Sharma, C. K., & Barua, P. (2017). Small tea plantation and its impact on the rural landscape of contemporary Assam. *International Journal of Rural Management*, *13*(2), 140–161.

Sharma, N., Madhusudan, M. D., & Sinha, A. (2012). Socio-economic drivers of forest cover change in Assam: A historical perspective. *Economic and Political Weekly*, 64–72.

Sun, M., Jia, X., Yang, D., Lu, B., Han, F., & Shi, F. (2024). Life cycle environmental impact assessment of green tea production in China. *Journal of Cleaner Production*, *434*, 140377.

Taher, M. (2022). Historical evolution and peopling. In A. K. Bhagabati, A. K. Bora, & B. K. Kar (Eds.), *Geography of Assam* (pp. 13−30). Rajesh Publications.

Tea Board of India. (1961–2023). *Tea statistics, 1961–2023*. Under Ministry of Commerce and Industry, Govt. of India.

Thushara, S. C. (2015). Sri Lankan tea industry: Prospects and challenges. *Proceedings of the second Middle East conference on global business, economics, finance and banking*.

Yu, Z., Jiao, W., & Min, Q. (2024). Carbon footprints of tea production in smallholder plantations: A case study of FenghuangDancong tea in China. *Ecological Indicators*, *158*, 111305.

Part II

Environmental and social challenges in agricultural land use

7 Agriculture's response to diverse environmental and other challenges in the driest state in the driest continent

Guy M. Robinson and Bingjie Song

Introduction

Agriculture dominates land use in those parts of Australia where broadacre and/or intensive farming, such as horticulture, dairying, and pig/poultry production, are possible. In 2022–2023, agricultural products accounted for 13.6% of goods and services exports by value, 2.7% of the country's total value added and 2.2% of employment (ABARES, 2024, p. 2). It produces enough food to feed over 60 million people but has a domestic population of only 27 million, so 72% of production is exported. In part, this reflects the range of different conditions under which agriculture operates – from tropical and sub-tropical conditions, which enable production of sugarcane, mangoes, and bananas, to more temperate and Mediterranean-type climate areas supporting wine, horticulture, dairying, oilseeds, and grains, to semi-arid to arid conditions, where extensive grazing on rangelands dominates over large areas, and there are 20 cattle stations over 10,000 km^2. In effect, agricultural activities and land management vary hugely across Australia, reflecting the diversity of climates and environmental conditions, accessibility to the major domestic markets and ports for export, the impact of the purchasing power of the big cities, notably Melbourne (5.3 million population), Sydney (5.2 million), South-East Queensland (4 million), Perth (2.1 million), and Adelaide (1.4 million), and market orientation. The labour force in agriculture is just over 250,000, not including short-term overseas workers who play a big role in the annual fruit harvest. In 2021–2022, there were 87,800 agricultural businesses with an estimated value of operations exceeding A$40,000, though the number of businesses has fallen for over five decades (FA, 2023).

One of the characteristics of farming in Australia is the high level of risk involved in operating farm businesses because of the heavy reliance on overseas markets and the harsh climatic conditions faced in many regions, exacerbated by human-induced climatic change. The risks vary greatly across locations, farm types, and sizes. For example, cropping farms tend to face the greatest challenges from climate change, as crops can struggle to deal with the hotter and drier conditions now occurring (Huai, 2017), whilst livestock producers tend to face a wider range of price fluctuations, and in the short-term they can partially offset the impacts of drought impacts by de-stocking, that is through increasing sales of livestock

DOI: 10.4324/9781032672779-10

(Hughes et al., 2019). Because they tend to be more diversified and possess more options for centralised marketing and storage, cropping farms may be able to withstand price fluctuations better, but loss of overseas markets can be devastating. A good recent example is the effect of the lack of access to the lucrative Chinese market for Australian wine when tariffs were raised in March 2021, contributing to a significant downturn in the wine industry.

Human-induced climate change presents increasing risks to the agricultural sector, with broadly a trend towards warmer and drier conditions across much of the continent, which is already very arid: with 33% classified as arid (<250 mm rainfall per annum) and 75% classified as semi-arid (<500 mm) (Hughes, 2011, p. 190). Moreover, around 80% of the continent has at least three months of the year lacking any effective precipitation. In the last century, overall Australia has warmed by about 1 °C whilst rainfall trends have seen a decline across the western half of the continent and some increases elsewhere, though often through damaging heavy rainfall events. Increases in potential evaporation are predicted across large areas, but increased CO_2 may favour forest growth and more danger of destructive bushfires. For agriculture, more severe storm events such as tropical cyclones will cause more destruction and flooding, major heat events and longer periods of drought will make adaptation more difficult, and, potentially, agriculture faces numerous serious climatic threats but distributed unevenly across both space and time.

It is predicted that by 2050, climate change will have halved the irrigated agricultural output of the Murray–Darling Basin, which currently accounts for 50% of Australia's irrigated agricultural output by value (around A$7.2 billion per year). Indeed, the Millennium Drought in the 2000s reduced yields from the Basin by about 20%, cutting the growth of national gross domestic product (GDP) by 1%, and such severe droughts may become more common. In the 2000s and 2010s, federal drought assistance cost A$6 billion (Steffen et al., 2019, p. 15). Wheat yields in the main producing area in Western Australia, which accounts for half of Australia's wheat crop, are projected to fall by >40% by 2090 if greenhouse gas emissions remain high (Steffen et al., 2019, p. 15).

This chapter examines how Australian agriculture is addressing some of the main environmental challenges, especially those from climate change. Emphasis is placed on the responses by farmers to risk, relating not only to the physical conditions of production but also to market fluctuations, which often reflect changes in overseas markets given Australia's high reliance on exporting agricultural produce, and threats to farming posed by urban sprawl and lack of government protection for the agricultural sector. The latter reflects the long-term application of a neoliberal approach favouring free trade and lack of substantive subsidies for agriculture (Rosewarne, 2019). This chapter will focus on two examples from South Australia, the driest of the Australian states and territories, with only one-fifth of the state receiving more than 250 mm (10 inches) per annum and, even in this more well-watered area, less than half exceeds 400 mm (16 inches) (McCaskill & Richards, 1998). One consequence is that substantial areas of the state do not support agriculture. In 1865, the state's Surveyor-General, George W. Goyder, produced a map showing those areas to the south of a line (subsequently termed Goyder's

Line) running roughly east–west that received annual rainfall of 250 mm and which he deemed suitable for cropping (Meinig, 1961) (Figure 7.1). However, this "frontier" of settlement fluctuated in the second half of the 19th century, with the area under cultivation being pushed northwards after growing seasons with good rainfall, only for subsequent retreats when the rains failed (Meinig, 1962). North of the line the land is suitable only for extensive grazing, though mineral deposits have supported a substantial mining industry. The grazing land is held as pastoral lease by graziers, but, especially in the north-western part of the state, there is land reserved for indigenous groups, notably A̲ngu Pitjantjatjara Yankunytjatjara formerly the North-West Aboriginal Reserve (covering 102,650 km²). The harsh physical conditions there limit economic opportunities for the around 2,500 population, and on the extensive grazing lands of the cattle stations to the east and south careful management of the rangelands is required to support livestock in the harsh conditions (Hacker et al., 2020; Waudby et al., 2012).

Towards the south of Goyder's Line there are two areas where a combination of irrigation and winter rainfall have helped to produce highly productive horticultural specialisation, making major contributions to the state's output of wine, fruit, and vegetables. These are the focus of this chapter, namely the Adelaide Hills, lying between 10 km and 45 km east of the state capital Adelaide, and the Riverland, immediately west of the state's border with New South Wales and Victoria, where River Murray provides precious water for irrigation (Figure 7.1). The chapter draws upon our research in these two areas, which face different environmental and other issues. So, although we focus especially upon environmental concerns, we also discuss other challenges faced by farmers relating to market volatility, the changing economics of farm businesses, and land use competition. This research employed sequential mixed methods in which questionnaire surveys to farmers were followed by interviews and focus groups with a purposively selected sample drawn from respondents to the questionnaire. Interviews were also conducted with representatives of farmers' organisations, and some participant observation was incorporated into the research via visits to farm tourism businesses and farmers' markets.

In the Adelaide Hills (see second section), a 35.4% response rate (n = 75) was obtained from a list of 212 farmers compiled from members of professional growers' and farmers' associations and agricultural businesses listed in trade directories and the Yellow Pages. They completed a questionnaire, supplying information about the farmer and the farm, agricultural production and marketing, farm support from government and advisers, on-farm tourism, risks, factors affecting decision-making, and the future of the farm. In the Riverland (see third section), the response rate was much smaller, just 12% (n = 69) from members of an irrigation trust. Whilst this is a small proportion of the membership, in terms of the crops grown by the sample respondents, the balance of crop types closely resembled agricultural census data for the study area in terms of the proportion of farms with different types of crops (Robinson & Song, 2023, p. 5). The questions in the survey covered similar topics to that for the Adelaide Hills but with additional information sought on irrigation and water-use practices. The Riverland respondents were

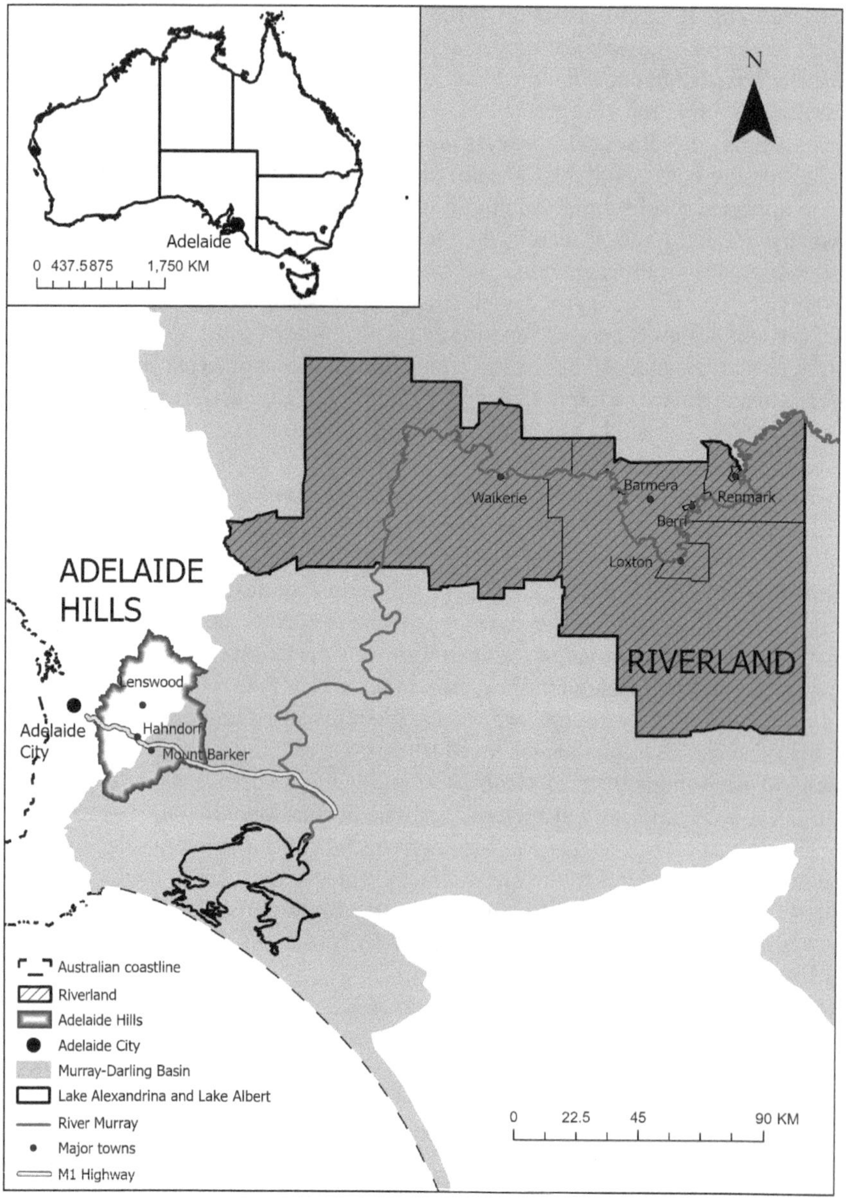

Figure 7.1 The study areas.

Source: Digital boundary files supplied from Australian Bureau of Statistics, Australian Statistical Geography Standard (ASGS) Edition 3; content prepared by the authors.

asked about their views of pro-environmental actions being taken by the irrigation trust, and in both study areas farmers were requested to provide information about their own on-farm pro-environmental actions.

Interactions between farming and urban development: the Adelaide Hills

The administrative districts of Adelaide Hills and Mount Barker, herewith collectively referred to as the Adelaide Hills, cover around 1,400 km^2, with a population of over 80,000 and an elevation above 300 m. A motorway constructed in the 1990s links the Hills to Adelaide (population 1.3 million) and has encouraged commuting to work from the Hills to the city as well as flows of tourists originating in the city, especially day-trippers, contributing to the growth of tourist enterprises in the Hills. Since the arrival of European settlers in the 1830s, the area has been dominated by horticulture and viticulture, with long-term specialisations in apples, pears, and cherries, and more recent expansion of wine grapes and a range of other fruit and vegetables, including strawberries, flowers, alpacas, nuts, olives, and brussels sprouts. There are also enterprises raising beef cattle, sheep, and horses on land suitable for pasture.

The annual rainfall varies from between 1,000 mm in the highest areas to just over 500 mm on the Hills' eastern slopes which are in rain shadow. More than 75% of the rain falls between April and October. Summer daily average temperatures are between 25 °C and 30 °C, with winter equivalents between 12 °C and 18 °C. In the higher and well-watered reaches of the central Hills (>800 m), around Lenswood, the higher rainfall has supported family farming based on production of apples and pears, occupying around 1,500 ha. There are some fourth- and fifth-generation farms in the area, but the relatively small size of farms (<50 ha) and steep slopes render mechanisation difficult and limit economies of scale on individual holdings. However, some advantages of scale have been developed from the 1930s in cooperative storage and processing.

A clear management strategy on many farms is to diversify into multiple functions, referred to as multifunctional agriculture (MFA), including activities additional to the traditional production of food and fibre. This is manifest not only in the growth of farm-based tourism and processing but also in an emphasis on environmental initiatives on some farms. Multifunctionality extends beyond diversification into a wider range of crops and livestock on individual farms to help minimise risk, but it can serve the same objective of providing multiple income streams. Wine production lends itself well to this through the production of wine on many farms that grow grapes, from small 'boutique' wineries to larger but largely family-owned businesses to a small number of large-scale producers on farms owned and run by corporate producers, such as Australian Vintage and Penfolds. While there are some grape growers who simply supply grapes to wine producers, most growers produce their own wines and have developed tourism businesses through offering cellar-door visits, restaurants/cafes, and high-quality accommodation. The Adelaide Hills have become well-known as a cool climate wine region with over 90 wine labels and 50 cellar doors. Its main wines are Chardonnay, Sauvignon Blanc, Pinot Noir, and Pinot Gris, but other cool climate wines such as Grüner Veltliner are growing in popularity. Several fruit producers in the Hills have also diversified into tourism via pick-your-own (PYO) fruit, food outlets, farm shops, and bed-and-breakfast whilst some of the larger operations have added processing activities, such as fruit juices, jams, chutneys, and preserves.

Tourism has grown since the 1960s, with day trips and overnight stays by city residents who participate in an array of farm-based activities, such as PYO, farm-gate retailing, cellar-door visits, cycling and walking trails, farmers' markets, and bed-and-breakfast accommodation. This reflects the proximity of the city, which has also given rise to the growth of hobby or lifestyle farming, in which urban-derived capital is invested into smallholdings, primarily as desirable lifestyle and retirement properties. A broad spectrum of farming is represented, ranging from commercial horticulture to less- and non-commercial farming, with many holdings diversifying into tourism-related activities and farm-based processing. The Adelaide Hills attracted 233,000 overnight visitors in 2023 (one-quarter from inter-state; 7% from overseas) and 1.4 million day trips, contributing $293 million to the local economy.

A decline in full-time commercial farms has been accompanied by a growth in part-time and hobby farms. Hobby farms accounted for 26.7% of Song et al.'s (2022a, 2022b) survey of farms in the Hills, compared with 16% for part-time farms. Urban sprawl and land subdivision have gradually decreased the amount of agricultural land in the peri-urban fringe, but farm fragmentation is creating opportunities for urban–rural migrant "lifestylers" (Spataru et al., 2020). This trend is leading to the growth of "amenity farming," in which landholders are motivated by lifestyle factors and environmental attributes, consuming both land and resources while producing little or nothing of value (Buxton & Butt, 2020, pp. 16–17). Amongst the hobbyists in the Hills, Song et al. (2022b) recognise two distinct groups, both investing urban-derived capital in the rural–urban fringe. One group consists of hobbyists buying land to establish horse-related properties, contributing to the growth of 'horseyculture.' These properties, often former dairy farms, maintain a pastoral landscape but feature new well-fenced paddocks and manèges (riding arenas). The second group is primarily interested in improving the natural environment on their smallholding. They may produce vegetables and fruit for home consumption and farm-gate or farmers' market sales, but their main concern is to establish native vegetation and maintain a healthier ecosystem. Some of these hobbyists as well as more commercially orientated farmers work with Landscape Boards, statutory bodies in South Australia established under the Landscape South Australia Act 2019. The Boards collaborate with communities and stakeholders to help manage landscapes (Bardsley et al., 2024; Robinson et al., 2018). They provide strategic leadership and address regional priorities for land, water, animals, pests, and biodiversity. In addition, a voluntary landcare programme has operated for over three decades, often involving local farmers, but it has delivered relatively piecemeal and small-scale environmental benefits (Robins, 2018).

Climate change is presenting major problems to viticulture and horticulture in the Hills, notably through higher temperatures and longer periods of extreme heat; more variable rainfall, and more extreme and longer droughts; increased storm volatility; and more intense, widespread bushfires (Rohr et al., 2020, p. 36). For the apple, pear, and cherry growers, the likelihood of hotter winters poses a threat to vernalisation and fruit set, while hotter conditions can also inhibit fruit development. Declining spring rains threaten the crops, while government-imposed Water Allocation Plans for the region have restricted water resource availability for orchards, partly to secure metropolitan water supplies but also to protect ecosystem functions. Research by Houston and Bardsley (2018) indicates that areas adjacent to the current apple and pear production zones could become suitable for cultivation under projected climate change conditions. However, "by 2030, climate change will shrink the bio-climatic envelope suitable for high quality apple production from 20,300 ha to about 9,000 ha" (p.35). This potential area is further reduced to 4,200 ha (omitting current production areas), after accounting for urban areas, rural residential development, public land, remnant native vegetation, and zoning restrictions likely to limit commercial-scale horticulture. Pursuing opportunities for new orchards in climatically suitable areas would likely intensify land use conflicts, both with other agricultural activities and with growing demands for residential development.

Bushfires pose an increasing threat, although the Hills have not experienced a destructive bushfire on the scale of the Ash Wednesday fires of 1983 in over 40 years. Nevertheless, two substantial fires have occurred in the last decade, including the Cudlee Creek fire in December 2019, which burned >5,000 ha, destroyed around one-third of vineyard production (c1,100 ha), worth A$20 million, and left other vineyards affected by smoke and ash taint. The fire also killed more than 3,750 cattle and horses, destroyed 98 homes, and damaged >540 outbuildings and other properties (SAIBRT, 2020). The threat from bushfires has intensified with hotter and drier conditions, further exacerbated by weather extremes, for example, very hot days and higher storm intensities, leading to more severe hailstorms. In response, expensive netting, part-funded by government grants, has been increasingly adopted as a protective measure.

The farmers themselves recognise the increasing environmental challenges they face. Hobby and part-time farmers acknowledged climate change as the chief risk to the farm, followed by pests and diseases, also partly related to the changing climate, with soil-borne fungal disease and fruit fly being the main threats identified. Full-time farmers placed risk from pests and diseases only fractionally ahead of climate change as a major threat to their business (Table 7.1). The group most readily acknowledging the threat posed by climate change comprised part-time farmers, with all but one citing this as a current or potential risk. These are mainly cherry, apple, and pear growers nearing retirement, often using PYO as a labour-saving means of cutting costs, but exposing them more to the vagaries of weather when essentially relying on tourists to harvest their crops. Over two-thirds of hobby farmers, the most well-educated farmers in the sample, recognised that the climate is gradually warming, with spring coming earlier, the length of time with high

Table 7.1. Principal current or potential risks on the farm, Adelaide Hills

Risky	*Hobby farmers*		*Part-time farmers*		*Full-time farmers*	
	n = 20	*%*	*n = 12*	*%*	*n = 43*	*%*
Climate change	14	70	11	92	27	63
Pests/diseases	12	60	5	42	28	65
Urban development	9	45	2	17	7	16
Excessive regulation	7	35	5	42	15	35
Government policy	6	30	4	33	3	7
Living costs	6	30	1	8	4	9
Biosecurity	5	25	1	8	10	23
Lack of customers unstable markets	5	25	0	0	3	7
Labour costs	3	15	5	42	27	63
Economic conditions	3	15	2	17	17	40
Succession	1	5	4	33	10	23
Oversupply of your produce	1	5	4	33	10	23

Note: Each respondent was asked to list up to five main risks.

temperatures increasing, and that there is a rising prevalence of extreme weather events (notably drought).

Using netting to protect fruit from inclement weather is just one example of ways in which farmers are not only adapting to perceived climate threats but are engaged in pro-environmental actions. When asked what the most important outcomes of their farming business were, the highest priorities from hobby farmers were protecting the environment and maintaining the variety of flora and fauna. Food production and economic profit were the most important for part-timers, whilst full-time farmers listed economic profit first but with environmental protection in second place.

For many of the farmers these desired outcomes were translated into specific actions (Table 7.2), with several activities relating directly to pro-environmental management. These included providing habitat for native plants and animals (including beneficial insects), which was undertaken by half of the farmers in the sample, and creating wildlife corridors (48%). Just over 40% actively restored native vegetation. These activities were not just confined to pro-environmental hobbyists and part-time farmers but were pursued by some full-time farmers: 42% providing habitat and 37% creating wildlife corridors. These farmers liked the idea of leaving the farm healthy and in good condition for their successors. This was closely related to their concept of what constituted a good farmer and how they might be viewed by their neighbours. It was also translated into direct actions to maintain the soil in good health, with 71% of all farmers acting to prevent soil erosion and 73% minimising the use of fertilisers and pesticides to prevent harmful run-off. Just over one-third preferred to use organic fertilisers whilst ten farmers claimed to farm organically, though not all had formal organic certification.

Table 7.2. Farm-based activities protecting the environment and biodiversity, Adelaide Hills

Activity	*Hobby farmers* (*n* = 20)		*Part-time farmers* (*n* = 12)		*Full-time farmers* (*n* = 43)	
	n	%	*n*	%	*n*	%
Managing nature reserve	7	35	5	42	7	16
Providing habitat	12	60	8	67	18	42
Improving management of farm wastes	4	20	4	33	14	33
Irrigation	4	20	5	42	33	77
Minimising fertilizers/ pesticides	11	55	7	58	33	77
Prevent soil erosion	12	60	9	75	32	74
Creating wildlife corridors	12	60	8	67	16	37
Protecting riparian environment	8	40	4	33	12	28
Restoring native vegetation	13	65	7	58	12	28
Covering unproductive land	5	25	4	33	14	33
Sequestering carbon	6	30	1	8	8	19
Using organic fertilizer	6	30	3	25	17	40
Organic farming	2	10	3	25	5	12

Labour costs, economic conditions, and excessive regulations featured prominently in the risks identified by full-time farmers. Some of the orchardists stressed the increasing difficulties of staying in business. Indeed, the problems faced more broadly by this sector in the Hills were illustrated in July 2024 by the demise of a fifth-generation apple producer after his family had grown pome fruit there for over 120 years. Despite being one of the major local producers, the 80-ha farm was put on the market after facing years of diminishing returns, with the owners citing rising costs and an inability to obtain higher prices for apples from the supermarkets, with apples selling for about the same price as in 2000. This is not the first such sale and reflects the relatively small scale of production under difficult physical conditions (steep slopes) compared with competitors in Victoria.

Many of the small horticulturalists are facing a price-cost squeeze in which the cost of inputs, notably labour, machinery, and oil-based fertilisers, are rising faster than the returns from sales. The growers are adopting various strategies to combat this problem, notably by reducing labour and avoiding restrictive contracts with the major supermarkets by selling direct at the farm gate and in farmers' markets (Fielke & Bardsley, 2013). Some fruit producers are operating PYO schemes, but many are constrained by agreements with supermarkets, a problem being addressed in 2024 by a federal government enquiry. Nevertheless, the number of producers is falling as part of a national decline in farms accompanied by a rise in average farm size, with a nationwide decline in farm numbers of >43% since 1980 (ABARES, 2024).

Economic concerns were uppermost in the farmers' responses to questions about the main factors affecting their decision-making (Table 7.3). For the full-time farmers the need to remain efficient and productive to maintain incomes was apparent, with 81% referring to 'maintaining profitability' and 63% to 'maintaining efficiencies.' More than half also highlighted the need to meet market requirements, but the same proportion referred to climate change playing a role in their decision-making. Similar proportions of hobbyists and part-time farmers referred to climate change. For hobbyists, maintaining their way of life was the most important decision-making factor (85%) closely followed by environmental concerns (75%) and maintaining biodiversity (55%). Part-timers regarded meeting market requirements as the chief factor (75%) with profitability third (50%).

Managing water supplies under climatic uncertainty: the Riverland

The Riverland covers an area of 9,386 km^2 in the valley of River Murray from where it flows westwards into South Australia to where it heads south past Swan Reach towards Lakes Alexandrina and Albert (Figure 7.1). With a population of around 35,000, it is overwhelmingly an agricultural area dominated by wine and horticulture, with a few small market towns. It is about one and three-quarters to three hours' drive north-east of the state capital Adelaide. The climate has Mediterranean characteristics, with hot dry summers (daily average maxima around 33 °C), mild winters (daily average maxima around 16 °C), and annual rainfall around 260–270 mm, concentrated mainly in the winter months. The long hours of summer sunshine ensure fruit ripens fully, and low relative humidity helps restrict crop disease. It is Australia's largest wine-producing region, with just over one-fifth of the national wine grape crush from around 21,000 ha on >1,000 holdings. It is also a large producer of citrus, almonds (20% of the national output), and stone fruit (notably nectarines, peaches, and apricots), in addition to Asian vegetables (such as Bok Choi), avocados, persimmons, pistachios, figs, cherries, and dates.

The Riverland is part of the largest river system in the country, the Murray–Darling Basin (MDB), which accounts for around one-third of the food and fibre produced by Australian farmers and 70% of the country's agricultural water use, with half of its irrigated agricultural production value, and contributing A$22 billion to the Australian economy (MDBA, 2021). It is of vital significance to Australia's rural and export economies whilst also containing 16 internationally recognised and protected wetlands, 120 species of waterbirds, over 50 native fish species, and 35 endangered species, plus 40 different groups of indigenous peoples reliant on its water to maintain traditional practices. However, its management has been controversial, reflecting competing interests of the four states (New South Wales, Queensland, South Australia, Victoria) and one territory (Australian Capital Territory) in which the MDB is located, and the various stakeholders seeking different outcomes from the management of the land use and river system, for example, irrigators, broadacre farmers, environmental groups, and indigenous

Table 7.3. Most important factors affecting decision making, Adelaide Hills

Factors	*Hobby farmers* (n = 20)		*Part-time farmers* (n = 12)		*Full-time farmers* (n = 43)	
	n	*%*	*n*	*%*	*n*	*%*
Market requirements	2	10	9	75	24	56
Climate change	12	60	7	58	24	56
Improving efficiencies	8	40	4	42	27	63
Food quality	4	20	5	42	9	21
Attracting tourists	3	15	3	25	11	26
Poor returns	2	10	2	17	15	35
Maintaining biodiversity	11	55	2	17	5	12
Access to technology	1	5	2	17	2	5
Environmental concerns	15	75	2	17	10	23
Maintaining way of life	17	85	5	42	13	30
Profitability	3	15	6	50	35	81
Enabling part-time work off-farm	1	5	1	8	2	5
Input from advisors	0	0	2	17	4	9
Lack of information	0	0	1	8	1	2
Maintain for future generations	5	25	0	0	12	28
Government policy	4	20	0	0	5	12
Following neighbours	2	10	0	0	0	0
Other	1	5	0	0	3	7

Note: Each respondent was asked to list up to five main factors.

peoples (Bischoff-Mattson et al., 2018). The desire to find a balance between competing uses for water, with sufficient water for irrigation whilst maintaining thriving ecosystems (Grafton et al., 2020) has also had to address the impacts of climate change, which is creating drier and hotter conditions, with more extreme weather events (Alexandra, 2021; Prosser et al., 2021). Indeed, the MDB may lose half of its irrigated agricultural output by 2050 and over 90% by 2100 through drought and decreased rainfall and run-off (Garnaut, 2008).

The area was not settled by Europeans until the 1880s when the Chaffey brothers were granted land around what is now Renmark. This became the first irrigation settlement in the country, encompassing 4,700 ha, originally in 6-ha lots (Tripney, 1993). Further small settlements followed in the 1890s, mainly as irrigation schemes funded by state government as a response to economic depression. More lands for irrigators were granted to returning servicemen from the First World War. In the Renmark area, irrigators received water rights with part-ownership of the water supply infrastructure, becoming ratepayers liable to pay annually to the

Renmark Irrigation Trust (RIT) for their water supply. Today, the Trust manages more than 140 km of water-related infrastructure, with around 570 members (irrigators). A larger group of irrigators are members of the Central Irrigation Trust (CIT), with members in Berri, Loxton, and Waikerie, which manages irrigation water for 1,600 growers irrigating 14,000 ha.

The majority of holdings in the region (termed 'blocks') are owner-occupied by families. The two irrigation trusts supply water from River Murray to their members through an infrastructure of pumps and pipes owned by the trusts, as well as supplying drinking water to the small towns. Although the average holding is small, at under 10 ha per farm, horticulture has remained the mainstay of the Riverland economy, with the Trusts developing substantial efficiencies in water usage by improving infrastructure and adopting either drip irrigation or under-tree sprinklers and installing and running new pipes from the river to individual holdings.

To regulate competition for water between different users Basin-wide, integrated catchment management has been pursued since 1986, with the creation of the Murray–Darling Basin Commission. However, many conflicts remained unresolved, and it was replaced in 2008 by the Murray–Darling Basin Authority (MDBA) to oversee and regulate water use within the Basin. The MDBA (2022) launched a Plan in 2012 to balance the needs of competing sectoral interests and the ecology of the river system. The Plan sets the amount of water to be taken from the Basin each year, through local water resource plans, including "water for the environment" (Colloff & Pittock, 2019), "whereby federal and state environmental water holders decide when, where and how much water is released for environmental purposes that have measurable outcomes" (Robinson & Song, 2023, p. 2).

The Plan embraces the operation of a water market, first introduced in the 1990s, aimed at meeting multiple objectives via application of a strong neoliberal market-oriented policy (Pahl-Wostl, 2019). Prior to 1994, water licences were attached to particular plots of land, but subsequently they have been "unbundled" or traded, so the market would ensure it found its highest-value use (Simons, 2020). This was extended to South Australia in the SA Irrigation Act 2009, separating a water right from the land (i.e. unbundling), which has then encouraged various forms of water trading, often with a 'middleman,' a water broker, handling the trade. During the Millennium Drought, from about 1997 to 2009, Riverland irrigators imported substantial volumes of water allocations from New South Wales to maintain high-value permanent crops, such as wine grapes. However, many small irrigators regard the operation of the market as being disadvantageous, because of fluctuating water prices and concerns about unbundling, with water traded outside the region. The RIT has traded out more water entitlement than it has bought in but mainly by trading water to the federal government in exchange for efficiency improvement funding (Bull, 2018, p. 266). It is possible for irrigators to make money via water trading, but at times of water shortage the high costs of purchasing water to supplement entitlements can be highly problematic for small irrigators.

A key idea behind the concept of 'water for the environment' is to maintain the health of the river, its wetlands, and floodplains, so that a balance is preserved between a viable and thriving agricultural industry based on irrigation and the ecology of the river system. The RIT has been an active participant in recent attempts to restore environmental health, strongly supported by its membership. Starting in 2013, the RIT used federal environmental water to inundate a local wetland. Deemed to be a success, this prompted expansion of the Trust's floodplain rehabilitation programme (Bull, 2018, p. 235). The RIT also participated in MDB Plan water efficiency programmes, with individual members returning 10.9 GL of water to the Commonwealth by mid-2018 (Bull, 2018, p. 236). In 2016, the RIT signed an agreement with the Commonwealth Environmental Water Holder (CEWH) enabling some of its water to be allocated to floodplain rehabilitation over a five-year period, largely outside the main growing season for irrigators. In the five years up to June 2022, the Trust delivered >2 GL to 12 sites via its own infrastructure, inundating 120 ha (RIT, 2024). In 2020, in recognition of its work in rehabilitating the Murray floodplain, the RIT was awarded platinum certification (the highest such honour) by the international Alliance for Water Stewardship.

The survey of 69 irrigators, all members of the RIT, revealed less interest in pro-environmental actions than was shown by farmers in the Adelaide Hills. This reflects two characteristics of farms in the Riverland. First, the size of holdings is appreciably smaller than in the Hills: an average holding size of 19.7 ha compares with 41.7 ha, and second, the Riverland has a much smaller proportion of farmers who had been to university. There was a statistically significant link ($P = 0.05$) that experience of higher education is associated with greater environmental awareness and willingness to engage in pro-environmental activities on the farm. Just under one-third stated they did not engage in any farm-based pro-environmental actions, but 19% engaged in five or more activities (Figure 7.2). No statistically significant variation occurred in the interest shown by respondents in on-farm environmental actions across full-time, part-time, and hobby farmers, though the latter, with their smaller blocks, were most likely to have no environmental activities. The most common actions were minimising the use of pesticides and fertilisers (45%), preventing soil erosion (26%), covering unproductive land (19%), and using organic fertiliser (16%). Much smaller numbers than in the Hills engaged in more active attempts to care for the land, such as providing habitat for native plants and animals ($n = 14.5\%$), restoring native vegetation (10%), and creating wildlife corridors (8.5%).

Of the respondents to the questionnaire, 58.8% were either part-time farmers (38.2%) or hobbyists (20.6%). The former had an average holding size of 9 ha compared with just 3.1 ha for hobbyists and 37.9 ha for full-time farmers. As in the Adelaide Hills, the hobbyists tended to be former urban residents, mainly from Adelaide, who had purchased land from exiting irrigators to create 'lifestyle blocks,' adding both to landscape diversity and fragmentation. Three kept horses whilst half engaged in pro-environmental actions. Most grew fruit and vegetables for their own consumption or occasional farm-gate sales and possibly at the Riverland farmers' market. No hobbyist derived even as much as 10% of their income from the farm and the majority generated no farm-based income. There was a higher proportion of female-headed households amongst the hobbyists, and only one-quarter

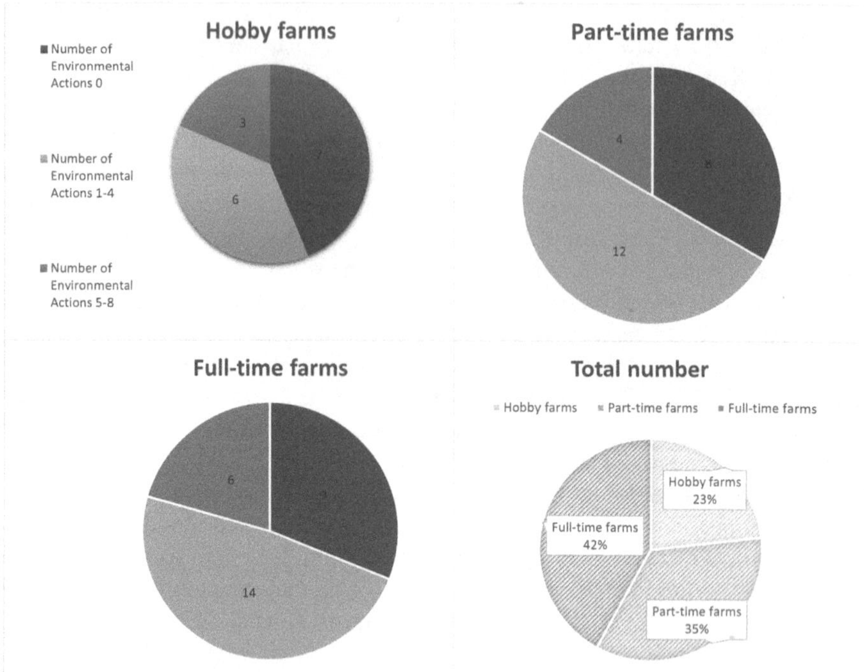

Figure 7.2 Number of on-farm environmental actions by farm type, Riverland.

Source: Prepared by the authors based on data collected via questionnaire survey.

were retirees, with the rest being part of a small-scale 'back-to-the-land' movement. Only 40% of part-time farmers derived less than one-quarter of their income from the land. Those with smaller farm-based incomes were largely nearing retirement, and their income came largely from non-farm sources.

Not surprisingly, given the reliance on irrigation water, the number one risk factor identified was reliable water supply, acknowledged by 77% of the respondents and reinforced by 45% who referred to insufficient water supply. In third place came climate change (38%) just ahead of oversupply of their produce in the market (36%). The latter referred primarily to growers of wine grapes, the majority of whom supply to major wine producers, and who comprised 42% of the sample. Wine producers in the Riverland have struggled to deal with the combined effects of climate change, water scarcity, and market volatility, the latter partly reflecting high import duties imposed on Australian wine between 2021 and 2024 by China, previously the largest importer of Australian wine (Tingey-Holyoak et al., 2024). Oversupply has also contributed to falling prices and featured prominently in interviews and focus groups with respondents. There is the prospect of grapes being grubbed up and replaced by crops more profitable at present, such as citrus or almonds, although the latter require more water and tend to be grown on larger holdings.

Water supplies also featured prominently in the decision-making factors of the irrigators, with ensuring sufficient water supply being a key factor for 70% of respondents. The economic viability of the farm featured prominently too with 68% referring to this. The threat to water supply from climate change was raised by irrigators in the focus groups and was cited as a decision-making factor by 49%. In effect, the irrigators engage in water trading when necessary to maintain sufficient water, purchasing additional water beyond entitlements in times of shortage and/or selling to make additional income if supplies are plentiful. However, during the Millennium Drought several irrigators gave up their entitlements under a federal government compensation scheme. These small block irrigators' exit grants removed 276 ha from receiving irrigation water in the Renmark area in the early 2010s, or 7% of the irrigated area under the RIT, though 44 ha were subsequently reinstated (RIT, 2022). Similar packages may be offered in future if serious drought restricts supplies, though removal of irrigation leaves stranded assets, with dead trees, vines, and bare ground alongside cultivated blocks, maybe with negative environmental impacts on the latter.

Profit was the most desired outcome from respondents' farming businesses, followed by producing more food, fibre, and beverages. Protecting food security also figured highly, just behind supporting the local community but ahead of protecting the natural environment and the landscape. Whereas diversification and multifunctionality were commonplace in the Adelaide Hills, partly as a means of spreading risk, it was less so in the Riverland, where holdings are smaller, and sunk costs associated with many of the crops currently grown limit the ability to switch from one crop to another. So diversification or wholesale change tends to be the prerogative of well-capitalised larger operators, which helps explain recent moves into almond production. In focus groups, several respondents explained that the size of their block constrained irrigators' ability to diversify, with 71% having a single crop specialisation, and wine grapes and citrus are dominant. Despite the limited potential for diversification the Millennium Drought had prompted some changes, with moves towards organic and "unconventional" production, that is into crops new to the area, especially those offering more water security (Wheeler & Marning, 2019). However, there had been few moves into on-farm tourism or processing, with the distance from Adelaide to Renmark, around 260 km, limiting the potential to attract day-visitors.

Conclusions

The two study areas illustrate different aspects of challenges facing Australian farmers more broadly across the continent. There are common issues relating to the vagaries of the market, which are threatening the survival of smaller producers unable to benefit from economies of scale in their production. Smallholders are also often disadvantaged when endeavouring to secure contracts with supermarkets. In both areas, concerns about the impacts of climate change featured prominently in risk assessments and decision-making. In the Riverland, this was manifest via the strong emphasis the irrigators placed on having access to sufficient water, with smallholders expressing concerns regarding the operation of the water market, where they felt disadvantaged compared with landholders with bigger water entitlements.

One distinctive characteristic of Australian agriculture, manifest with respect to both managing environmental and economic concerns, is the neoliberal approach by successive federal governments. This has meant that direct assistance to farmers is minimal compared with that enjoyed by farmers in many other countries. For example, direct agricultural support in Australia comprises just 0.24% of GDP (OECD, 2023), mostly invested in research and development and public goods. So the agricultural sector is encouraged to pursue high productivity and competitiveness in a commercial paradigm. This has favoured the largest producers to the extent that nationwide 50% of total output is derived by the largest 10% of broadacre farms (ABARES, 2023).

There is little direct support for farmers and graziers to develop on-farm pro-environmental measures or actions to ameliorate the effects of climate change. There are localised impacts from the voluntary Landcare programme and work by Landscape Boards, but the federal focus has primarily featured the promotion of carbon sequestration, introduced in 2011, which from 2023 has become the Australian Carbon Credit Unit Scheme. This post-dates the research reported here, with none of the farmers surveyed engaging in sequestration, though many reported changing farming practices to adapt to climate changes. Moreover, especially in the Adelaide Hills, farmers engaged in a range of pro-environmental actions to improve agro-ecology and biodiversity. With their smaller size, the scope for such actions on many Riverland farms was restricted, but the environmental watering projects overseen by the RIT align well with both farmers' concerns and environmental issues of interest to the public and government, which have contributed to its success. This is in contrast to the Adelaide Hills, where much of the environmental management in the landscape heavily relies on individual farmers' initiatives.

The future of agriculture in the two study areas is uncertain but for different reasons. Climate change adds uncertainty, threatening water supplies in the Riverland and raising risks of damaging bushfires in the Adelaide Hills. In both areas, full-time farmers will need flexibility to respond to changing market conditions. Currently, this is expressed in the Riverland in moves away from wine grapes, and in the Hills through diversification, including into tourism, on-farm processing, and direct marketing initiatives. The Hills faces greater threats from urban development and landscape fragmentation, whilst the small size of many holdings in the Riverland may favour more cessation of farm businesses, bringing about both a greater influx of hobbyists through land parcellation and consolidation into larger holdings.

References

Alexandra, J. (2021). Navigating the Anthropocene's rivers of risk–climatic change and science-policy dilemmas in Australia's Murray-Darling Basin. *Climatic Change*, *165*(1), 1–21.

Australian Bureau of Agricultural and Resource Economics and Sciences (ABARES). (2023). *Disaggregating farm performance statistics by size*. https://www.agriculture.gov.au/abares/research-topics/surveys/disaggregating-farm-size

Australian Bureau of Agricultural and Resource Economics and Sciences (ABARES). (2024, March). Snapshot of Australian agriculture 2024. *ABARES Insights*, *1*.

Bardsley, D. K., Winsborough, S., Skinner, W., & Drew, G. (2024). The governance of hydrosocial risk in peri-urban South Australia. *Geographical Research*, *62*(4), 553–568.

Bischoff-Mattson, Z., Lynch, A. H., & Joachim, L. (2018). Justice, science, or collaboration: Divergent perspectives on Indigenous cultural water in Australia's Murray–Darling Basin. *Water Policy*, *20*(2), 235–251.

Bull, G. (2018). *One hundred and twenty-five years of the Renmark Irrigation Trust No. 1, 1893 to 2018*. Renmark Irrigation Trust.

Buxton, M., & Butt, A. (2020). *The future of the fringe: The crisis in peri-urban planning*. CSIRO Publishing.

Colloff, M. J., & Pittock, J. (2019). Why we disagree about the Murray–Darling Basin Plan: Water reform, environmental knowledge and the science-policy decision context. *Australasian Journal of Water Resources*, *23*, 88–98.

Fielke, S. J., & Bardsley, D. K. (2013). South Australian farmers' markets: Tools for enhancing the multifunctionality of Australian agriculture. *GeoJournal*, *78*, 759–776.

Freshlogic Analytics (FA). (2023). *Understanding the Australian farming industry in 2023*. https://freshlogic.com.au/articles/australian-farming-industry-in-2023/

Garnaut, R. (2008). *The Garnaut climate change review*. Cambridge University Press.

Grafton, Q., Colloff, M., Marshall, V., & Williams, J. (2020). Confronting a 'post-truth water world' in the Murray-Darling Basin, Australia. *Water Alternatives*, *13*(1), 1–27.

Hacker, R. B., Sinclair, K., & Pahl, L. (2020). Prospects for ecologically and socially sustainable management of total grazing pressure in the southern rangelands of Australia. *The Rangeland Journal*, *41*(6), 581–586.

Houston, P., & Bardsley, D. K. (2018). Climate change adaptation for peri-urban horticulture: A case study of the Adelaide hills apple and pear industry. *South Australian Geographical Journal*, *114*(1), 29–42.

Huai, J. (2017). Dynamics of resilience of wheat to drought in Australia from 1991–2010. *Scientific Reports*, *7*(1), 9532.

Hughes, L. (2011). Climate change and Australia: Key vulnerable regions. *Regional Environmental Change*, *11*, 189–195.

Hughes, N., Galeano, D., & Hattfield-Dodds, S. (2019). The effects of drought and climate variability on Australian farms. *ABARES Insights*, *6*.

McCaskill, M., & Richards, E. S. (1998). *Climate of South Australia*. Britannica, https://www.britannica.com/place/South-Australia

Meinig, D. W. (1961). Goyder's line of rainfall: The role of a geographic concept in South Australian land policy and agricultural settlement. *Agricultural History*, *35*(4), 207–214.

Meinig, D. W. (1962). *On the margins of the good earth: The South Australian wheat frontier, 1869–1884*. Rand McNally.

Murray-Darling Basin Authority (MDBA). (2021). *Murray-Darling Basin authority annual report 2020–21*. MDBA.

Murray-Darling Basin Authority (MDBA). (2022). *A plan for the Murray-Darling Basin*. MDBA. https://www.mdba.gov.au/basin-plan/plan-murray-darling-basin

Organisation for Economic Co-Operation and Development (OECD). (2023). *Agricultural policy monitoring and evaluation 2023*. OECD.

Pahl-Wostl, C. (2019). The role of governance modes and meta-governance in the transformation towards sustainable water governance. *Environmental Science and Policy*, *91*, 6–16.

Prosser, I. P., Chiew, F. H., & Stafford Smith, M. (2021). Adapting water management to climate change in the Murray–Darling Basin, Australia. *Water*, *13*(18), 2504.

Renmark Irrigation Trust (RIT). (2022). *Water stewardship plan*. RIT. https://rit.org.au/wp-content/uploads/2022/12/RIT-Water-Stewardship-Plan-2022-23.pdf

Renmark Irrigation Trust (RIT). (2024). *Water for the environment*. https://rit.org.au/environmental-watering/

Robins, L. (2018). More than 30 years of 'landcare' in Australia: Five phases of development from 'childhood' to 'mid-life' crisis or renewal? *Australasian Journal of Environmental Management*, *25*(4), 385–397.

Robinson, G. M., Bardsley, D. K., Raymond, C. M., Underwood, T., Moskwa, E., Weber, D., Waschl, N., & Bardsley, A. M. (2018). Adapting to climate change: Lessons from farmers and peri-urban fringe residents in South Australia. *Environments*, *5*(3), 40.

Robinson, G. M., & Song, B. (2023). Managing water for environmental provision and horticultural production in South Australia's Riverland. *Sustainability*, *15*, 11546.

Rohr, S., Mounter, S., Fleming, E., & Griffith, G. (2020). The Australian apple industry – trends and challenges. *Australasian Agribusiness Review*, *28*(2), 28–46.

Rosewarne, S. (2019). The structural transformation of Australian agriculture: Globalisation, corporatisation and the devalorisation of labour. *Journal of Australian Political Economy*, *84*, 175–218.

Simons, M. (2020). Cry me a river: The tragedy of the Murray-Darling Basin. *Quarterly Essays*, *77*, 1–114.

Song, B., Robinson, G. M., & Bardsley, D. K. (2022a). Multifunctionality and path dependence: Farmer decision-making in the peri-urban fringe. *Journal of Rural Studies*, *96*, 64–77.

Song, B., Robinson, G. M., & Bardsley, D. K. (2022b). Hobby and part-time farmers in a multifunctional landscape: Environmentalism, lifestyles, and amenity. *Geographical Research*, *60*(3), 480–497.

South Australian Independent Bushfire Review Team (SAIBRT). (2020). *Independent review into South Australia's 2019–2020 bushfire season*. Government of South Australia.

Spataru, A., Faggian, R., & Docking, A. (2020). Principles of multifunctional agriculture for supporting agriculture in metropolitan peri-urban areas: The case of Greater Melbourne, Australia. *Journal of Rural Studies*, *74*, 34–44.

Steffen, W., Mallon, K., Kompas, T., Dean, A., & Rice, M. (2019). *Compound costs: How climate change is damaging Australia's economy*. Climate Council of Australia.

Tingey-Holyoak, J. L., Wheeler, S. A., Seidl, C., & Zuo, A. (2024). Understanding viticultural financial returns: A case study from the Riverland, South Australia. *Journal of Rural Studies*, *110*, 103334.

Tripney, D. L. (1993). *A brief history of the Renmark Irrigation Trust. The first hundred years, 1983–1993*. Renmark Irrigation Trust.

Waudby, H. P., Petit, S., & Robinson, G. M. (2012). Pastoralists' perceptions of biodiversity and land management strategies in the arid Stony Plains region of South Australia: Implications for policy-makers. *Journal of Environmental Management*, *112*, 96–103.

Wheeler, S. A., & Marning, A. (2019). Turning water into wine: Exploring water security perceptions and adaptation behaviour amongst conventional, organic and biodynamic grape growers. *Land Use Policy*, *82*, 528–537.

8 Shaping agriculture and agri-food system futures in Aotearoa New Zealand

The role of values and relationships with land and self

Daniellé A. du Plooy and Sean Connelly

Introduction

It is widely agreed that systemic change is required for the long-term sustainable transformation of agri-food systems (e.g. see Johnston et al., 2009; Levkoe, 2011; Widener & Karides, 2014; Allen, 2014; Haynes-Maslow & Salvador, 2015; Moragues-Faus & Marsden, 2017). Global food production is a major contributor to climate change and environmental damage, driving issues like deforestation, habitat loss, water contamination and depletion, and burning hill and high-country tussock grasslands (MEA, 2005; Willett et al., 2019). Additionally, the heavy use of chemicals and pesticides harms soil health while agriculture remains a significant source of greenhouse gas emissions. As a result, a radical transformation of the global, industrialized agri-food system is urgently needed. Academic debate has focused on the nature of agri-food system change and how to understand and create more positive and sustainable outcomes (Seymour & Connelly, 2022). However, the politics of change are complex and challenging and are closely linked to how land is used.

Agri-food networks are a convoluted entanglement of relations between people, environments, and land (Sonnino & Milbourne, 2022), some of which are visible, transparent, and reciprocal such as agri-food value chains, while others are hidden, hierarchical, and contradictory. It is the latter that has provided a focus for academic debate on agri-food network transformation towards sustainability, which, building on Patterson et al. (2017), is defined as a "process of radical change in the structural, functional and relational aspects of a food network that leads to more just socio-ecological relationships, patterns of interactions and outcomes" (Sonnino & Milbourne, 2022, p. 915). This chapter aims to discuss the intertwined complexity of the human–environment relational aspect of agri-food networks, which has the potential to influence decision-making regarding the future of agri-food systems.

This chapter is structured into three interrelated subsections. The first section examines the agricultural history of Aotearoa New Zealand (AoNZ), providing essential context for understanding how the productivist agri-food model has shaped both the current and potential future narratives of the country's agricultural system – what we refer to as the growth mindset. The second section invites the

DOI: 10.4324/9781032672779-11

reader to consider an alternative worldview centered on interconnection, values, and relationships – capacities that have the potential to drive transformative change in the agri-food system – what we refer to as relational values. This leads into the third section, which explores the present-day realities and experiences of diverse stakeholders, illustrating how farmers are grappling with relational values associated with the land in light of environmental challenges and the deeply embedded productivist logic that has shaped agricultural land use. We then reflect on how this might shape the future trajectory of AoNZ's agri-food landscape. Together, these sections offer a framework for understanding the complex dynamics that influence agricultural transformation in AoNZ.

Situating AoNZ's Agricultural Land Use: the Past and Present

AoNZ's export-oriented agriculture sector has historically been underpinned by a strong productivism ideology and a moral imperative to 'feed the world' (Rosin, 2012). With the arrival of Europeans and colonialism, Māori adopted European farming techniques. Māori produce was crucial for the survival of the new colonial settlements, while Māori agricultural products were also exported to Australia in the 1850s and 1860s. As colonial structures took root, Māori were rapidly alienated from land in general, and highly productive land in particular, alongside attempts to assimilate agriculture aligned with private landholdings and modernist principles that contributed to AoNZ being seen as Britain's 'farm in the South pacific' (Campbell, 2020, p. 2).

After the World Wars, advances in agricultural science and technologies, as well as the plunder of phosphate from island nations like Nauru, allowed for the 'grasslands revolution' to occur (Brooking et al., 2002). Government-funded research to support extensive pastoral farming through selective planting of exotic rye grasses, aided by substantial fertilizer inputs, provided the basis to extend agriculture beyond previous limits of the land, fully entrenching the growth mindset. By the 1970s, AoNZ's landscape had been transformed, with 51% of the country's surface area turned into exotic pasture and sheep numbers peaking at 70 million in 1982 (Brooking et al., 2002). There was a glorification of this 'success' of modernist agricultural science that 'tamed' the wild lands of Aotearoa and rendered invisible the ecological damage and colonizing nature of such a system (Brooking et al., 2002; Brooking & Pawson, 2010; Campbell, 2020).

However, this growth paradigm system hit a crisis when Britain began to unwind from reliance on its colonial trade connections in the 1970–1980s (Campbell, 2020) and necessitated searching out for new markets. Government provided a financial buffer to the loss of preferential market access through subsidies and industry support programmes in agriculture. The restructuring of the economy in response to inflationary pressures and growing foreign debt, in the mid-1980s, and the associated removal of trade barriers, meant that the agricultural sector became fully exposed to global market forces to respond to the financial crisis. The government saw neoliberalizing agriculture as the solution, rolling back its previously strong supportive ties to the industry. This process had major negative impacts on farm incomes and debt levels and drove significant stress into rural communities. This period became known as the 'rural downturn' (Haggerty et al., 2009), with many farmers getting out of farming entirely, ushering in a period of increased intensification.

In response to these shocks, AoNZ's agriculture entered a new phase of intensification where stocking units, fertilizer, pesticide use, food stock inputs, and animal productivity have all increased substantially (MacLeod & Moller, 2006). Dairy farming epitomized this shift (Tall & Campbell, 2018). As an example, in the Southland Region of AoNZ, the number of farm holdings declined from 5,379 in 1990 to 3,540 by 2013 while the total area of farms stayed relatively constant (1.27 million hectares in 1990 to 1.17 million hectares in 2013). Over that same period, dairy cattle increased from 37,772 to 615,428 and sheep declined from 8.9 million to 4.3 million (Connelly & Nel, 2017). The dairy boom was facilitated by land use change to accommodate an 84% increase in the dairy herd nationally from 3.4 million in 1990 to 6.3 million by 2019 (Stats NZ, 2021). As a result of this growth, dairy product exports were valued at $25.7 billion in 2023, accounting for a quarter of the value of all exports and 2.8 times the value of meat exports (Sense Partners, 2023). This growth-oriented mindset, rooted in historical productivism, contributes to the agricultural crises we are facing today.

Alongside the dairy boom, there were also growing environmental concerns, which led to dramatic shifts in approaches to environmental management from state-centered to consumer-and industry-driven regulation (Haggerty et al., 2009). New tensions were created between emerging environmental aspirations and 'green' claims and the need for agricultural producers to use land more intensively to generate the financial returns needed to support investments in agricultural infrastructure, such as water storage, irrigation, and artificial inputs, required in the pursuit of increased production. The intensification of agriculture resulted in fewer people working more land, decreased on-farm employment, and the decline and consolidation of agricultural services in fewer rural communities and towns (Joseph et al., 2001). The early 2000s also saw public debate develop over the environmental impacts of intensification. In particular, the 'dirty dairying' debate emerged to counter the country's 'clean green' reputation (Blackett & Le Heron, 2008). Notably, some of the most contentious environmental issues have been those concerning water quality and allocation.

The scope of the environmental crisis in agriculture is significant and increasingly well-documented. Agriculture accounted for over 53% of AoNZ's greenhouse gas emissions in 2022 (MfE, 2024). The State of Environment: Our Freshwater (2023) report states that while the health of our freshwater ecosystems is central to our well-being and to our economy, many freshwater ecosystems have been degraded by the effects of excess nutrients and other contaminants from our land-based activities in rural and urban environments, with agriculture being a key driver. The Environment Aotearoa (2022) report, identifies increased pressures on species and ecosystems, impacts on land and soil, loss of biodiversity, degradation of freshwater systems, declining health of our marine environment, changing climate, and air pollution. It calls for processes to make more responsible decisions about our environment, by better integrating the environment into policy. More recently, the Parliamentary Commissioner for the Environment reported on land use change (2024) and states, 'we need to change the way we use our land if we want to hold the line on environmental quality, let alone improve it,' and that we need to have hard conversations and address the hard questions about future land use decisions. The increased intensification and resulting environmental outcomes are leading to questions about agriculture's social license to operate, with both urban and rural residents expressing low levels of trust in the agricultural sector and a growing disconnect between farmers and the public (Beban et al., 2023). These disconnections are contributing to a sense of uncertainty over the future of agriculture in AoNZ.

Transforming agricultural land use will require consideration of the values that underpin land use decisions. While there is agreement across government and industry that a significant transition process in the agri-food industry will have to be undertaken, the current approach to deal with the crisis uses mostly reformative change over addressing the root cause of environmental, social, cultural, and economic issues. These histories of the past and stories of the present agricultural crisis in AoNZ create a context for the system into which we start to rethink the future of agricultural land use and agri-food system transformation.

In the section that follows, we outline how greater attention to relational values, grounded in care, could provide a foundation for rethinking future narratives of agriculture and might challenge the dominant growth mindset. We then explore the extent to which agricultural producers are engaging with the tensions between the growth mindset and more relational values.

Rethinking Future Agricultural Land Use: the Value of Interconnection and Human–Environment Relationships

Values are fundamentally important for the interconnected care of humans and the environment, affecting how we experience the world and our interconnected place within it. Values are ways in which people assign importance to human and more-than-human nature and their relations (Muraca, 2016; Pascual et al., 2017). Values reveal and shape people's experiences and ontological conceptions about what is morally right and about what entities exist in the world (Muraca, 2016). Moreover, they may also inform how human–environment relations form and interact with agri-food systems, shaping land use decisions.

To gain a better understanding of human–environment connections and to respond to the sustainability crises, there is a call for greater attention to relational approaches in sustainability science (Walsh et al., 2020; West et al., 2020). This shifts attention from growth mindsets based on 'perceiving more-than-human nature as a resource to be exploited to a soulful family member that holds legal personhood, agency, dignity, and intrinsic value to speak with an own voice' (Artmann, 2023, p. 12). Relational approaches require blurring of the artificial boundaries between nature and culture, economy and society, humans and environment that are central to the growth paradigm (Ives et al., 2020).

However, the creation of dualisms between nature and culture is not universal – as many indigenous cultures have never formed such divisional constructs. Within indigenous culture, the sustainable use of natural resources is connected to traditional knowledge, which is often embedded in a cosmology that reveres the oneness of life, considers nature as sacred, and acknowledges humanity as a part of it (UN Environment, 2020). Muller et al. (2019, p. 5) emphasize that 'many Indigenous worldviews have the power to radically transform environmental management by inscribing values of connectivity, reciprocity, and trust in ethical relationships with all species.' An ethos of care that embraces reciprocal relationships and interconnectedness stands in contrast to human–environmental divides and can form the basis for renewed ethical responsibilities in these communities (Whyte & Cuomo, 2016). Thus, restructuring our relationship with nature, and removing the hierarchical separations of humans from the environment, aims to deeply unsettle the assumption of the separation and superiority of humans in agri-food systems and highlights the importance of these interconnected relationships as critical components of transformative change.

Indigenous scholars (Battiste, 2002; Cajete, 2000b; Deloria, 2005; LaDuke, 1994; McGregor, 2004) argue that traditional knowledge is place-based, and therefore it can't be confined to a particular source of knowledge. Anishinaabe scholar Deborah McGregor points out that to understand where traditional knowledge comes from one must start with indigenous people and their own understanding of the world (McGregor, 2004). In AoNZ, this means deepening our understanding of Te Ao Māori (the Māori worldview) and listening to our indigenous communities' wisdom in how principles such as Mātauranga Māori (Māori knowledge), mauri ora (life force), whakapapa (genealogy), and kaitiakitanga (guardianship) might shape future agri-food systems and our relationship with Papatūānuku (Earth Mother) (Roberts, 2013; Durie, 2001; Marsden, 2003; Salmond, 2012). These values and principles are ways of being and inform the way Māori relate to the world and to others; it is a holistic worldview acknowledging the sacred relationships humans have with the environment and the interdependence of those relationships (Cheung, 2008). In caring for the well-being of nature, Māori values demonstrate how they ensure that the environment itself is a stakeholder with a voice and listen directly to nature rather than the voice of intermediaries such as institutions (Spiller et al., 2011). We have seen this reflected in the shifting legal recognition of the more-than-human world in AoNZ through Te Awa Tupua (Whanganui River), Te Urewera (a former National Park), and Taranaki Maunga (Mount Taranaki) as each of these were granted legal identity/personhood status (Winter, 2019), in which Māori are guardians of these legal landscapes, which may provide lessons for how we might think of our relationship with the land differently – in that we are not separate from the land and might recognize our interconnected existence. Western and Mātauranga Māori knowledge streams may coexist – sometimes separately, other times coming together – but when they are both acknowledged and given respect, opportunities for innovation and solutions to environmental challenges emerge that would not be possible if only one knowledge stream is considered (MacFarlane, 2024; Neumeier, 2017; Bock, 2012).

To bridge the human–environment divide, integrating relational values and more-than-human relationships is key. This perspective emphasizes our coexistence with the natural world and recognizes the dynamic interrelationships between humans, physical elements, materials, immaterial entities, and other beings (Ortiz-Przychodzka et al., 2023). Embracing these relational approaches requires rethinking our values and practices related to land use. Gosnell et al. (2019) argue that industrial farmers can transition from traditional methods to more holistic, regenerative practices by understanding complex, interrelated living systems. They cite regenerative farmer Charles Massy, who advocates for 'listening to the land' to foster deeper connections with nature and community, aligning with broader stewardship values (2019, p. 9). Evidence of the blurring of the human–environment divide is documented in the caring practices among regenerative farmers (Seymour & Connelly, 2023), in farmer peer-to-peer learning exchanges that build on linkages between food sovereignty and indigenous knowledge (Mann, 2019), and in calls for more careful reflexive practices to guide agri-food system transformation (Sharp, 2019). Regenerative farmers have shifted from intensive land management to low-stress livestock handling (Gosnell et al., 2019) and provided an alternative to industrial, productivist agricultural models (Gordon et al., 2021).

In the following section, we present some of the results from 62 semi-structured interviews that were conducted in 2023 to gather insights from key stakeholders within AoNZ's agri-food system. Participants included members of alternative food networks (AFNs), industrial farmers, and local and regional council members. The interviews aimed to explore perspectives on the challenges and opportunities for transforming the agri-food system. The transcribed data was analyzed thematically to identify key trends, divergent viewpoints, and underlying values that contribute to the ongoing discourse within agri-food system transformation. The outcomes of these conversations were used to construct the following narrative of how agri-food stakeholders are navigating tensions between growth mindsets and relational values.

Current Narrative of What Is (Not) Happening in AoNZ

In making land use decisions, farmers must navigate multiple sets of values that shape their practices. At times, the growth mindset and structural constraints prevent alignment of personal values with practices due to the broader system's limitations. This section outlines the challenges and barriers experienced and voiced by agri-food system stakeholders while emphasizing the relational potential between industrial farmers, AFN producers, and local and regional council members. It highlights the importance of recognizing our interconnectedness with each other and the environment, exploring the role of policy and regulations, the dominance of productivism, how our values can foster transformative change, as well as examples of transformative practices on the ground.

Role of Policy and Regulations

When discussing sustainable and regenerative practices, most producers agreed there was no single definition, making it difficult to understand what was expected of them. Industrial farmers generally supported government policies around land use regulation but emphasized that these policies needed to be 'sensible.' As one industrial sheep and beef farmer put it:

> We just need common sense, workable plans.

All participants concurred that 'blanket' solutions from Central Government were inadequate, and advocated for closer relationships with catchment groups, as well as local and regional councils. They suggested that granting more autonomy to councils in land management decisions could foster a more responsive and 'understood' agri-food future. Council members highlighted that place-based, localized decision-making would enable them to better address specific community needs, while industrial farmers expressed frustration with national policies that fail to account for regional differences, as solutions need to be adaptable to local contexts, including the diverse farming systems across varying climates, soil types, and cultural norms in AoNZ. An industrial sheep farmer suggests that:

> This is where legislation must be sensible and understand regional differences. The soil and the topography in Southland are completely different from the soil and topography in Otago . . . Our policy has to reflect each region for the best outcome.

While the call for place-based regulations can be interpreted to align with relational values of connection to the land, soil, and waterways, they are constrained by productivist logics that take priority. In this context, 'sensible' policy is policy that does not infringe on production. This is particularly evident in the context of intensive winter grazing (IWG) regulations. IWG is a practice that concentrates stock in a limited area over winter to ensure that high-density forage crops provide the necessary feed requirements for animals while pasture growth is limited. Due to concerns with run-off, erosion, and freshwater pollution, rules to regulate IWG activities were introduced in 2020 as part of National Environmental Standards for Freshwater and came into effect in 2022. IWG regulations restrict the area, slope, and vegetation for IWG and provide a requirement for farms to have freshwater management plans. However, the variability of soils and topography raises questions about whether uniform IWG regulations can effectively improve freshwater outcomes (Bresnell, 2023). The assumption that IWG is accepted as common practice, despite the negative environmental impacts, illustrates the power of the growth mindset.

Aligning policy with local realities can create a more caring and responsive agri-food system. By emphasizing place-based autonomy and recognizing diverse farming systems, there is potential for a deeper understanding of the interconnectedness between individuals, communities, and the environment. The call for localized, context-specific decision-making reflects an ethic of care that considers the unique needs of industrial farmers, local communities, and the land. However, for these values to take root, they need to be able to unsettle the growth mindset to promote sustainable and transformative change within the agri-food system.

Challenge of Productivism

Industrial farmers are making efforts to adopt nature-positive practices but are constrained by AoNZ's historical agricultural production markets. There is a consensus that shifting away from the current production system is difficult because of the significant investments in global, industrialized agriculture, including high upfront capital costs, cultural legacies, and large debt loads. These 'path dependencies' generate considerable resistance to embracing alternatives. Additionally, a risk-averse culture rooted in short-term interests was seen as a major barrier to adopting alternatives. One regional council member explained:

> The thought of change itself can be a barrier. A lot of investment has gone into current farming systems, and companies and banks have contributed to this. That investment locks farmers into an intensive agricultural model to sustain their finances. Recognizing this is essential to understanding the challenge.

The fear of change, uncertainty about the future, and the difficulty of transitioning away from intensive farming systems are significant challenges. As the council member continued:

> Fear of what's coming is a huge barrier. Change can be scary, especially if it means altering land use or shifting away from intensive dairy farming. Providing some certainty about the future, even if change is required, would help farmers plan and adapt.

Many participants felt that the agri-food system was 'broken' and that it would need to change, but no one could elaborate on what that change would be. It was suggested that the path forward was not through fighting against the status quo but rather by creating something new that makes the status quo obsolete.

While industrial farmers are open to discussions about shifting agri-food relations, they remain locked into export markets and large-scale processing, which further entrenches path dependency. Powerful corporate actors play a significant role in shaping and constraining efforts to transform agriculture, often prioritizing their own interests and values, which are misaligned with the goal of creating a more sustainable and inclusive agri-food system (Conti et al., 2021).

Meanwhile, AFNs are emerging nationally, as models for a more values-driven agri-food system. These stakeholders vary from market gardeners selling at their farm gate, small-scale alternative producers pushing their food through privately owned greengrocers, individuals participating in farmers markets, community gardens, 'pick your own' operations, and several more unique food avenues. However, many face structural constraints, including stringent health and safety regulations and rising compliance costs. This points to the challenge of promoting values-based approaches to transformation in isolation from wider systems and structures that shape agri-food systems. To remain financially viable, they would need to scale up to fit into AoNZ's productivist agri-food model, despite it going against their values. As one AFN producer notes:

> It seems to be a 'go big or go home' kind of industry. There's little evidence that small-scale operations can succeed financially.

Another AFN producer added:

> When we met with the Ministry for the Environment about supporting local food production, they said it wasn't scalable or replicable.

Despite these challenges, AFN producers who are succeeding believe that a significant mindset shift is needed in AoNZ to prioritize local and seasonal eating. They feel that education could play a central role in driving this change. Interestingly, many of the barriers these producers face are also the motivations that drive them to continue pushing for change within the agri-food system.

It is evident to see that both industrial farmers and AFN producers in AoNZ are constrained by a historical, growth-oriented mindset rooted in the productivist agri-food system model. Overcoming entrenched path dependencies, corporate interests, and regulatory barriers is crucial to creating an inclusive and sustainable agri-food future. Agri-food transformation is not simply about policy shifts; it requires a deeper shift in mindset and values, where both individual actions and systemic changes reflect a more caring, compassionate, and environmentally attuned approach to agriculture and food. Tangible practices for transformation may come from collaborative, place-based solutions that support the values of producers while addressing the broader societal needs for a more just and inclusive agri-food future.

The Role of Shared Values

The role of shared values in reshaping the agri-food narrative is a recurring theme voiced by a diverse range of academic and industry leaders, particularly at conferences focused on the future of AoNZ's agri-food industry. As one speaker noted:

> Policies mean very little if they are not grounded in values and wisdom – relationality, reciprocity, reverence, respect, trust, and place – and that people need to be connected to the land in order to restore the land and restore ourselves.
>
> Principal Strategist of a Trust and Independent Board Member at an Ecological Conservation Project (2023)

Another speaker emphasized the collective power of shared values:

> Only when we realize that we all have value together can we come together and share our values and narratives to make systemic, transformative change.
>
> Executive Director at an Indigenous Tertiary Provider (2023)

Despite different motivations, stakeholders across sectors expressed shared values. AFNs were driven by a vision for a transformed agri-food system, grounded in community, environmental stewardship, and a planetary systems approach. They valued the well-being of people and the planet and felt a deep sense of responsibility toward building resilient, local communities. Both AFNs and industrial farmers shared a commitment to generational care for the land, recognizing the importance of 'place.' However, generational care for the AFN was driven by greater care for the general public, while the industrial farmers' generational care was driven by familial farm succession.

Despite the call for attention to relational values, there is significant focus on land use practices, with little attention given to the politics of the agri-food system. AoNZ's approach to land management is to focus on what individual farmers are doing on their land, without paying attention to the larger agri-food system that constrains their options. Health and safety regulations and compliance costs are geared toward being a volume player, placing extreme pressures on the individual to maximize productivity from their land for export markets. Although the system in place values scalability, participants suggest that if AoNZ had more of a domestic market into which they could funnel some of their products, they would take that opportunity to improve the resilience of AoNZ's highly centralized agri-food system. This led participants to emphasize the need for AoNZ to support the development of a domestic agri-food system alongside the existing global/industrial system. While there is recognition of the benefits of embracing a diversity of values to drive agricultural practices, relational values such as trust, reciprocity, and care continue to sit uncomfortably alongside values of efficiency, scale, and profit. Participants identified opportunities for greater transparency in the agri-food system as being crucial for unsettling the growth mindset, with collaboration among policy, agriculture, and consumers, along with collective action and social license playing key roles in transforming the system.

Toward Caring Practices

The potential for shifts toward more caring practices is evident in interviews with industrial farmers, where the focus is shifting from traditional, modernized farming practices to nature-positive approaches – more closely aligned with regenerative farming principles. Rethinking values has the potential to drive changes in agricultural land use, leading to better social and environmental outcomes, such as improved soil health, biodiversity, and water quality. The importance of integrating diverse values and deepening human connections with the environment in transforming agri-food systems is exemplified by a Central Otago beef farmer who, by connecting with her soil, transformed her relationship with the land, herself, and her sense of purpose. Through this more-than-human relationship, she transitioned from intensive land management to nature-positive, regenerative practices, deepening her connection with nature and aligning her actions with her personal values.

This conceptual framework encourages a rethinking of diverse economies through the interconnected existence of humans and more-than-humans. These shifts in values are also reflected in the practices of a sheep farmer and a beef farmer who describe the purpose of their farm being 'in service to life.' Many AFN participants discussed their purpose as having a strong emphasis on reducing food miles, adopting organic land management practices, and ensuring affordability and accessibility to healthy food, which reflects a shift toward a more just, inclusive, and sustainable agri-food future. A notable example is a collaborative AFN from the Wairarapa, established to provide the local community with organic, locally grown produce through a box scheme. This initiative not only diversified food options but also created opportunities for the community to support local farmers directly.

At the heart of this initiative was the founder, who, driven by her personal experience of struggling to access organic produce for her family during COVID-19, prompted her to start her own garden, which later turned into the collaborative box scheme model. Driven by an ethic of care, the founder spent hours on social media, promoting the importance of eating seasonally and supporting local agriculture, cultivating an environment of mutual care, respect, and trust within the community which she claims aided in the success of her business model.

By educating the community on where their food originated and how to prepare it, the initiative helped bridge the gap between humans and nature through a strengthened relationship with food. It fostered a deeper understanding of seasonal produce, and the labor involved in food production, fostering a more mindful and sustainable relationship between people, their environment, and the agri-food system. This initiative not only supported local farmers but also nurtured a sense of belonging and responsibility, reconnecting individuals with the land and strengthening ties within their community – reshaping the agri-food system into a more localized, just, and ecologically conscious model that extends well beyond the growth paradigm and extends beyond just agricultural land use practices by strengthening relations between producers and consumers.

These shifts, rooted in a relational ontological transformation, demonstrate how personal change can contribute to broader societal and environmental change. By reflecting on what we value, we can recognize that these internal shifts may be the change reflected in the world around us. Cultivating reciprocal relationships involves an internal healing process (Warber et al., 2020; Woiwode et al., 2021) through care, compassion, reciprocity, self-acceptance, and listening – beyond the self – contributing to a cosmological shift toward a more relational ontology. Whatmore (2018) argues that this shift redefines materiality from an external, indifferent world 'out there' to the intimate, embodied experience of human existence 'in here.' Structurally, this requires bridging the divide between internal and external dimensions, shifting focus from technological solutions to addressing inner dimensions to overcome the fearful barriers that constrain fundamental changes toward sustainability.

Toward a Relational Shift in Agri-Food Systems

AoNZ has a long history of deliberate decisions and practices that have intensified land use, which are increasingly recognized as incompatible with both local and global environmental outcomes. These decisions have largely been shaped by values that prioritize a growth paradigm to serve export markets. However, our interviews with a range of industrial farmers, AFN producers, and local and regional council members reveal how diverse stakeholders are now discussing alternative relational values. Some are even engaging in practices that challenge the deeply embedded productivist logic that has historically driven land use transformation. These findings suggest that adopting more relational values, grounded in care, could help reframe and reconstruct new narratives around our relationships with the land.

However, moving beyond individual shifts requires collective responsibility – particularly through investment in education and awareness about the challenges and opportunities of agri-food transformation. This awareness, combined with inclusive policymaking that amplifies marginalized voices (including indigenous leadership and farmers' perspectives), can help promote sustainable, resilient, and decentralized agri-food systems. Ultimately, transforming the agri-food system requires a relational shift in how we understand our interconnectedness, recognizing that both personal and collective responsibility, rooted in values and reciprocal moralities, will drive the necessary changes within individuals and society (Moore, 2017). Through this shift, we can redefine our relationship with the environment and foster more sustainable practices that care for both people and the planet.

References

Allen, P. (2014). Divergence and convergence in alternative agrifood movements: Seeking a path forward. In *Alternative agrifood movements: Patterns of convergence and divergence* [Online] (pp. 49–68). Research in Rural Sociology and Development. Emerald Group Publishing Limited. Retrieved March 27, 2019, from https://www.emeraldinsight.com/doi/full/10.1108/S1057-192220140000021008

Artmann, M. (2023). Human-nature resonance in times of social-ecological crisis – a relational account for sustainability transformation. *Ecosystems and People*, *19*(1). https://doi.org/10.1080/26395916.2023.2168760

Battiste, M. (2002). *Indigenous knowledge and pedagogy in first nations education: A literature review with recommendations*. Retrieved October 31, 2016, from http://www.usask.ca/educa tion/documents/profiles/battiste/ikp_e.pdf

Beban, A., Korson, C., Reid, J., Procter, J., Halley, J., & Mackenzie, K. (2023). *Building a place-based social licence to operate*. Our Land and Water NZ. https://ourlandandwater.nz/wp-content/uploads/2023/10/Building-a-Place-based-Social-Licence-to-Operate-Report-Oct-2023.pdf

Blackett, P., & Le Heron, R. B. (2008). Maintaining the 'clean green' image: Governance of on-farm environmental practices in the New Zealand dairy industry. In R. B. Le Heron & C. Stringer (Eds.), *Agri-food commodity chains and globalising networks* (pp. 89–102) [Online]. Routledge. Retrieved May 11, 2021, from https://www.taylorfrancis.com/, https://www.taylorfrancis.com/chapters/edit/10.4324/9781315566276-11/maintaining-clean-green-image-governance-farm-environmental-practices-new-zealand-dairy-industry-paula-blackett-richard-le-heron

Bock, B. (2012). Social innovation and sustainability; How to disentangle the buzzword and its application in the field of agriculture and rural development. *Studies in Agricultural Economics*, *114*, 57–63. https://doi.org/10.7896/j.1209

Bresnell, J. (2023). *Investigating the relationship between the loss of social license to farm and the development of regulation* [Unpublished Masters dissertation]. University of Otago.
Brooking, T., Hodge, R., & Wood, V. (2002). The Grasslands revolution reconsidered. In E. Pawson & T. Brooking (Eds.), *Environmental histories of New Zealand* (pp. 169–182). Oxford University Press.
Brooking, T., & Pawson, E. (2010). *Seeds of empire: The environmental transformation of New Zealand* [Online]. I. B. Tauris & Company, Limited. Retrieved May 11, 2021, from http://ebookcentral.proquest.com/lib/otago/detail.action?docID=688257
Cajete, G. (2000b). *Native science: Natural laws of interdependence*. Clear Light Pub.
Campbell, H. (2020). *Farming inside invisible worlds: Modernist agriculture and its consequences* [Online]. Bloomsbury Publishing Plc. Retrieved May 11, 2021, from http://ebookcentral.proquest.com/lib/otago/detail.action?docID=6268901
Cheung, M. (2008). The reductionist-holistic dilemma. *MAI Review*, *3*, 1–7. Research Note 5.
Connelly, S., & Nel, E. (2017). Restructuring of the New Zealand economy: Global-local links and evidence from the West Coast and Southland regions. In G. Halseth (Ed.), *Transformation of resource towns and peripheries* (pp. 112–135). Routledge.
Conti, C., Zanello, G., & Hall, A. (2021). Why are agri-food systems resistant to new directions of change? A systematic review. *Global Food Security*, *31*, 100576. https://doi.org/10.1016/j.gfs.2021.100576
Deloria, V. (2005). Indigenous peoples. In W. Schweiker (Ed.), *The Blackwell companion to religious ethics* (pp. 552–559). Blackwell Publishing.
Durie, M. (2001). *Mauri Ora: The dynamics of Māori health*. Oxford University Press.
Gordon, E., Davila, F., & Riedy, C. (2021). Transforming landscapes and mindscapes through regenerative agriculture. *Agriculture and Human Values*, *39*(2), 809–826. https://doi.org/10.1007/s10460-021-10276-0
Gosnell, H., Gill, N., & Voyer, M. (2019). Transformational adaptation on the farm: Processes of change and persistence in transitions to 'climate-smart' regenerative agriculture. *Global Environmental Change*, *59*, 101965. https://doi.org/10.1016/j.gloenvcha.2019.101965
Haggerty, J., Campbell, H., & Morris, C. (2009). Keeping the stress off the sheep? Agricultural intensification, neoliberalism, and 'good' farming in New Zealand. *Geoforum*, *40*(5), 767–777.
Haynes-Maslow, L., & Salvador, R. (2015). The food system should unite us, not divide us. *Journal of Agriculture, Food Systems, and Community Development*, *5*(4), 105–108.
Ives, C. D., Freeth, R., & Fischer, J. (2020). Inside-out sustainability: The neglect of inner worlds. *Ambio*, *49*(1), 208–217. https://doi.org/10.1007/s13280-019-01187-w
Johnston, J., Biro, A., & MacKendrick, N. (2009). Lost in the supermarket: The corporate-organic foodscape and the struggle for food democracy. *Antipode*, *41*(3), 509–532.
Joseph, A., Udgard, J. M., & Bedford, R. (2001). Dealing with ambiguity: On the interdependence of change in agriculture and rural communities. *New Zealand Geographer*, *57*(1), 16–26. https://doi.org/10.1111/j.1745-7939.2001.tb01591.x
LaDuke, W. (1994). Traditional ecological knowledge and enviromental futures. *Journal of International Environment and Policy*, *5*, 127–135.
Levkoe, C. Z. (2011). Towards a transformative food politics. *Local Environment*, *16*(7), 687–705.
MacFarlane, A. (2024). Prologue. In A. Macfarlane, M. Derby, & S. Macfarlane (Eds.), *He Awa Whiria – braiding the knowledge streams in research, policy and practice* (pp. 15–19). University of Canterbury Press.
MacLeod, C. J., & Moller, H. (2006). Intensification and diversification of New Zealand agriculture since 1960: An evaluation of current indicators of land use change. *Agriculture, Ecosystems & Environment*, *115*(1), 201–218.

Mann, A. (2019). Education for food sovereignty as transformative ethical practice. *Policy Futures in Education*, *17*(7), 862–877.

Marsden, M. (2003). *The woven universe: Selected writings of Rev. Māori Marsden*. Estate of RevMāori Marsden.

McGregor, D. (2004). Coming full circle: Indigenous knowledge, environment, and our future. *American Indian Quarterly*, *28*(3/4), 385–410. http://www.jstor.org/stable/4138924

Millennium Ecosystem Assessment. (2005). *Ecosystems and human Well-being: Synthesis*. Island Press.

Ministry for Environment. (2022). *Environment aotearoa 2022*. https://environment.govt.nz/publications/environment-aotearoa-2022/

Ministry for Environment. (2023). *State of environment: Our freshwater*. https://environment.govt.nz/publications/our-freshwater-2023/

Ministry for Environment. (2024). *New Zealand's greenhouse gas inventory 1990–2022: Snapshot*. https://environment.govt.nz/publications/new-zealands-greenhouse-gas-inventory-19902022-snapshot/

Moore, K. D. (2017). *Great tide rising: Towards clarity and courage in a time of planetary change*. Counterpoint Press.

Moragues-Faus, A., & Marsden, T. (2017). The political ecology of food: Carving 'spaces of possibility' in a new research agenda. *Journal of Rural Studies*, *55*, 275–288.

Muller, S., Hemming, S., & Rigney, D. (2019). Indigenous sovereignties: Relational ontologies and environmental management. *Geographical Research*, *57*(4), 399–410. https://doi.org/10.1111/1745-5871.12362

Muraca, B. (2016). Relational values. *Balkan Journal of Philosophy*, *8*(1), 19–38. https://doi.org/10.5840/bjp2016813

Neumeier, S. (2017). Social innovation in rural development: Identifying the key factors of success. *The Geographical Journal*, *183*(1), 34–46. https://doi.org/10.1111/geoj.12180

Ortiz-Przychodzka, S., Benavides-Frías, C., Raymond, C. M., Díaz-Reviriego, I., & Hanspach, J. (2023). Rethinking economic practices and values as assemblages of more-than-human relations. *Ecological Economics*, *211*, 107866. https://doi.org/10.1016/j.ecolecon.2023.107866

Parliamentary Commissioner for the Environment. (2024). *Going with the grain: Changing land uses to fit a changing landscape*. https://pce.parliament.nz/publications/going-with-the-grain-changing-land-uses-to-fit-a-changing-landscape/

Pascual, U., Balvanera, P., Díaz, S., Pataki, G., Roth, E., Stenseke, M., Watson, R. T., Başak Dessane, E., Islar, M., Kelemen, E., Maris, V., Quaas, M., Subramanian, S. M., Wittmer, H., Adlan, A., Ahn, S., Al-Hafedh, Y. S., Amankwah, E., Asah, S. T., . . . Yagi, N. (2017). Valuing nature's contributions to people: The IPBES approach. *Current Opinion in Environmental Sustainability*, *26–27*, 7–16. https://doi.org/10.1016/j.cosust.2016.12.006

Patterson, J., Schulz, K., Vervoort, J., van der Hel, S., Widerberg, O., Adler, C., Hurlbert, M., Anderton, K., Sethi, M., & Barau, A. (2017). Exploring the governance and politics of transformations towards sustainability. *Environmental Innovation and Societal Transitions*, *24*, 1–16.

Roberts, M. (2013). Ways of seeing: Whakapapa. *Sites: New Series*, *10*, 93–120. https://doi.org/10.11157/sites-vol10iss1id236

Rosin, C. (2012). Food security and the justification of productivism in New Zealand. *Journal of Rural Studies*. *29*(2), 50–58.

Salmond, A. (2012). Back to the future: First encounters in Te Tai Rawhiti. *Journal of the Royal Society of New Zealand*, *42*, 69–77.

Sense Partners. (2023). *Solid foundations: Dairy's economic contribution to New Zealand*. Dairy NZ. https://www.dairynz.co.nz/media/0oibxesz/solid-foundations-4-september-2023.pdf

Seymour, M., & Connelly, S. (2023). Regenerative agriculture and a more-than-human ethic of care: A relational approach to understanding transformation. *Agriculture and Human Values*, *40*, 231–244. https://doi.org/10.1007/s10460-022-10350-1

Sharp, E. L. (2019). The role of reflexivity in care-full food systems transformations. *Policy Futures in Education*, *17*(7), 761–769.
Sonnino, R., & Milbourne, P. (2022). Food system transformation: A progressive place-based approach. *Local Environment*, *27*(7), 915–926. https://doi.org/10.1080/13549839.2022.2084723
Spiller, C., Pio, E., Erakovic, L., & Henare, M. (2011). Wise up: Creating organizational wisdom through an ethic of Kaitiakitanga. *Journal of Business Ethics*, *104*(2), 223–235. https://doi.org/10.1007/s10551-011-0905-y
Stats NZ. (2021). *Livestock numbers*. Retrieved July 15, 2024, from https://www.stats.govt.nz/indicators/livestock-numbers
Tall, I., & Campbell, H. (2018). The 'dirty dairying' campaign in New Zealand: Constructing problems and assembling responses. In J. Forney, C. J. Rosin, & H. Campbell (Eds.), *Agri-environmental governance as an assemblage: Multiplicity, power, and transformation. Earthscan food and agriculture*. Routledge.
UNEnvironment. (2020, June 8). *Indigenous peoples and the nature they protect*. Retrieved October 14, 2020, from https://www.unenvironment.org/news-and-stories/story/indigenous-peoples-and-nature-they-protect
Walsh, Z., Böhme, J., & Wamsler, C. (2020). Towards a relational paradigm in sustainability research, practice, andeducation. *Ambio*. https://doi.org/10.1007/s13280-020-01322-y
Warber, S. L., Irvine, K. N., Quinn, B. F., Hansen, A. L., Hypki, C., & Sims, E. (2020). Methods for integrating transdisciplinary teams in support of reciprocal healing: A case study. *Ecopsychology*, *12*(3), 222–230. https://doi.org/10.1089/eco.2020.0008
West, S., Haider, L. J., Stålhammar, S., & Woroniecki, S. (2020). A relational turn for sustainability science? Relational thinking, leverage points and transformations. *Ecosystem People*, *16*(1), 304–325. https://doi.org/10.1080/26395916.2020.1814417
Whatmore, S. (2018). Materialist returns: Practicing cultural geography in and for a more-than-human world. *Culture and Society*, 481–490. https://doi.org/10.4324/9781351160360-21
Whyte, K. P., & Cuomo, C. (2016). Ethics of caring in environmental ethics: Indigenous and feminist philosophies. In S. M. Gardiner & A. Thompson (Eds.), *Ethics of caring in environmental ethics* (pp. 234–248) [Online]. Oxford University Press. Retrieved May 17, 2021, from http://oxfordhandbooks.com/view/10.1093/oxfordhb/9780199941339.001.0001/oxfordhb-9780199941339-e-22
Widener, P., & Karides, M. (2014). Food system literacy. *Food, Culture & Society*, *17*(4), 665–687.
Willett, W., Rockström, J., Loken, B., Springmann, M., Lang, T., Vermeulen, S., Garnett, T., Tilman, D., DeClerck, F., Wood, A., Jonell, M., Clark, M., Gordon, L. J., Fanzo, J., Hawkes, C., Zurayk, R., Rivera, J. A., De Vries, W., Majele Sibanda, L., . . . Murray, C. J. (2019). Food in the anthropocene: The eat–lancet commission on healthy diets from sustainable food systems. *The Lancet*, *393*(10170), 447–492. https://doi.org/10.1016/s0140-6736(18)31788-4
Winter, C. J. (2019). Does time colonize intergenerational environmental justice theory? *Environmental Politics*, *29*(2), 278–296.
Woiwode, C., Schäpke, N., Bina, O., Veciana, S., Kunze, I., Parodi, O., Schweizer-Ries, P., & Wamsler, C. (2021). Inner transformation to sustainability as a deep leverage point: Fostering new avenues for change through dialogue and reflection. *Sustainability Science*, *16*(3), 841–858. https://doi.org/10.1007/s11625-020-00882-y

9 Recent changes of land use in Southern Moravia's lowland and highland regions

Milada Šťastná and Antonín Vaishar

Introduction

Czechia is mainly a country of highlands. Lowlands, suitable for intensive agriculture, occupy only a small part of the area. At the same time, agricultural lands cover more than half of the territory. Until the Second World War, all available areas were used, including less fertile land. Collectivization meant, on the one hand, the consolidation of lands and the creation of large ropes, suitable for deploying heavy machinery. Large-scale land reclamation was also carried out to make the maximum amount of land available for agricultural production. On the other hand, unsuitable plots were grassed over or afforested (Bičík et al., 2015).

The mentioned trends, together with the intensive mechanization and chemicalization of agricultural production, had serious environmental consequences, which increased especially in the 1980s in connection with the orientation toward the quantity of production at the expense of its quality. There was soil degradation, significant reduction of biodiversity, pollution of underground and surface water, especially residues of chemical substances and other consequences. The water regime, which is a key condition for agriculture, was significantly disrupted. Vast areas were open to water and wind erosion.

In the former Czechoslovakia, a land reform took place after the First World War, when the land of large estates over 150 hectares was distributed to small owners. After the Second World War, there was a confiscation of German land (2.4 million hectares) followed by a revision of the land reform, which distributed estates over 50 hectares to small owners. However, this stage has almost merged with collectivization (Homoláč & Tomšík, 2016). At the end of the 1980s, almost all agricultural land was managed by unified agricultural cooperatives, state estates, or other state organizations (the army, schools).

After 1990, restitution of private land (to owners who acquired it based on the land reform, its revision, and confiscation of German land), privatization (parts of state land) began. At the same time, the right to dispose of the land, that is to manage, sell, or rent it, was restored to the cooperative members. Transformed cooperatives, corporations, and larger private owners rented land and gradually tried to buy it at the market price (Swain, 1999).

DOI: 10.4324/9781032672779-12

The result of this process is that Czechia has an extremely fragmented structure of landowners and, on the other hand, a high organizational concentration of agriculture. Agricultural enterprises thus often operate on land belonging to tens or even hundreds of owners. Family farms have resumed but to a lesser extent than expected. In addition, in many cases, they are not traditional farmers, but often originally residents of cities, who are often also highly qualified, although not necessarily in agricultural fields. The traditional relationship of the peasants to the land was interrupted by collectivization. The majority of farmers are employees of large corporations. The structure of large agricultural enterprises was preserved – transformed cooperatives, joint-stock companies, and limited liability companies. This structure was supplemented by individual peasants (family farms). The result is by far the largest average area of agricultural land per farm in the European Union. The average family farm manages an area of 39 hectares. Another consequence is the extremely high proportion (73%) of farming not on own land but on leased land. Agriculture contributes 1.79% to the gross domestic product of the Czech Republic, and its share of jobs is 1.99%.

Feranec et al. (2017) state that the main driving forces of land use changes in the period 1996 to 2006 (i.e. in the period preceding the entry of the Czech Republic into the European Union) were urbanization, agricultural intensification, agricultural extensification, afforestation, deforestation, construction, and management of water bodies. Štych et al. (2019) state that in the 20 years before 2010, agricultural intensification took place in the fertile regions of Czechia, accompanied by an increase in the share of arable land, while in the less fertile areas grass and afforestation took place. Built-up areas were concentrated.

In 2004, the Czech Republic joined the European Union, and its agriculture was implemented within the framework of the Common Agricultural Policy. Subsidies have become decisive for the management of agricultural land. Support is directed, for example, to the development of organic agriculture, greening, and growing energy crops. Czech agriculture occupies a unique position within the European Union. It ranks first in the average area of agricultural land per holding, first in the share of entities over 500 hectares (6.6%), first in the number of employees per entity (4.9 people), and second place in the share of organically managed entities. In contrast, it ranks twenty-eighth in the share of family workers (37.2%), twenty-fourth in the share of own agricultural land (26.9%), and twenty-fourth in the number of workers per 100 hectares of agricultural land (3.8 persons). In 2016, a hypothetical agricultural holding in Czechia utilized 132 hectares of agricultural land. Out of this area, it owned 35 hectares, and the remaining 97 hectares were rented. On average, seven persons worked in one agricultural holding; out of them, five persons were fully employed.

Today, the Czech countryside is very little dependent on agriculture in terms of both job opportunities and the rural economy. Agriculture is particularly important for the creation and maintenance of the landscape, among other things, to support its attractiveness for tourism and housing. In terms of the workforce, there has been a significant decline since 2000. However, the situation is completely different for family farms and legal entities. While the number of agricultural workers on family farms has hardly changed, on legal entities it has fallen by a quarter since 2000. In 2016, 36.9% of owners, 39.3% of their family members, and 10.9% of employees worked on family farms (the rest were agency workers, contract workers etc.). For legal entities, 60.9% were employees, 35.2% were irregular (mostly seasonal) workers, and only 2.3% were owners and their family members.

In the age structure of people engaged in agriculture, 15,215 are younger than 30 years, 70,938 are middle-aged (30–60), and the rest are older than 60: 12,646 persons. The share of older workers has been decreasing recently but remains higher than in other sectors of the national economy. In terms of education, 9.7% of workers in agriculture have a basic (even incomplete) education, 52.2% have an apprenticeship, 25.6% have a high school diploma, and 10.7% have a university education. Unemployment in agriculture is increasing and has reached 2.2%. This results in a high proportion of foreigners (22,600 people). However, unemployment in the countryside is lower than in the cities.

Currently, Czech agriculture has to cope with new challenges. Climate change brings increased temperatures, the extension of the growing season, and also the danger of drought, which can also be reflected in the use of land. The most fertile lowland regions of the state, especially in southern and central Moravia, are at risk of drought, while the higher upland locations will not be affected as much. Other challenges include the energy crisis or immigration, which paradoxically helps to solve the labor problem in agriculture. Rural sustainability is the main political challenge (Zagata et al., 2020). The research question is how the current changes are reflected in land use in the lowland and mountainous parts of the state.

Theory

From the point of view of land use, we distinguish landscape macrostructure and landscape microstructure (Lipský, 2007). The macrostructure of the landscape includes ways of its use, which are usually recorded in statistics. Agricultural land usually consists of arable land, meadows and pastures, and special crops such as gardens, orchards, vineyards, hop fields. From the point of view of the primary sector, forest land is a complement to agricultural land. Other components of land use are built-up areas, water areas, and other areas that hide a variety of territory, from natural (e.g. rock formations, swamps) to various infrastructural or undefined ways of use.

The landscape microstructure relates to the mosaic (in our case, agricultural) landscape, that means, for example, the size of the fields, the mosaic of alternation of individual uses, small landscape elements such as ditches, hedge and tree lines, grass margins borders, field roads, elements of anti-erosion protection, elements to increase biodiversity, possible small technical and infrastructural facilities of the point, linear or planar in nature. This microstructure is sometimes measured in terms of its influence on the ecology of the landscape (Mõisja et al., 2016) using methods of remote sensing.

Moreover, we distinguish between land use and land cover (Piao et al., 2023). Land use relates to what purpose the land is utilized, especially different forms of agricultural or non-agricultural use. In contrast, land cover states specific landscape patterns and characteristics. Land use and land cover are important aspects of land resources (Nedd et al., 2021). Significant land use changes are taking place on a global scale. Winkler et al. (2021) estimated up to 32% change over the past 60 years.

Land use and land cover depend on several factors. Natural conditions, or their suitability for specific ways of use, can be characterized as primary. These conditions are relatively stable in the short term but may change in the long term. Currently, it is mainly climate change that can affect the use of the land through geographical redistribution of plant and animal species (Pecl et al., 2017), or vice versa (Searchinger et al., 2018).

Socio-geographical factors represent another group. Originally, it was the conversion of non-agricultural land into agricultural land due to the increasing consumption of food for the growing population of the Earth and regions. Later it was about commercial interests, for example, plantations for growing lucrative monocultures for export from developing countries. A contemporary example is Amazonia (Cardoso da Silva et al., 2017). This category also includes experiments with, for example, the ploughing of virgin soils in Kazakhstan (McCauley, 1976) or the drying up of ponds within the framework of socialist economies.

In recent times, significant changes have also taken place under the influence of urbanization (Nuissl & Siedentop, 2021). It is not just about the area of built-up and paved areas but often about larger areas of infrastructure and similar facilities. These changes are significant not only in terms of quantity, but also because they significantly change the soil, hydrological and biogeographic properties of the territory, and their eventual reversal is not easy.

The Czech landscape has gone through long-term development, during which its macrostructure has more or less stabilized, and the ongoing changes, including the consequences of changes in political regimes, are manifested relatively little. On the other hand, almost complete collectivization and nationalization in former Czechoslovakia and the USSR led to significant changes in the landscape microstructure in the second half of the last century (Lipský, 1995). The organizational concentration of agriculture led to the creation of large chains of fields, the ploughing of borders, the elimination of smaller islands of greenery, and melioration changes. The original mosaic of different types of use has almost disappeared and a uniform field landscape has emerged. This makes Czechia different from some other post-socialist countries, especially Poland (Stacherzak et al., 2019).

Large-scale farming has significant environmental consequences in the form of the opening of large areas to water and wind erosion, a decrease in biodiversity, a decrease in natural resistance to extreme climatic influences, and also a decrease in the aesthetic value of the landscape concerning other potential uses – tourism and residential function (Tittonell et al., 2020). It seems that even the transformation after 1989 did not significantly improve the situation, as the organizational structure of large agricultural enterprises remained (Fukalová & Mašíček, 2018).

In recent times, the use of land has been influenced by the Common Agricultural Policy of the European Union through subsidy measures and legislative frameworks. This policy, which places increasing emphasis on the ecological pillar of sustainability through standards of good agricultural and environmental status soil (Pe'er et al., 2019), comes into conflict with purely market principles.

Currently, there are conflicting opinions on further changes in the landscape. While the landscape adapted to intensive agriculture gives higher yields and more efficient use, the traditional mosaic landscape is considered more environmentally and ecologically favorable. This is where some of the conventional farmers' attacks against the Green Deal originate (van der Ploeg, 2020). It would probably not be entirely optimal to repeat the mistake from the socialist period, when pressure was exerted on intensive agricultural production even in mountainous locations, where it was not appropriate, and to introduce identical organic farming rules everywhere (Guyomard et al., 2023).

It can be assumed that regional land use is influenced by the overall character of the region. In fertile lowlands, we expect a high proportion of arable land; in mountainous areas, we will probably find a higher proportion of forests; while in the vicinity of large cities a higher proportion of built-up and so-called other areas. But what is the development in different types of landscape? The existing works focused on the transformation period, when the Czech market with agricultural commodities opened up to the international market, and at the same time the previous subsidy policy was abolished. It is a question of whether the trends have changed since the beginning of the new millennium. The research question of our contribution, which focuses on the differences in land use in fertile and less fertile areas of South Moravia, is aimed at this. The South Moravian Region will be the case territory, which includes various types of territory.

Methods

The region under study

The South Moravian Region is situated in the south-east of the Czech Republic. It can be described as a rural region with diverse natural conditions. This includes the warmest and most fertile lowlands in the basin of the Morava, Dyje, and Svratka rivers. Most of the Czech vineyards are concentrated there, among others. A relatively significant part of the territory is covered by highlands, especially the Bohemian-Moravian Highland, which represents a natural border against Bohemia and the Drahanská Highland with the most significant karst in Czechia. The eastern border opposite Slovakia is formed by the White Carpathian mountains. The urbanized landscape is represented by the country's second-largest city, Brno (about half a million inhabitants), and its immediate surroundings.

Despite being a predominantly rural region, the average population density is 169 persons per km^2, which is the third highest among the NUTS 3 regions. The structure of land use at the end of 2023 is in Table 9.1. In the South Moravian Region, there is a larger share of agricultural land, but the share of cropping is far more significant, where the share of arable land from agricultural land in the region reaches 82%, while nationally it is only 69%. On the other hand, the region records

Table 9.1 Structure of land use 2023 [%]

Area	*Including*				
	Agricultural land	*Arable land*	*Permanent crops*	*Grasslands, meadows*	*Forests*
Czechia	53.2	36.9	3.2	13.2	34.0
South Moravian Region	58.6	48.0	6.2	4.3	28.1

Source: Public database. Czech Statistical Office Praha.

significantly lower shares of permanent grasslands and forest areas (even though the forests of South Moravia are of high quality). The shares of other areas[1] do not differ significantly. Due to the high visibility, the coefficient of ecological stability of the territory of the South Moravian Region is one of the lowest (0.69 compared to the national average of 1.11).[2]

Workflow

A certain methodological problem is the fact that, while data is available for administrative units, natural conditions respect these units only to a limited extent. Therefore, it is necessary to go to the smallest possible territorial units and then group them without regard to superior units. The lowest statistical unit in the Czech Republic is municipality. However, due to the fragmented structure of municipalities, where the majority are small and very small municipalities, we decided on municipal authority districts with extended powers.[3] The data comes from the public database of the Czech Statistical Office in Prague, which receives it from the Czech Geodetic and Cartographic Office. The data is updated every year (always as of December 31) and is available for municipalities, districts of municipal authorities with extended powers, regions, and the state. Methodologically, they are completely comparable in the period from 2001 to 2023.

The next step was the selection of areas representing individual types of territory. We divided the territory of the region completely according to the use of the territory. These types are fertile lowlands with intensive agriculture (micro-regions of Břeclav, Hodonín, Hustopeče, Kyjov, Mikulov, Pohořelice, Znojmo), highlands with mixed agriculture (Boskovice, Bučovice, Ivančice, Moravský Krumlov, Slavkov u Brna, Rosice, Tišnov, Vyškov, Židlochovice), mountain areas focused almost exclusively on animal production (Veselí nad Moravou) and suburban areas around Brno (Blansko, Brno, Kuřim, Šlapanice). Since there is only one microregion of the South Moravian Region in the mountain range, we expanded the analysis to include two micro-regions of a similar nature in the neighboring Zlín Region: Uherský Brod and Luhačovice (see Figure 9.1).

The problem is that physical-geographical regions are fundamentally homogeneous, while administrative regions are heterogeneous. Therefore, it is impossible to find objective criteria for their intersection. In such a case, the experience of the

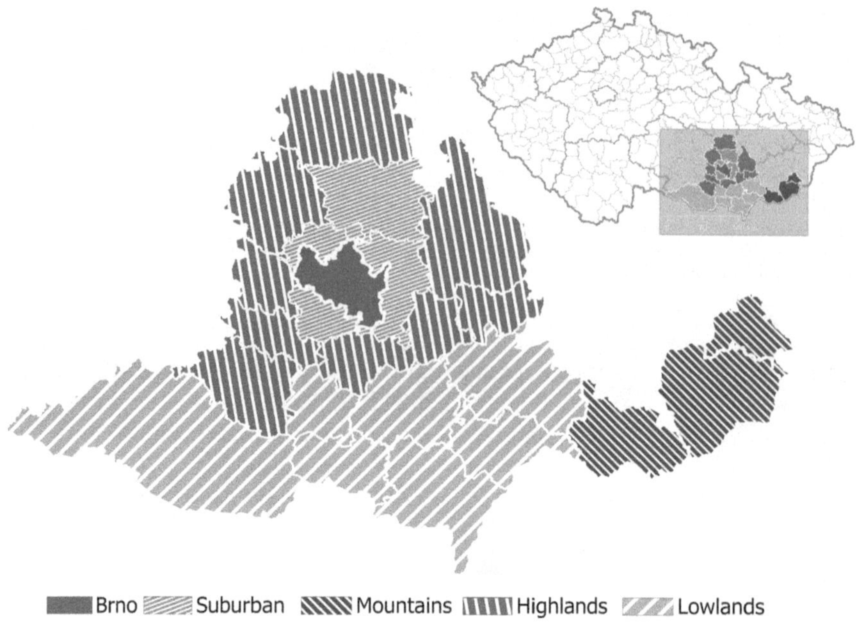

Figure 9.1 Area under study. Areas of municipal offices with extended powers.

Source: Drawn by P. Zálešák.

researchers is decisive. The territories of individual types form continuous regions. The suburbanized region was singled out as Brno and neighboring micro-regions, while the other typologies related to the character of the landscape. It is not a uniform typology, but the suburbanized landscape is obviously different from the other types, even though it is similar in phase-geography.

In addition to stability against changes in land use, we also monitored the ecological stability of the landscape. We calculated the coefficient of ecological stability[4] for individual types of regions, which is defined as the proportion of ecologically stable and ecologically unstable areas (Bažík et al., 2014). We considered arable land, built-up, and other areas to be ecologically unstable areas and all other forms of use to be ecologically stable.

Subsequently, we recalculated the development of the use of the main types of land use since the beginning of the millennium (2001–2023) to reveal trends in the latest development of land use in individual types of territory. Five groups of land use were analyzed: arable land, special crops (vineyards, gardens, orchards), permanent grasslands, forests, and others, where built-up and so-called other areas were included. Water bodies that play a minimal role in the region, their area is very small, and it is difficult to assign them to another type of use were omitted from the analysis. In this way, we gained knowledge about the development of landscape macrostructure.

Results

Development of land use according to the types of territory

In the fertile micro-regions, arable land completely predominates. However, its area decreased by more than 9%, which is also the main observed trend. The areas that are the result of greening (permanent grasslands and forests) have grown only slightly. The decrease in arable land is mainly due to the area of so-called other areas. The increase in areas of special crops by almost 20% is quite exceptional. This is due to the feverish increase in the area of vineyards, but also of gardens, while the area of orchards is decreasing.

Land use in highland micro-regions shows similar main trends as in other regions, that is a reduction in the area of arable land and an increase in the area of built-up and other areas. Here, too, the largest part of the territory is used as arable land, although the proportion of forests is higher than in the lowlands. The areas of permanent grassland are quite low.

The main type of land use in Brno and its surroundings are forest areas, followed by arable land. This result is not surprising if we consider that Brno lies in the contact zone between the lowlands in the south and east and the highlands in the north and west. In the city itself, so-called other areas make up 28% of the area, as well as forests; arable land occupies 21% and built-up areas 9%. The most striking trend is the continuous decrease in arable land (a decrease of 6.7% in 20 years). This is in line with general trends unless we realize that this is largely the most fertile black soil. The second obvious trend is the constant increase in the area of built-up and other areas (an increase of 11.1%). At the same time, the so-called other areas significantly outweigh the built-up land itself.

At the beginning of the millennium, arable land occupied the highest share of the area even in mountainous regions. Since then, it has decreased, so that its area has equaled the area of forests. The areas of permanent grassland have grown relatively significantly in connection with the development of subsidized organic agriculture. It appears that the land use structure in the mountainous regions is beginning to approach their natural assumptions.

An interesting finding may be that the highest changes in land use were experienced by the mountain and lowland regions, while the highland regions and the suburbanized zone were subject to significantly smaller changes in land use. Roughly half of this is due to the loss of arable land in lowland and mountain areas. Does this mean that upland and suburbanized areas are closer to the optimal state?

The ecological stability coefficient values are in Table 9.2. The data shows that the ecological stability coefficient of the landscape has improved in all types of regions. It is also clear that the ecological stability coefficient is higher with increasing altitude. Perhaps unexpectedly, the value of the coefficient of ecological stability in Brno and its surroundings has the second-best value. In the city of Brno itself, the value of the ecological stability coefficient increased from 1.02 to 1.03 in the monitored period. This confirms the knowledge that the rural landscape is not always ecologically more stable than the urban landscape.

Table 9.2 The ecological stability coefficient values

Region	*Year*	*Non-stable [ha]*	*Stable [ha]*	*Coefficient*
Fertile	2001	214,245	108,982	0.51
	2023	208,413	114,805	0.55
Highland	2001	137,419	94,047	0.68
	2023	141,758	101,920	0.72
Mountain	2001	45,520	53,945	1.19
	2023	41,996	57,417	1.37
Suburban	2001	49,631	50,523	1.02
	2023	48,930	51,250	1.05

Source: Public database. Praha: Czech Statistical Office. Own elaboration.

Comparison of relative data according to the types of land use

The most common way of land use in the South Moravian Region is arable land (Figure 9.2). Its area is decreasing in all types of territory. However, both absolutely and relatively, arable land is being lost in the fertile lowlands. This is not optimal, as it is usually the highest quality soil, which is, however, threatened by drought and erosion. Arable land is also decreasing relatively quickly in mountainous areas, where the share of its area has fallen to the level of a suburbanized region.

The shares of areas of special crops are growing in all types of regions, which can be described as a more or less positive trend. In lowland areas, vineyards are the main driver; in other regions, gardens are the main drivers, while the area of fruit orchards is mostly decreasing. We observed an increase in the area of vineyards before the Czech Republic joined the European Union, which limited the further expansion of vineyards.

The increase in the area of grasslands is supported by the Common Agricultural Policy of the European Union as one of the environmentally friendly measures. The largest area and the increase in the area of permanent grasslands can be found in mountainous areas due to the natural conditions and the orientation of agriculture to animal production in a pastoral way, often in subsidized organic agriculture (Figure 9.3). In the other types of regions, the area of permanent grasslands is increasing slowly, in the lowlands almost not at all.

The forest cover of the South Moravian Region is considered insufficient. It seems that the areas of forests are increasing in all areas of the territory only slightly in the order hundredths of a percent of the year. In the suburbanized zone, the forest areas are practically unchanged. The minimum areas of forests are of course found in the lowlands, whereas the biggest share can be found in the city and its suburban zone.

The ever-increasing trend of land use is recorded by built-up and other areas as a response to urbanization. At the same time, the so-called other areas are growing significantly more (probably mainly logistical and infrastructural), while the growth of directly built-up areas is not so great; in some types of regions and periods, they

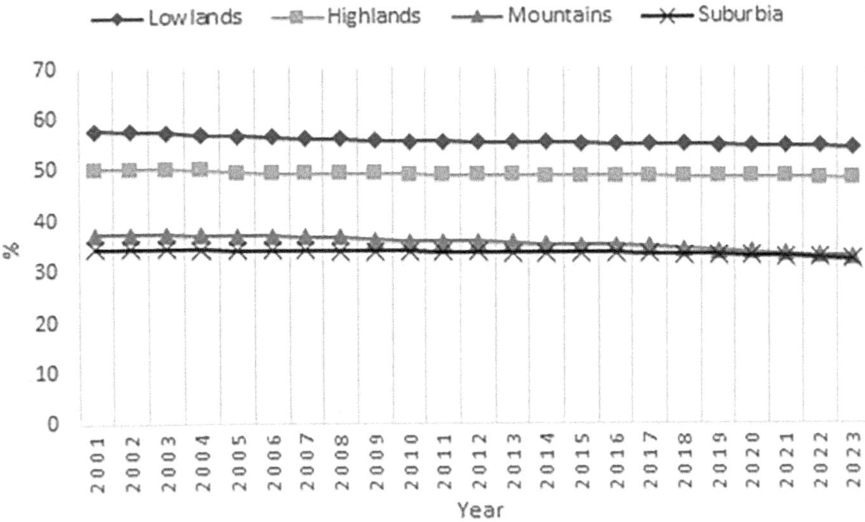

Figure 9.2 Development of arable land in different types of regions during 2001–2023 [%].

Source: Public database, Czech Statistical Office Praha. Own elaboration.

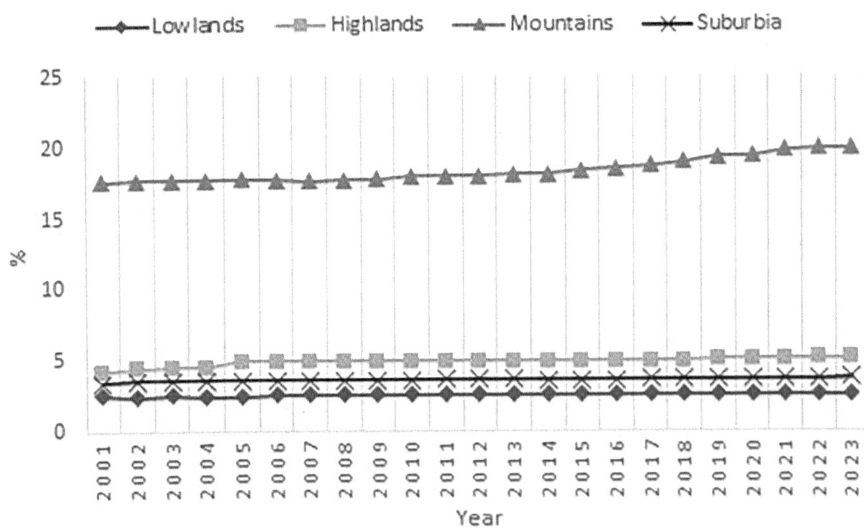

Figure 9.3 Development of permanent grasslands in different types of regions during 2001–2023 [%].

Source: Public database, Czech Statistical Office Praha. Own elaboration.

even decrease. Brno and its surroundings understandably have roughly double the share of these areas than other regions.

If we combine the ecological stability of the landscape with stability to changes, Brno and its surroundings appear to us as an ideal area. This would mean denying the usual notion that big cities are unstable while the countryside is stable. This fact cannot be generalized in full. Brno is known for its forested northern half arc around the built-up area, which other cities may not have. On the other hand, it is a reminder that big cities are not necessarily on the wrong side in terms of their landscape in comparison with the countryside.

Another factor can be the load on the landscape, which can be based on population density. The population density values per km^2 reach 106 people in the lowlands, 107 people in the highlands, 108 people in the mountains, and 549 people in Brno and its surroundings. The high burden of the landscape in the regional metropolis and its surroundings is not surprising. The landscape there is burdened not only by residents but also by people temporarily present and various logistical and infrastructure facilities. What is surprising, however, is the same population density in lowland, upland, and mountain areas. This may be partly because our regions have a functional character and include both the peak areas and the foothills, where cities and villages are quite densely concentrated.

Discussion

Land use trends in four types of territory of the South Moravian Region (with the addition of two mountainous territories of the neighboring Zlín Region) were evaluated. Our research has some limitations. The first of them is the nature of the data that is taken from the statistical service. This means that they characterize official data, which, despite annual updates, may not always correspond to reality. Sometimes, there can be a certain time discrepancy between the real change in land use and its statistical recording.

The second limitation, based on the statistical nature of the data, is that they are presented as administrative units. Even if we went down to the level of micro-regions, there are different types of landscapes at this level as well. If we went down to the level of municipalities, we could get more detailed data. On the other hand, such a mosaic would probably be more difficult to evaluate.

Conclusion

The following conclusions follow from the performed analysis:

The most significant trend in land use over the last quarter of a century is the decrease in arable land. The negative thing is that the most fertile soil in the lowlands in particular is decreasing. Although there are currently no special requirements to increase agricultural production, high-quality soil is a fundamental building block of food security, and its current decline could have a negative impact in the future.

Another significant trend is the increase in built-up and other areas. However, the area of the so-called other areas is growing even faster. Various purposes and functions are hidden in this category, but the areas for infrastructural facilities, especially transport and logistics areas and the like, are probably growing the most. Ultimately, the decrease in the share of ecologically unstable arable land is compensated by the increase in the share of unstable built-up areas.

Greening, which is among the preferred trends of the EU's common agricultural policy, is manifested but to a relatively small extent. The growth in the share of forests and permanent grasslands does not correspond to their importance for increasing ecological stability, nor for its aesthetics and recreational use. The increase in the share of gardens is also positive, but this is somewhat offset by the decrease in the area of fruit orchards.

The comparison of individual types of landscapes did not give quite the expected results. From this comparison, the landscape of the fertile lowlands, which was previously considered rich, is the worst. A relatively rapid decrease in the share of arable land is typical, but this is not sufficiently compensated by the development of green areas. The coefficient of ecological stability is very low. Moreover, the macrostructure of the landscape, where large stretches of fields still prevail, is not attractive for alternative uses. The only positive is the expansion of the vineyard area, which does not exist in other regions of Czechia.

The upland area showed the least land use changes. However, it is a question whether this fact can be interpreted in such a way that their landscape is approaching the current optimal state. Its ecological stability is relatively low. On the other hand, the microstructure of the upland landscape forms a mosaic in which fields, meadows, woods, ponds, and villages alternate. The soil is of lower quality, but due to the ongoing climate change, this region will have more favorable climatic conditions for agriculture.

Mountain areas understandably have the largest shares of ecologically stable areas of forests and meadows. However, the mosaic of the landscape is not so varied, which can reduce the attractiveness of the landscape (at least for some types of tourists).

The landscape of Brno and its surroundings turned out surprisingly well. It has a relatively high coefficient of ecological stability, which is mainly due to the existence of a forest belt in the northern semi-arc. On the other hand, the landscape of the city is heavily burdened by human activities.

It can be considered positive that all types of landscapes in the South Moravian Region offer certain opportunities or benefits in terms of land use. Surprisingly, the average population density in different types of landscapes (with the understandable exception of the city of Brno itself) is relatively high and generally balanced.

It is, of course, a question of to what extent the results are transferable to other territories in Czechia or Europe. Here it is necessary to draw attention to some specifics that may play a role. In the South Moravian Region, there are no distinctly lagging areas at the level of micro-regions (e.g. compared to the border micro-regions of northern and western Bohemia and the Czech part of Silesia).

The territory is covered by a dense network of small towns, which are supplemented (with exceptions) by medium and large villages. The coverage of the territory by public transport is above standard from a European point of view, which enables a high commute to work and thus the independence of rural municipalities (including peripheral ones) from local employment opportunities in agriculture.

Nevertheless, it will be necessary to monitor the further development of land use and supplement the obtained results with the study of landscape microstructure – preferably with the use of remote sensing. It will be also especially interesting to monitor the development in the previously fertile lowlands which are undergoing structural economic changes (Goddard et al., 1993).

Acknowledgment

The contribution was prepared as part of unspecified research of the Department of Applied and Landscape Ecology, Faculty of AgriSciences, Mendel University in Brno.

Notes

1 This type of use is defined as such use, which cannot be included in other types of land. This can include, for example, roads, paths, highways, railways, various paved areas for handling (e.g. with cargo), sports fields, cemeteries, and so on but also rocky areas or wetlands without agricultural or other use, landfills, brownfields, tailings dumps, quarries.
2 The proportion of ecologically stable (forests, permanent grasslands, water bodies) to ecologically unstable areas (arable land, built-up and other areas).
3 Due to the fact that very small municipalities are not able to fulfill their state administrative function in full, some of competences of very small municipalities have been taken over by larger, whose offices have the necessary professional apparatus. Their territorial database was used.
4 The coefficient of ecological stability is a dimensionless quantity introduced by Míchal (1994), which was adopted by the Czech Statistical Office for calculating the carrying capacity of the territory.

References

Bažík, J., Muchová, Z., & Petrovič, F. (2014). Assessment of ecological situation in a landscape based on calculation of ecological stability coefficient. In J. Fialová (Ed.), *Public recreation and landscape protection-with man hand in hand* (pp. 36–43). Mendel University in Brno.

Bičík, I., Kupková, L., Jeleček, L., Kabrda, J., Štych, P., Janoušek, Z., & Winklerová, J. (2015). *Land use changes in the Czech Republic 1845–2010*. Springer. https://doi.org/10/1007/978-3-319-17671-0

Cardoso da Silva, J. M., Prasad, S., & Felizola Diniz-Filho, J. A. (2017). The impact of deforestation, urbanization, public investments, and agriculture on human welfare in the Brazilian Amazonia. *Land Use Policy*, *65*, 135–142. https://doi.org/10.1016/j.landusepol.2017.04.003

Feranec, J., Soukup, T., Taff, G. N., Stych, P., & Bičík, I. (2017). Overview of changes in land use and land cover in Eastern Europe. In G. Gutman & V. Radeloff (Eds.), *Land-cover and land-use changes in Eastern Europe after the collapse of the Soviet Union in 1991* (pp. 13–33). Springer. https://doi.org/10.1007/978-3-319-42638-9_2

Fukalová, P., & Mašíček, T. (2018). Impacts of intensive agriculture on current rural landscape – a case study for South Moravian landscape, Czech Republic. *Acta Universitatis Agriculturae et Silviculturae Mendelianae Brunensis*, *66*(5), 1099–1109. https://doi.org/10.11118/actaun201866051099

Goddard, E., Weersink, A., Chen, K., & Turvey, C. G. (1993). Economics of structural change in agriculture. *Canadian Journal of Agricultural Economics*, *41*(4), 475–489. https://doi.org/10.1111/j.1744-7976.1993.tb03772.x

Guyomard, H., Soler, L. G., Détang-Dessendre, C., & Réquillart, V. (2023). The European Green Deal improves the sustainability of food systems but has uneven economic impacts on consumers and farmers. *Communication Earth and Environment*, *4*, 358. https://doi.org/10.1038/s43247-023-01019-6

Homoláč, L., & Tomšík, K. (2016). Historical development of land ownership in the Czech Republic since the foundation of the Czechoslovakia until present. *Agricultural Economics/Zemědělská Ekonomika*, *62*(11), 528–536. https://doi.org/10.17221/250/2015-AGRIECON

Lipský, Z. (1995). The changing face of the Czech rural landscape. *Landscape and Urban Planning*, *31*(1–3), 39–45. https://doi.org/10.1016/0169-2046(94)01034-6

Lipský, Z. (2007). Methods of monitoring and assessment of changes in land use and landscape structure. *Journal of Landscape Ecology*, 105–118.

McCauley, M. (1976). *Khrushchev and the development of Soviet agriculture*. The Macmillan Press.

Míchal, I. (1994). *Ekologická stabilita*. Veronica.

Mõisja, K., Uuemaa, E., & Oja, T. (2016). Integrating small-scale landscape elements into land use/cover: The impact on landscape metrics' values. *Ecological Indicators*, *67*, 714–722. https://doi.org/10.1016/j.ecolind.2016.03.033

Nedd, R., Light, K., Owens, M., James, N., Johnson, E., & Anandhi, A. (2021). A synthesis of land use/land cover studies: Definitions, classification systems, meta-studies, challenges and knowledge gaps on a global landscape. *Land*, *10*, 994. https://doi.org/10.3390/land10090994

Nuissl, H., & Siedentop, S. (2021). Urbanization and land use change. In T. Weith, T. Barkmann, U. Gaasch, S. Rogga, Ch. Strauß, & J. Zcheischler (Eds.), *Sustainable land management in a European context: A co-design approach* (pp. 75–99). Springer.

Pecl, G., Araújo, B., Bell, J. D., Blanchard, J., Bonebrake, T. C., Chen, I.-C., Clark, T. D., Colwell, R. K., Danielsen, F., Evengård, B., Falconi, L., Ferrier, S., Frusher, S., Garcia, R. A., Griffis, R. B., Hobday, A. J., Janion-Scheepers, C., Jarzyna, M. A., Jennings, S., . . . Williams, S. E. (2017). Biodiversity redistribution under climate change: Impacts on ecosystems and human well-being. *Science*, *355*, eaai9214. https://doi.org/10.1126/science.aai9214

Pe'er, G., Zinngrebe, Y., Moreira, F., Sirami, C., Schindler, S., Müller, R., Bontzorlos, V., Clough, D., Bezák, P., Bonn, A., Hansjürgens, B., Lomba, A., Möckel, S., Passoni, G., Schleyer, Ch., Schmidt, J., & Lakner, S. (2019). A greener path for the EU Common Agricultural Policy. *Science*, *365*, 449–451. https://doi.org/10.1126/science.aax3146

Piao, Y., Xiao, Y., Ma, F., Park, S., Lee, D., Mo, Y., Jeong, S., Hwang, I., & Kim, Y. (2023). Monitoring land use/land cover and landscape pattern changes at a local scale: A case study of Pyongyang, North Korea. *Remote Sensing*, *15*, 1592. https://doi.org/10.3390/rs15061592

Searchinger, T. D., Wirsenius, S., Beringer, T., & Dumas, P. (2018). Assessing the efficiency of changes in land use for mitigating climate change. *Nature*, *564*, 249–253. https://doi.org/10.1038/s41586-018-0757-z

Stacherzak, A., Hájek, L., & Heldak, M. (2019). Changes in the use of agricultural land in Poland and the Czech Republic. *Journal of Ecological Engineering*, *20*(7), 211–221. https://doi.org/10.12911/22998993/109869

Štych, P., Kabrda, J., Bičík, I., & Laštovička, J. (2019). Regional differentiation of long-term land use changes: A case study of Czechia. *Land*, *8*, 165. https://doi.org/10.3390/land8110165

Swain, N. (1999). Agricultural restitution and co-operative transformation in the Czech Republic, Hungary and Slovakia. *Europe-Asia Studies*, *51*(7), 1199–1219. https://doi.org/10.1080/09668139998507

Tittonell, P., Piñeiro, G., Garibaldi, L. A., Dogliotti, S., Olff, H., & Jobbagy, E. G. (2020). Agroecology in large scale farming – a research agenda. *Frontiers in Sustainable Food Systems*, *4*, 584605. https://doi.org/10.3389/fsufs.2020.584605
van der Ploeg, J. D. (2020). Farmers' upheaval, climate crisis and populism. *The Journal of Peasant Studies*, *47*(3), 589–605. https://doi.org/10.1080/03066150.2020.1725490
Winkler, K., Fuchs, R., Rounsevell, M., & Herold, M. (2021). Global land use changes are four times greater than previously estimated. *Nature Communications*, *12*, 2501. https://doi.org/10.1038/s41467-021-22702-2
Zagata, L., Hrabák, J., & Lošťák, M. (2020). Post-socialist transition as a driving force of the sustainable agriculture: A case study from the Czech Republic. *Agroecology and Sustainable Food Systems*, *44*(2), 238–257. https://doi.org/10.1080/21683565.2019.1585400

10 Transformation of cultural landscapes

Case studies from Slovenia

Daniela Ribeiro and Matej Gabrovec

Introduction

This chapter deals with the changes in land use in Slovenia in a 22-year period between 2002 and 2024 and, the consequences of these changes. The land use changes reflect the changes in socio-economic conditions in Slovenia in this period.

In the context of natural conditions in Slovenia, the diverse interweaving of forested and agricultural land, which is possible due to complex cropping patterns, can be considered as a quality landscape. From an economic perspective, the fragmentation of land for agricultural production is not desirable; but from a cultural landscape perspective, the diversity of landscape patterns and the interweaving of different land use types in a sustainable landscape ensure greater biodiversity (Pinto-Correia, 2000; Blondel, 2006; Blondel et al., 2010; Bugalho et al., 2011; Lieskovský et al., 2015); it also represents natural and cultural heritage and identity (Golobič & Lestan, 2016; Perko & Ciglič, 2020). This landscape diversity is a typical Slovenian characteristic (Perko & Ciglič, 2020), and therefore the preservation of such land use patterns (interweaving of fields, meadows, forests, and semi-natural landscape) can be seen as a positive indicator for spatial development.

In recent history, changes and development of the Slovenian landscapes have been mainly influenced by changes in the technology of agriculture and social conditions (Žiberna & Ivanjšič, 2022). In the last decades, in addition to the accelerated settlement development (expansion of urban areas and dispersed construction) and infrastructure systems (e.g. motorways, transmission lines, hydropower plants) (Bole, 2015), one of the most perceptible processes of changing the cultural landscape is forest expansion (approximately, 1.65%), which is associated with favourable natural conditions for the rapid growth of the forest and above all with the withdrawal of agricultural activities and finding employment in other sectors.

In terms of environmental issues, the accelerated urbanization of agricultural land is the biggest one. This is the most destructive process for the soil as a natural resource, as it has far-reaching and irreversible effects – a permanent destruction of the soil. The conversion of agricultural land into forests, on the other hand, does not lead to permanent damage of the soil as it is a reversible process. From a nature conservation perspective, this change is often seen as beneficial for biodiversity, climate mitigation, and ecosystem restoration. However, if we look at land use

DOI: 10.4324/9781032672779-13

changes and their proportions in Slovenia in general, the expansion of forests can be seen as negative. Around 60% of Slovenia's area is forested, making it one of the most densely forested countries in Europe (Gabrovec et al., 2020). From this perspective, the further accelerated expansion of forests of predominantly agricultural land is a negative process that threatens the functional diversity of cultural landscapes and impoverishes the diverse habitats that are created in areas where agricultural and forestry use alternate (Ribeiro & Šmid Hribar, 2019). Industrialization and changes in transport infrastructure and urban planning are reflected in ever larger areas being used for these activities. The urbanization of land is increasing and with it the irreversible destruction of natural resources.

The chapter is divided into three thematic sections: the first addresses the transformation of agricultural land in Slovenia over the last 20 years, the second shows case studies of different processes of change and, the third discusses the consequences of these changes to biodiversity and landscape sustainability.

Current transformation of agricultural land in Slovenia

Data and methods to assess land use changes

Land use is an indicator of both the natural elements and the social conditions in a given area (Mander & Uuemaa, 2010). Quantifying the dynamics of land use change is critical in tackling global environmental challenges such as climate change and biodiversity loss.

In Slovenia, at the national level, the Land Use Database from the Ministry of Agriculture, Forestry and Food (MKGP) is the only database that was produced according to a consistent methodology on the plot level and can be used to identify the current situation in the country. The first data for the whole Slovenia was established in 2002. We compared two data sets from MKGP Land Use Database, of 2002 and 2024, to provide a more detailed and appropriate description of changes in land use. The land use change map was produced by overlaying the two data sets (Figure 10.1). Several studies on land use changes have been prepared using this database (Gabrovec & Kumer, 2019; Horvat & Žiberna, 2020; Lisec et al., 2013; Žiberna & Ivanjšič, 2022).

Typology of land use change in Slovenia between 2002 and 2024

We defined seven processes of land use change: Agricultural land conversion (neutral), Agricultural land conversion (decreasing land use), Agricultural land conversion (increasing land use), Agricultural land abandonment, Deforestation, Land take for urban use, and Wetland reclamation (Table 10.1). In the neutral conversion of agricultural land, the function of an area is changed while the intensity of use remains the same; for example the conversion of arable land into a vineyard. Agricultural land conversion (decreasing land use) means that the function of an area changes but with less intensity, such as the conversion of an intensive orchard into

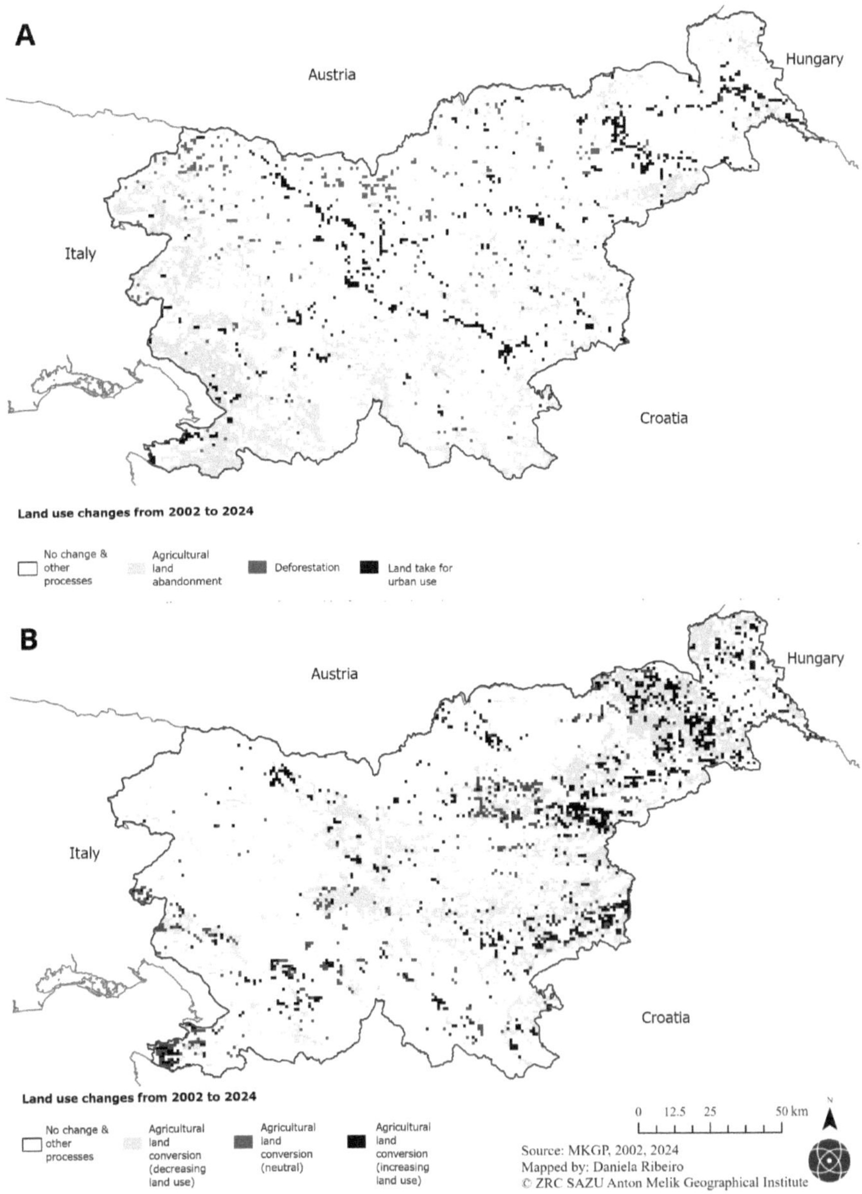

Figure 10.1 Processes of land use change in Slovenia between 2002 and 2024, shown on a 1 km grid: (A) Geographical distribution of abandonment of agricultural land, deforestation, and land take for urban use; (B) Geographical distribution of conversion within agricultural land.

Source: MKGP, 2002, 2024. Daniela Ribeiro is the author of the figure map.

an extensive orchard. Agricultural land conversion (increasing land use) occurs when the function of the land changes with increasing intensity; for example the conversion of permanent grassland into arable land. Abandonment of agricultural land occurs when agricultural use is reduced or abandoned completely, resulting in a transition to non-agricultural uses, with the exception of urban uses. An example is the conversion of temporary grassland to trees and shrubs. Deforestation occurs when forests are converted to land for other purposes, excluding urban uses. For example forest areas can be replaced by other permanent crops. Land take for urban purposes means that non-urban land is converted to urban uses, such as the conversion of temporary grassland to built-up areas and related surfaces. Wetland reclamation is the conversion of wetlands, such as swamps, bogs, and fens, into dry land for various purposes, such as agriculture and forestry. One example is the conversion of a swamp into a forest plantation. No change was mapped when there was no change from one land use type to another. When changes in land use were most likely attributable to a redefinition or refinement of land use categories, rather than a physical transformation of the landscape itself, they were defined as no clear trend.

Land use changes didn't show dramatic changes over the last two decades. Slightly more than 84% of Slovenia's territory has not changed since 2002 (Table 10.1). The remaining 16% has been subject to land use changes, with the abandonment of agricultural land being the most important process in the last two decades. This process is typical in western and southern Slovenia in karst and hilly areas. Increasing land use predominates in land conversion within agricultural land, indicating an intensification of agricultural cultivation. Around 1.6% of the national land area was taken up for urban use, which is a very dispersed process. Major changes occur primarily on the outer edges of settlements and along major infrastructure facilities like motorways. Deforestation occurs only in small isolated areas, mostly in the Alps in northern Slovenia, where individuals have cleared small areas of forest for grassland or pastures (Figure 10.1).

Table 10.1 Surface area of land use change processes

Process of land use change	*Total area (ha)*	*Total share (%)*
Agricultural land conversion (neutral)	36,048.14	1.85
Agricultural land conversion (decreasing land use)	28,490.02	1.46
Agricultural land conversion (increasing land use)	57,155.47	2.94
Agricultural land abandonment	72,004.33	3.70
Deforestation	23,204.17	1.19
Land take for urban use	30,989.22	1.60
Wetland reclamation	2,107.38	0.11
No change	1,637,206.71	84.15
No clear trend	58,409.70	3.00

In Figure 10.1, we mapped land use change processes using a 1 km grid resolution. To ensure that only significant changes were considered, we applied a threshold, taking into account only those land use changes where the difference exceeded 5% of the total area within each 1 km grid cell. This threshold was selected to eliminate minor variations, which could result from small-scale fluctuations or mapping inaccuracies, and to focus on substantial and meaningful changes in land use patterns.

The Land Use Database has been collected for more than 20 years, and in the meantime the methodology of land use data collection has evolved and improved. These changes have therefore affected the current land use data and therefore the results of this analysis. The changes include, among others, changes in built-up area and related surface and reed. We highlight these land use types because they influenced our results. The mapped minimum area of built-up area and related surface was 10 m^2 in 2002 and increased to 25 m^2 from 2004 (Foški, 2018).

What also influenced the total built-up area and related surface was the changes that occurred in other types of land use, and over time these types came to be regarded as built-up areas. Water area decreased over the years, but the largest change occurred in 2005, when the salt pans were considered as built-up area rather than water area. This change can be observed in the south-western part of Slovenia (Figure 10.1).

The grasslands around the airport runways were mapped as permanent grassland in 2002 and as built-up areas in 2024 (Klar, 2016).

These three changes in mapping methodology led to an increase in land take for urban use.

The reed areas have been mapped since 2002, and the mapped minimum area has not changed. The description of this land use type has not changed either. In 2004, however, a small addition was made to the description – that no agricultural production takes place in these areas. This was probably the reason for the large change between reed areas and permanent grassland between 2002 and 2005. Most of these areas are located in Lake Cerknica; this shows how much changes in mapping methodology can affect the results (Klar, 2016). The changes we see here in the context of land use change in wetland reclamation are largely due to these changes in mapping methodology, as can be seen in Figure 10.1.

Case studies of different processes of change

The analysis of land use changes at the local level (settlement) involved a detailed examination of how different types of land use evolve over time within the selected study areas of Bojanci and Krušni Vrh. Both Bojanci in Bela Krajina and Krušni Vrh in Suha Krajina are small settlements located in south-eastern Slovenia, and despite their distinct regional identities, they share several commonalities. The regions where these settlements are located – Bela Krajina and Suha Krajina – are characterized by a karst landscape, with limestone formations, caves, and rocky terrain, which influences local agriculture, water availability, and natural resources. Both Bojanci and Krušni Vrh are rural settlements, with a focus on traditional village life, agriculture, and forestry.

Using ArcGIS Pro 3.3, we compared land use data from 2002 (MKGP, 2002) with current land use data from 2024 (MKGP, 2024) to detect changes in land use, such as expansion of urban areas, shifts in agricultural practices, and deforestation. This analysis highlighted two opposite trends, on the one side abandonment of agricultural practices and on the other the intensification of agricultural use and expansion of urban areas.

The Bojanci case study

Table 10.2 shows shifts in land use categories over the 22-year period in Bojanci. Arable land has experienced a sharp decline from 3.79% in 2002 to just 0.44% in 2024; vineyards have decreased from 0.17% to 0.03%; and permanent grassland has also seen a notable reduction, from 14.72% in 2002 to 7.90% in 2024. These reductions reflect land abandonment due to reduced agricultural activity, driven by rural depopulation and economic migration, as stated by Ribeiro (2020). The decline in grassland might also be linked to a reduction in livestock farming in the area. The share of extensive orchards has slightly increased from 0.28% to 0.99%, while intensive orchards disappeared completely (0.07% in 2002 to 0% in 2024). This trend suggests a more traditional or subsistence-oriented orchard practices. The increase in overgrown agricultural areas (from 0.98% to 2.49%) and the appearance of a new category – trees and shrubs (3.90%) – suggest significant natural succession processes. As agricultural activities decline, previously cultivated areas are becoming overgrown with wild vegetation, leading to reforestation. The expansion of forest cover from 78.34% to 82.47% further reinforces this trend, highlighting the dominance of forest regeneration in areas where farming has receded. The built-up area has remained relatively stable (from 1.22% to 1.04%), this small difference between 2002 and 2024 could be due to map errors.

As it can be seen in Figure 10.2, a comparative analysis of the orthophotos from 1998 and 2021 reveals a significant expansion of forested area in the surroundings

Table 10.2 Surface area of land use categories in Bojanci over a 22-year period (2002–2024)

Land use category	*Total area (ha)*		*Total share (%)*	
	2002	*2024*	*2002*	*2024*
Arable land	28.64	3.32	3.79	0.44
Vineyard	1.27	0.26	0.17	0.03
Extensive orchard	2.10	7.45	0.28	0.99
Intensive orchard	0.54	/	0.07	/
Permanent grassland	111.34	59.75	14.72	7.90
Overgrown agricultural area	7.41	18.80	0.98	2.49
Forest plantation	3.28	0.22	0.43	0.03
Trees and shrubs	/	29.46	/	3.90
Uncultivated agricultural land	/	5.42	/	0.72
Forest	592.46	623.72	78.34	82.47
Built-up area and related surface	9.23	7.86	1.22	1.04

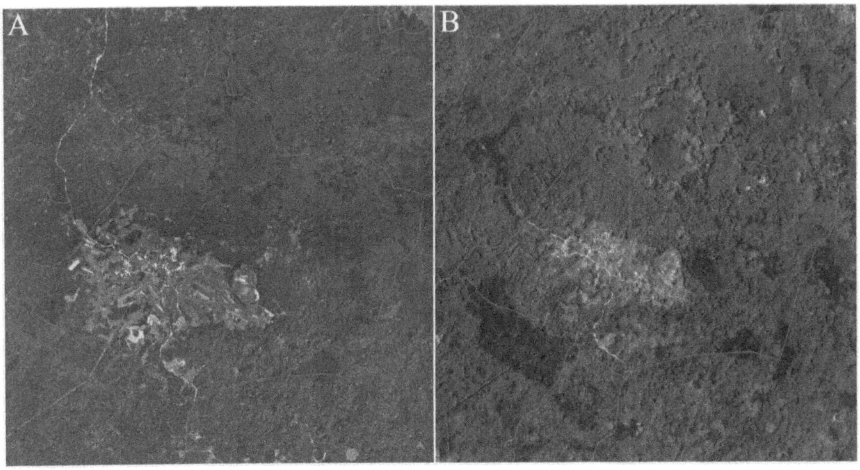

Figure 10.2 Landscape changes in the settlement of Bojanci: (A) an orthophoto from 1998 and (B) a present-day orthophoto from 2021.

Source: Geodetska uprava Republike Slovenije, 1998, 2021.

of the village. In the 1998 image, Bojanci was already characterized by a close landscape, consisting of forests. However, by 2021, the orthophoto shows signs of continuous forest expansion.

A reason for this forest expansion in Bojanci is the abandonment of agricultural land. A shift from traditional farming to urban migration or other economic activities leads to reduced maintenance of open landscapes, allowing natural forest regeneration to take place. This happens as agriculture is not economically viable in this karst landscape.

From an ecological point of view, the increase in forest cover can contribute to improved ecosystem services, such as carbon sequestration, and protection against soil erosion. However, from a socio-economic perspective, this transition reflects declining agricultural activity, which can impact local livelihoods and their self-sufficiency in food, as well as contribute to the loss of traditional cultural landscapes.

The Krušni Vrh case study

The comparison of land use data between 2002 and 2024, as shown in Table 10.3, reveals notable shifts in agricultural land use, forest cover, and urbanization in this case study. These changes suggest both agricultural intensification and natural succession processes, alongside a modest expansion of built-up areas. Arable land increased from 9.26% in 2002 to 10.20% in 2024, and permanent grassland expanded from 29.25% to 31.41%. This increase may indicate a revitalization of agricultural activities, potentially due to policy incentives. On the other hand,

Table 10.3 Surface area of land use categories in Krušni Vrh over a 22-year period (2002–2024)

Land use category	*Total area (ha)*		*Total share (%)*	
	2002	*2024*	*2002*	*2024*
Arable land	10.91	12.01	9.26	10.20
Vineyard	1.55	0.04	1.31	0.03
Extensive orchard	0.93	1.10	0.79	0.94
Other permanent crop	/	0.17	/	0.15
Permanent grassland	34.44	36.99	29.25	31.41
Overgrown agricultural area	1.87	7.40	1.59	6.28
Trees and shrubs	0.93	0.47	0.79	0.40
Uncultivated agricultural land	/	8.23	/	6.99
Forest	64.24	46.97	54.56	39.89
Built-up area and related surface	2.88	4.37	2.44	3.72

vineyard area saw a sharp decrease, from 1.31% in 2002 to just 0.03% in 2024. This reduction could be driven by several factors, such as a decline in the economic viability of viticulture in the area, shifts in agricultural focus, or indicate a movement away from labour-intensive crops like vineyards in favour of other forms of agriculture or land uses (e.g. grassland). Overgrown agricultural areas have increased considerably, from 1.59% to 6.28%, reflecting a trend towards land abandonment. However, forest cover has notably decreased from 54.56% in 2002 to 39.89% in 2024. This is one of the most significant land use changes and is due to deforestation for agricultural expansion and infrastructure development, as can be seen in the example given in Figure 10.3. Built-up areas and related surfaces have grown from 2.44% to 3.72%, reflecting modest urban expansion. This increase could be linked to infrastructural development (seen in Figure 10.3) or the spread of rural settlement.

A comparative analysis of the orthophotos from 1995 and 2021, as it can be seen in Figure 10.3, reveals agricultural expansion in Krušni Vrh, particularly in the upper left section of the selected area. In the 1995 orthophoto, this area was covered by forest. However, by 2021, a significant portion of the forest has been cleared, and the land has been converted to agricultural use (meadows). This change suggests an intensification of agricultural activities over the past few decades. The observed conversion of forest land to agriculture is likely driven by economic factors, most probably connected to policies incentivizing the expansion of agricultural activities.

The land use changes in Krušni Vrh between 2002 and 2024 reveal a complex interplay between agricultural intensification, land abandonment, and deforestation. The increase in arable land and permanent grassland suggests an intensification of agriculture, which may boost local food production or livestock farming. On the other hand, the rise in overgrown agricultural areas and uncultivated land reflects a trend of land abandonment, likely driven by socio-economic factors such as rural outmigration and ageing population. While the expansion of built-up areas

Figure 10.3 Landscape changes in a selected area of Krušni Vrh: (A) an orthophoto from 1995 and (B) a present-day orthophoto from 2021.

Source: Geodetska uprava Republike Slovenije, 1995, 2021.

is relatively modest, it points to ongoing rural development. This could bring both opportunities and challenges, such as improved infrastructure and services on the one hand but increased land competition and environmental pressures on the other.

Consequence of these changes to biodiversity/landscape sustainability – dilemmas and challenges

The sustainability of a landscape is intricately tied to how land is managed (Wascher, 2004) and requires the coordination of different practices to prevent potentially irreversible damage to the landscape.

While land use changes have contributed to significant net gains in human well-being and economic development, these gains have come at a high price in terms of biodiversity loss and ecosystem degradation (Millennium Ecosystem Assessment, 2005) as well as social impacts, such as increased financial burden on smallholders (Meyfroidt et al., 2018).

Meeting the future needs of a growing population while conserving natural habitats, halting biodiversity loss, and increasing the share of renewable energy will accelerate the competition for land (Lambin & Meyfroidt, 2011).

In Slovenia, in the period 2002–2024, agricultural land use declined, accompanied by a considerable land conversion within agricultural land (6.3%), the increasing land use intensification being the major process within the agricultural land conversion. Other two quite important land use change processes, each with a share higher than 1%, are land take for urban use and deforestation. Urbanization accounts for approximately 1.60%, while deforestation is occurring at a rate of 1.19%. At the same time, rural depopulation has driven land abandonment, enabling natural forest regeneration, as observed in Bojanci. While forest expansion contributes to ecosystem services like carbon sequestration and erosion control, it also reflects declining agricultural activity, jeopardizing local livelihoods, food security, and the preservation of Slovenia's traditional cultural landscapes. These trends highlight the duality of forest expansion – a dynamic often celebrated at the European level (e.g. Navarro & Pereira, 2012; Zethof et al., 2019) but potentially detrimental for marginal rural areas facing depopulation.

The example of Krušni Vrh, with its agricultural intensification, underscores another facet of Slovenia's land use dynamics. Agricultural expansion offers economic opportunities and enhanced food production but also raises concerns about its environmental sustainability and long-term viability. These contrasting trends illustrate the multifaceted and often conflicting nature of land use changes.

From a broader European perspective, the expansion of forests is often seen as a positive development as it contributes to climate goals, biodiversity, and ecosystem resilience. However, in marginal rural areas such as the depopulated regions of Slovenia, this trend can have unintended socio-economic consequences, including the erosion of traditional agricultural practices and rural identities that are an integral part of cultural heritage. This conflict between environmental aspirations at the European level and local socio-economic realities requires a differentiated approach to land management that balances environmental benefits with socio-economic vitality.

Understanding the balance between human activities and natural processes is essential for sustainable land management and policymaking. Insights from Slovenia's experiences provide valuable contributions to European scientific discourse on land use transformations, emphasizing the need for policies that integrate ecological, social, and economic dimensions. Effective local planning and conservation strategies must account for the competing pressures of human activity and natural processes, ensuring that land use decisions support both local resilience and broader sustainability goals. Such studies are indispensable for addressing the profound challenges of harmonizing European-level environmental priorities with the lived realities of marginal rural communities.

Acknowledgement

The research was financially supported by the Slovenian Research and Innovation Agency research core funding programme »Geography of Slovenia« (P6–0101).

References

Blondel, J. (2006). The "design" of Mediterranean landscapes: A millennial story of human and ecological systems during the historic period. *Human Ecology*, *34*, 713–729. https://doi.org/10.1007/s10745-006-9030-4

Blondel, J., Aronson, J., Boudiou, J. Y., & Bœuf, G. (2010). *The Mediterranean Basin – biological diversity in space and time*. Oxford University Press.

Bole, D. (2015). *Spreminjanje prometne rabe zemljišč v Sloveniji* [Transformation of transportation land use in Slovenia]. Založba ZRC. https://doi.org/10.3986/9789610503569

Bugalho, M. N., Caldeira, M. C., Pereira, J. S., Aronson, J., & Pausas, J. G. (2011). Mediterranean cork oak savannas require human use to sustain biodiversity and ecosystem services. *Frontiers in Ecology and the Environment*, *9*(5), 278–286. https://doi.org/10.1890/100084

Foški, M. (2018). The (non)usefulness of the register of existing agricultural and forest land use for monitoring the processes in urban areas. *Acta geographica Slovenica*, *58*(1), 69–82. https://doi.org/10.3986/AGS.1805

Gabrovec, M., & Kumer, P. (2019). Land-use changes in Slovenia from the Franciscean Cadaster until today. *Acta Geographica Slovenica*, *59*(1), 63–81. https://doi.org/10.3986/AGS.4892

Gabrovec, M., Kumer, P., Ribeiro, D., & Šmid Hribar, M. (2020). Land use in Slovenia. In Perko et al. (eds.), *The geography of Slovenia*, *World regional geography book series* (pp. 279–290). https://doi.org/10.1007/978-3-030-14066-3_18

Geodetska uprava Republike Slovenije. (1995). Orthophoto F2209.

Geodetska uprava Republike Slovenije. (1998). Orthophoto G1910.

Geodetska uprava Republike Slovenije. (2021). Orthophotos F0549 and G0250.

Golobič, M., & Lestan, K. A. (2016). Potential impacts of EU policies on cultural landscape diversity: Example of Slovenian coastal landscapes. *Annales*, *26*(2), 193–212. https://doi.org/10.19233/ASHS.2016.16

Horvat, U., & Žiberna, I. (2020). The correlation between demographic development and land-use changes in Slovenia. *Acta Geographica Slovenica*, *60*(2), 33–55. https://doi.org/10.3986/AGS.7611

Klar, T. (2016). *Preučevanje vpliva metodologije zajema dejanske rabe zemljišč na spreminjanje površin vrste rabe zemljišč med letoma 2002 in 2016 v Sloveniji*. University of Ljubljana.

Lambin, E. F., & Meyfroidt, P. (2011). Global land use change, economic globalization, and the looming land scarcity. *Proceedings of the National Academy of Sciences of the USA*, *108*(9). https://doi.org/10.1073/pnas.1100480108

Lieskovský, J., Bezák, P., Špulerová, J., Lieskovský, T., Koleda, P., Dobrovodská, M., Bürgi, M., & Gimmi, U. (2015). The abandonment of traditional agricultural landscape in Slovakia – analysis of extent and driving forces. *Journal of Rural Studies*, *37*, 75–84. https://doi.org/10.1016/j.jrurstud.2014.12.007

Lisec, A., Pišek, J., & Drobne, S. (2013). Suitability analysis of land use records of agricultural and forest land for detecting land use change on the case of the Pomurska Statistical Region. *Acta Geographica Slovenica*, *53*(1). https://doi.org/10.3986/AGS53104

Mander, Ü., & Uuemaa, E. (2010). Landscape assessment for sustainable planning. *Ecological Indicators*, *10*(1), 1–3. https://doi.org/10.1016/j.ecolind.2009.08.003

Meyfroidt, P., Chowdhury, R. R., de Bremond, A., Ellis, E. C., Erb, K.-H., Filatova, T., Garrett, R. D., Grove, J. M., Heinimann, A., Kuemmerle, T., Kull, C. A., Lambin, E. F., Landon, Y., le Polain de Waroux, Y., Messerli, P., Müller, D., Nielsen, J. Ø., Peterson, G. D., Rodriguez García, V., Schlüter, M., Turner II, B. L., & Verburg, P. H. (2018). Middle-range theories of land system change. *Global Environmental Change*, *53*. https://doi.org/10.1016/j.gloenvcha.2018.08.006

Millennium Ecosystem Assessment. (2005). *Ecosystems and human well-being: Synthesis*. http://www.millenniumassessment.org/documents/document.356.aspx.pdf

MKGP. (2002). *Land use database*. https://rkg.gov.si

MKGP. (2024). *Land use database*. https://rkg.gov.si

Navarro, L. M., & Pereira, H. M. (2012). Rewilding abandoned landscapes in Europe. *Ecosystems*, *15*, 900–912. https://doi.org/10.1007/s10021-012-9558-7

Perko, D., & Ciglič, R. (2020). Slovenia's landscapes. In Perko et al. (eds.), *The geography of Slovenia*, *World regional geography book series*. https://doi.org/10.1007/978-3-030-14066-3_14

Pinto-Correia, T. (2000). Future development in Portuguese rural areas: How to manage agricultural support for landscape conservation? *Landscape and Urban Planning, 50*(1–3), 95–106. https://doi.org/10.1016/S0169-2046(00)00082-7
Ribeiro, D. (2020). *Bela krajina – sustainability in a karst landscape.* Založba ZRC. https://doi.org/10.3986/9789610504627
Ribeiro, D., & Šmid Hribar, M. (2019). Assessment of land-use changes and their impacts on ecosystem services in two Slovenian rural landscapes. *Acta Geographica Slovenica, 59*(2), 143–59. https://doi.org/10.3986/AGS.6636
Wascher, D. M. (2004). Landscape-indicator development: Steps towards a European approach. In R. G. H. Jongman (ed.), *The new dimensions of the European landscape: Frontis Workshop on the future of the European Cultural landscape* (pp. 237–252) Wageningen UR Frontis series. Springer.
Zethof, J. H. T., Cammeraat, E. L. H., & Nadal-Romero, E. (2019). The enhancing effect of afforestation over secondary succession on soil quality under semiarid climate conditions. *Science of the Total Environment, 652*, 1090–1101. https://doi.org/10.1016/j.scitotenv.2018.10.235
Žiberna, I., & Ivanjšič, D. (2022). Land use changes by mesoregion in Slovenia between 2000 and 2022. *Revija za geografijo, 17*(2), 37–54. https://doi.org/10.18690/rg.17.2.2729

11 Agricultural land use and its climate change vulnerability in Asia

Moving from vulnerability to resilience

Sivapuram Venkata Rama Krishna Prabhakar

Introduction

Land serves as a vital natural resource for many Asian nations, with agriculture remaining the predominant sector utilizing this resource across the continent (FAO, 2016). Agricultural land is crucial for fulfilling the food, climate, and human security objectives of Asian countries, and any practices that lead to unsustainable land use could jeopardize the achievement of these essential goals. The agricultural landscape in Asia is experiencing profound changes driven by factors such as an increasing population, rapid urbanization, shifts in income levels, and evolving lifestyles, particularly in food consumption. These influences are exerting considerable stress on land resources, resulting in degradation that adversely affects the environment, the economy, and the livelihoods of communities reliant on these resources. As Asia's population is expected to rise soon, the strain on land resources is likely to intensify, leading to further negative repercussions. This situation underscores the urgent need to comprehend the emerging factors influencing land use and its transformations in Asia, enabling the development of effective policy responses.

Asian nations exhibit dynamic agrarian economies, with a considerable segment of their populations reliant on agriculture for their livelihoods. Over the years, these countries have gradually advanced their agricultural policy frameworks through coordinated efforts, resulting in increased food self-sufficiency and security. Nevertheless, the risks posed by climate change threaten to undermine this progress and jeopardize future advancements. While there have been strides in agricultural technology development and its dissemination, the challenges extend beyond mere scale. The new obstacles introduced by climate change necessitate more innovative solutions that have yet to be explored by these nations. It is imperative to enhance climate risk management within the agricultural sector, which requires a comprehensive approach that transcends traditional boundaries and sectoral limitations. Additionally, countries must explore new methods for generating high-quality data that can be effectively shared among various stakeholders, fostering data democracy in the region. As regional integration deepens, these nations must cultivate trust in collaborative efforts and reinforce regional adaptation strategies. This chapter has established a vital framework to facilitate such regional cooperation in Asia.

DOI: 10.4324/9781032672779-14

The heightened vulnerability of the agrarian population in Asia can be attributed, in part, to their significant reliance on rainfed agriculture, coupled with the inadequate development of reliable irrigation systems. This issue is particularly pronounced in South Asia, where freshwater resources are subjected to considerable stress. Countries within this region exhibit a marked susceptibility to fluctuations in climate and weather patterns. Alterations in rainfall distribution can severely impact agricultural output, thereby affecting both rural and urban communities. The repercussions of select historical weather-related disasters on agricultural productivity are evident. Notably, a drought event in India in 2002 led to a staggering loss of 32 million metric tonnes of paddy production, necessitating a recovery period of six years to return to pre-drought production levels. Given that agriculture serves as a fundamental input for the economy, such disruptions can have extensive ramifications across various sectors.

Climate change represents a significant and escalating threat, with varying degrees of vulnerability experienced by countries worldwide. The latest IPCC Assessment Report highlights agriculture as a critical sector that not only faces risks from climate change but also contributes to it through greenhouse gas emissions. The repercussions of climate change on agriculture could jeopardize the livelihoods and well-being of millions in South Asia.

Countries in Asia are currently navigating a pivotal stage in their development, where the significance of agriculture for national well-being is increasingly precarious. The agricultural sector in Asia is particularly susceptible to the adverse effects of climate change. This susceptibility is exacerbated by existing conditions that have not been adequately addressed, despite extensive developmental initiatives in the region. Nevertheless, Asian countries also possess the potential to effectively tackle climate change challenges in the future. A deeper comprehension of these vulnerabilities and capacities is essential for formulating collaborative strategies that nations can implement to confront climate change threats in the coming years.

Acknowledging the critical role of agriculture in national GDP and the well-being of populations, Asian countries are channelling significant resources into enhancing agricultural resilience across multiple dimensions. These investments encompass research and development aimed at creating new crop varieties and agro-technologies designed to adapt agriculture to the challenges posed by climate change while also working to reduce greenhouse gas emissions from agricultural and related sectors. These initiatives are still in their early stages, highlighting the necessity to assess the current vulnerabilities of countries to climate change and the measures undertaken to mitigate these risks. Land plays an important role in contributing to these initiatives and can sustain a development base for centuries in the future if managed sustainably.

The existing discourse predominantly focused on the effects of climate change on agricultural output. Nevertheless, there is a significant gap in the literature regarding the assessment of vulnerabilities and adaptive capacities of agricultural systems within Asia concerning how the land is managed, its contribution to the vulnerability and capacity of people and ecosystems to potential climate change impacts, and what opportunity it provides in further strengthening the adaptive capacity to people, societies, and nations. Agricultural systems in this context include not only cropping systems but also governmental institutions and policies, the farmers, supply chains, and production systems, which also encompass irrigation infrastructure. A comprehensive understanding of these agricultural systems is crucial for fully grasping the potential future impacts of climate change on agriculture.

In light of this context, this chapter describes the prevailing knowledge concerning land use changes and other land use-related issues in Asia. It evaluates how the use (or misuse) of land in general and agricultural land in particular can affect the vulnerabilities and capacities of national agricultural systems.

Vulnerability of agricultural systems in Asia

An examination of the climate change vulnerability of Asian countries necessitates a more comprehensive treatment than what is feasible within the confines of a standard book chapter. Nevertheless, this section aims to provide a concise overview of the current vulnerabilities associated with climate change in Asia. It helps us to see where the land and related factors play an important role in the overall vulnerability picture. Here, we discuss a typology of risks and vulnerabilities that Asian countries are experiencing and the possible connections with land use in a qualitative manner.

Climate change impacts in Asia are well documented. The IPCC AR6 highlighted that Asia has experienced significant warming and is projected to experience hotter summer climate with the potential for heat waves and droughts, with some regions experiencing increased precipitation in South Asia and North Asia (Shaw et al., 2022). South, Southeast, and East Asia could experience intense precipitation. This could result in large increase in flood frequency and related impacts. Extreme hot days could result in severe drought conditions in west and central Asia. Mountainous regions could experience elevation-dependent warming. Warming in the region is also projected to increase the winter precipitation. As a result, climate change will act as one of the important drivers of territorial ecosystems change in the region.

The literature from Asian countries indicate a prevalent issue of high poverty levels, particularly in rural regions, which are marked by low human development indices and other globally recognized developmental indicators (Klasen & Waibel, 2015; Tan et al., 2023). This situation establishes a foundational vulnerability for these populations, which in turn limits the agricultural sector's potential contributions to the region's economy. In turn, the prevailing poverty also affects the way the population use agricultural land. With sustainable farming and a lack of appropriate education and skills to manage land sustainably, agricultural land in Asia has been continuously subject to exploitation, denudation, and degradation.

The deficiencies in road infrastructure, restricted market access, inadequate educational facilities, limited availability of advanced agricultural technologies, and overall social underdevelopment render these countries exceptionally susceptible to various shocks, including those induced by climate change or global transformations (Thomas, 2023; Blondin, 2023). Furthermore, any climatic stressors may exacerbate the existing socio-economic challenges, intensifying the effects of climate change. The underlying vulnerabilities in these nations often remain obscured until they are triggered by climatic or non-climatic disturbances.

Broadly, farmers and agricultural systems in the region face two primary categories of risks: idiosyncratic and covariate risks (see Table 11.1). Idiosyncratic risks pertain to shocks that are unique to individual farming contexts and their specific vulnerabilities. These risks can be mitigated through income diversification and are typically insurable, as they do not correlate with external factors beyond the control of the affected individuals. Consequently, insurance providers encounter minimal risk when underwriting such policies. Conversely, covariate risks exert a considerable influence and frequently lie outside the realm of individual control.

How these risks are related to land is shown in Figure 11.1. The figure shows that both the idiosyncratic and covariate risks have feedback connections and that these risks can have compound impacts on the way the land is managed leading to overall high vulnerability of populations to climate change.

To deeply understand these dynamic connections between land use and climate change vulnerability, we also need to understand the trends in land use and how these trends are detrimental to environmental sustainability and social resilience. In the following section, an effort has been made to succinctly present the land use trends in the recent past.

Land use changes in Asia

Land in Asia, both agricultural and non-agricultural, are going through significant changes (Prabhakar, 2021). These changes have sociopolitical and economic push and pull factors. With a growing population and demand for more food production, change from non-agricultural land to agriculture is one set of changes we can observe. This includes changing land from productive forests to productive agriculture. These changes largely happen in major production centres. Another set of changes that we can observe is changing agricultural land back to non-agricultural

Table 11.1 Farmers in the Asia region encounter both unique and correlated risks

Unique risks		*Correlated risks*
Income risks	High cost of inputs Reduction of profits	Droughts Floods
	Loss of employment	High-temperature shocks
Asset risks	Damage to infrastructure Death of animals	Low-temperature shocks Forest fires
	Breakdown of equipment	Disease and pest outbreak
	Fire outbreak	Labor shortage
		Market fluctuations

Source: Author.

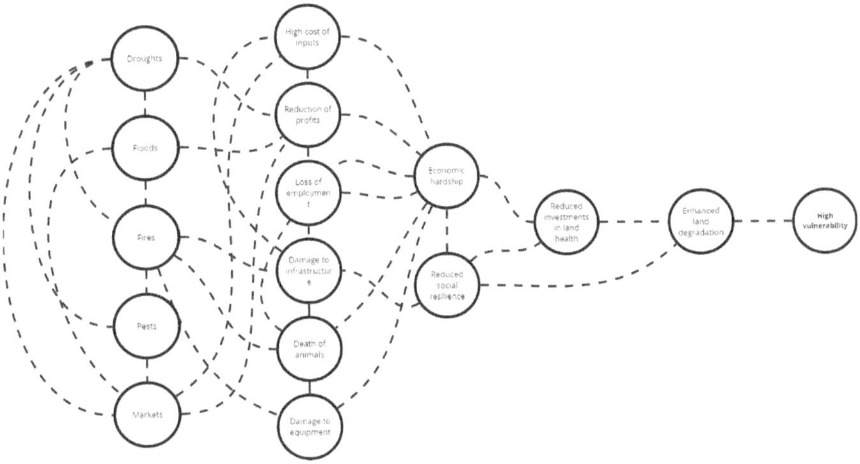

Figure 11.1 Feedback connections between covariate and idiosyncratic risks leading to compound impacts on the land and overall vulnerability.

Source: Author.

lands. In the same way, another set of productive lands in forests is being converted to productive urban, industrial and infrastructure use. These changes largely happen in major urban centres.

Depending on how one views land use changes, the changes are happening within the productive category and between productive and unproductive categories of land. Productive categories of land use are agricultural land; pristine and planted forests; and urban, industrial, and infrastructure lands. The major unproductive lands are ravines, gullies, rocky, eroded, salt, saline, acidic, degraded, and denuded lands. The bold arrows show the significant directional changes from one use category to another use category. The narrow-dotted arrows show the change of land in the reverse direction but at a much lower scale.

While the changes are happening between the major land use categories, changes are also happening within a land use category with many nuances. Examples of these changes include changes in cropping systems, that is changes from one crop to another crop; changes in forest vegetation due to both natural and human pressures; and changes in urban areas from a small town to a metropolis. The global discourse has been heavily focused on shifts from one land use category to another use category and very little has been explored on how land use is changing within a land use category.

Agricultural land use changes

Agricultural land use changes are the most significant land use change in Asia, and it deserves focused discussion (Prabhakar, 2021; Huang et al., 2023). Over the last several decades, Asia has experienced notable shifts in its agricultural land utilization. Between 1961 and 2015, data reveals that agricultural land remained stable in eight nations, while it saw a decrease in ten others and an increase in 29 (FAO, 2024). Among the countries that have faced substantial reductions in agricultural land are Brunei Darussalam, which saw a staggering decline of 66.7% since 1961, followed by Japan with a 48% decrease, the Republic of Korea at 35.3%, and Mongolia with a 24.6% drop. Conversely, the nations that have recorded the most significant growth in agricultural land include Malaysia, which increased by 49.5%, Vietnam at 47.8%, the United Arab Emirates with a 44.5% rise, and Bhutan, which saw a 40.7% increase.

China (153 million hectares), Indonesia (18.6 million hectares), and Saudi Arabia (87.2 million hectares) have transformed extensive tracts of land for agricultural purposes, while Mongolia (27.8 million hectares) and Iran (13.7 million hectares) have experienced a notable decline in their agricultural land during the same time frame (FAO, 2024). The origins of the land that has been converted to agriculture vary among these nations. In China and Saudi Arabia, a considerable portion of the newly cultivated land has come from areas classified as 'other land.' Conversely, Indonesia's agricultural expansion has primarily occurred at the expense of its forests. In the cases of Iran and Mongolia, the reduction in agricultural land can be attributed to either afforestation efforts or shifts to different land use categories. Several factors contribute to the dynamics of land loss or gain in agriculture. In regions where agricultural land has diminished, this may be linked to changes in land management practices aimed at enhancing productivity, the influence of global agricultural trade that favours sourcing food from the most cost-effective producers, and a trend of farmers transitioning away from full-time farming to pursue alternative income opportunities. On the other hand, in countries that have seen an increase in agricultural land, motivations such as the need for national food security, government subsidies, the pursuit of food exports to specialized markets, and employment regulations may play a significant role in these developments.

Commercial agriculture has played a crucial role in the transformation of land for agricultural use, as highlighted by the FAO in 2016 (FAO, 2016). The advent of mechanization has significantly facilitated this shift by decreasing reliance on manual labour. Recently, private sector investments and foreign direct investments have emerged as key factors driving the ongoing demand for agricultural land, as noted by Ravanera and Gorra in 2011. Additionally, the opportunities for food trade stemming from regional economic integration within the ASEAN region have further bolstered this demand.

Conversely, several factors have contributed to the loss of agricultural land, including urbanization, industrialization, and infrastructure development (Junaid & Gokce, 2024). The soaring land prices in urban areas have prompted the conversion of agricultural spaces into residential developments. Furthermore, the declining rural population in developed Asian nations has led to a shift of agricultural land towards non-agricultural uses. Various challenges such as decreasing farm profits, market inefficiencies, land degradation, unfavourable pricing policies, natural disasters, and rising education levels have also pushed individuals away from farming, making agricultural land more susceptible to conversion. Urbanization stands out as a significant emerging factor, particularly in countries like Lao PDR, Thailand, China, and Bhutan, which are experiencing rapid urban population growth. In contrast, some West Asian nations are witnessing a gradual deurbanization, exemplified by Lebanon's 12% decline in urban population over the past 15 years (FAO, 2024). Notably, countries exhibiting a negative correlation between agricultural land and urban population have seen a marked decrease in agricultural land, while those with a positive correlation have experienced an increase in agricultural land across Asia.

Predisposing vulnerability factors

Asia is marked by various vulnerability factors that significantly influence the condition of land, and how the land is managed, which in turn affects the resilience and capacity of its people and nations. These characteristics are likely relevant in other regions as well, but this chapter emphasizes Asia, allowing us to delve into our current comprehension of these factors. By examining the challenges faced in this region, we can glean valuable insights that may lead to effective solutions.

The significant reliance on climate-sensitive sectors like agriculture and livestock, combined with a scarcity of varied livelihood options, emerges as a primary factor contributing to vulnerability in Asian nations (Behera et al., 2024; Chandio et al., 2023). The primary dependency doesn't always translate into significant investments in safeguarding land resources, and this is an important issue. This is largely due to a lack of appropriate policies and sustainability education that is well-linked with the local cultures and values. While other developing countries may have a larger proportion of their populations dependent on land-based livelihoods such as agriculture, they often exhibit greater resilience due to a more diversified agricultural and livelihood base that mitigates the impact of environmental shocks on land and in turn on their livelihoods. How livelihoods are diversified is crucial; for instance, a farmer who grows multiple crops but relies solely on agriculture for income may face greater risks from severe droughts or floods compared to one who has diversified into non-agricultural activities. Additionally, the availability of support services is vital for enabling farmers to recover and resume agricultural activities when conditions improve after a climate-related disaster. Essential resources such as credit, seeds, and fertilizers are critical immediately following adverse weather events, allowing farmers to capitalize on favourable conditions. However, establishing these support systems requires years of dedicated effort from both governments and civil society organizations to ensure they are effectively developed and can provide timely assistance to farmers in need.

Figure 11.2 illustrates the connection between various income sources and the percentage decrease in household income following floods in Myanmar, with farm income being the sole focus. Insights drawn from household surveys (n=198) conducted by the author highlight the critical need to encourage diverse income streams. While the research does not provide strong evidence on the extent to which these multiple income sources impact land management, it stands to reason that families with adequate off-farm income are better positioned to invest

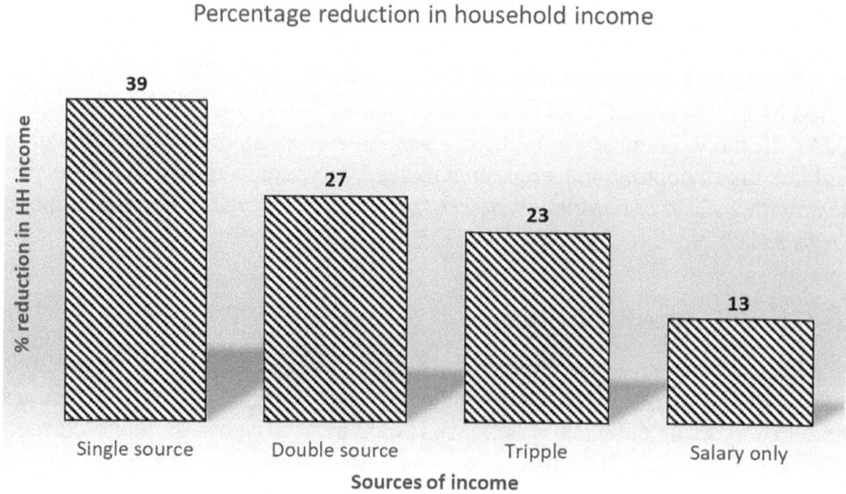

Figure 11.2 Diverse livelihoods observed in rural communities in Myanmar and its impact on overall income loss at the family level

Source: Author-based on questionnaire surveys conducted in Bago, Myanmar.

in sustainable land practices, thanks to the extra financial resources available for agricultural endeavours.

Another vulnerability factor that communities have is social inequality. Even though this is a social issue and is not directly addressed by individuals, it affects the way they have access to resources. Limited access to natural resources is an important issue undermining the resilience of communities in the face of climate change. The limited access to resources, combined with the swift deterioration of the natural resource base, represents a significant vulnerability for the Asian nations. Over recent decades, these countries have experienced a notable drop in factor productivity, indicating that crop yields are not keeping pace with the rising input quantities and associated costs. This imbalance has resulted in increased production costs and diminished overall farm productivity, including labour efficiency. A considerable portion of this decline in productivity can be linked to long-standing unsustainable agricultural practices, the overuse of land and water resources in certain regions, and the excessive application of agricultural chemicals. These detrimental practices, alongside the absence of effective policy interventions at the ground level, have contributed to ongoing natural resource degradation, manifesting in issues such as soil erosion, reduced soil fertility, and pollution of land and water. Over the years, many regions have also faced challenges like salinity and alkalinity, and the lack of consistent reclamation efforts has exacerbated these problems, leading to a deterioration in the socio-economic conditions of affected areas.

Land management requires substantial resources, necessitating significant investments in equipment, skill enhancement, and ongoing commitment over time.

As social living standards evolve and more individuals migrate from rural to urban areas, finding available labour for land management has become increasingly challenging. This labour shift and its implications for land management are often overlooked in existing literature, which tends to concentrate on the livelihood aspects of communities. The scarcity of agricultural labour presents a pressing issue that demands immediate policy intervention from governments. This labour shortage has led to soaring labour costs, which in turn inflate production expenses and hinder farmers from executing timely agricultural tasks, such as planting crops during favourable weather or harvesting before adverse conditions like cyclones. The trend of rural populations relocating to urban centres in search of better livelihoods has intensified over the years. Numerous studies have explored the factors driving this migration, revealing that unreliable livelihood opportunities, inadequate rural infrastructure, and limited access to essential services like healthcare and education are significant contributors. While many of these studies have examined the impact of climate change, particularly concerning droughts and their role in persistent crop failures, the broader implications of climate change on rural outmigration have not been extensively addressed.

Agricultural intensification refers to the extent of yield enhancement resulting from specific production decisions (Dietrich et al., 2012). While crop yields can serve as a measure of this intensification, it is essential to approach such interpretations with caution, as external factors like favourable weather conditions can also play a significant role in yield improvements. The degree of agricultural intensification can be assessed by examining the land utilized to generate a certain quantity of food relative to the inputs used. Irrigation is a key driver of this intensification, enabling farmers to grow multiple crops within a single year, cultivate high-yield varieties, and apply fertilizers, pesticides, and mechanization. Consequently, when data on other intensification methods is lacking, irrigation can be viewed as a primary indicator of agricultural intensification.

Data regarding chemical inputs and the percentage of irrigated agriculture reveals a notable increase in chemical usage across countries such as Bangladesh, China, India, Indonesia, and Vietnam. In contrast, Japan and the Republic of Korea have seen a decline in chemical inputs, primarily due to a reduction in agricultural land and a shrinking farming population. Among these nations, Indonesia has experienced the most significant rise in chemical usage, followed closely by India and Bangladesh.

The socio-economic challenges stemming from outmigration have garnered significant focus in recent research. In response to these challenges, various governments have enacted policies aimed at curbing outmigration and mitigating its socio-economic repercussions. These initiatives include skill development programmes, crop diversification efforts, enhancements to irrigation infrastructure, rural employment guarantee schemes, direct benefit transfers, and health and nutrition programmes for women and children. While these measures have proven effective in lessening the adverse socio-economic effects of outmigration, they have not entirely halted the trend, primarily due to the severe livelihood losses caused by unpredictable weather and the compelling allure of urban areas that offer considerably higher wages and better job opportunities than rural regions. A promising avenue for strengthening these initiatives and achieving a meaningful reduction in outmigration lies in investing in land management. In India, both the central and local governments have rolled out soil card programmes that allow farmers to periodically test their soil, providing them with essential guidelines for land, crop, and soil management based on the test results. Additionally, practices like land levelling are gradually gaining traction, particularly in the Indo-Gangetic Plain, where zero and reduced tillage technologies are being promoted alongside laser-assisted land levelling, often facilitated by farmer cooperatives and progressive farmers who offer these services for a fee. Such interventions have the potential to significantly stabilize crop yields and agricultural incomes, thereby alleviating the pressures that drive outmigration.

Asian nations have been grappling with inadequate governance across political, social, environmental, and economic spheres, which has hindered the advancement of agricultural systems in the region. Historically, these governments have prioritized technocratic approaches that emphasize the distribution of high-yield crop varieties aimed at achieving food self-sufficiency and security, often at the expense of fostering sustainable agricultural practices aligned with long-term social development objectives. This narrow focus on immediate solutions has led to missed opportunities for enhancing agricultural governance, particularly from the viewpoints of land management and sustainability.

Highland regions in various Asian countries have often been overlooked, and this neglect is increasingly raising alarms for multiple reasons. The combination of overexploitation of the land and inadequate slope stabilization has led to significant erosion, rendering these slopes incapable of supporting intensive farming practices. As a result of prolonged degradation and a lack of attention, many individuals are migrating from the highlands to the plains, which in turn places excessive strain on the natural resources in those areas. The newcomers often face inadequate livelihoods and poor living conditions, leading to marginalization and exposing them to various forms of social and political exploitation. Additionally, this outmigration depletes the highlands of the human resources necessary for effectively implementing any policy initiatives, creating a vicious cycle that further exacerbates the challenges faced by these regions. Despite the critical importance of highland agriculture, many agricultural systems, including national research and policies, have largely ignored its sustainability. However, recent efforts in countries like India and Nepal to leverage highland and hilly ecosystems for organic agriculture and food production represent a positive shift, albeit one that many consider to be long overdue.

Vulnerability impacts of land use

While social, economic, political, environmental, and institutional vulnerabilities affect the way land is managed sustainably, the condition of the land also affects the vulnerability of societies and countries in turn. In this section, the vulnerability impacts of land use in Asia have been discussed.

The transformations in agricultural land use previously examined reveal significant socio-economic and environmental trade-offs across Asia. In the Gangetic Basin, there has been a notable increase in agricultural areas, yet this has coincided with a decline in total factor productivity (Prabhakar & Elder, 2009). The implementation of food subsidies, particularly minimum support prices for food crops, has resulted in uneven growth of certain crops, which has subsequently led to heightened consumption of water and chemicals (Sharma, 2015). This intensive chemical usage has caused pollution in both surface and subsurface water sources, increased run-off, and contributed to the silting of freshwater bodies, ultimately resulting in eutrophication that adversely affects aquatic ecosystems (Wasantha et al., 2015). Furthermore, the detrimental effects of land use changes on soil and water ecosystems have been documented (Vlek et al., 2017).

The production of biofuels is another area of concern, as it requires substantial water resources; for instance, soy feedstock can demand between 47 litres and 215 litres of water for every megajoule of energy produced (O'Connor, 2009). The expansion of biofuel cultivation has led to a net increase in greenhouse gas emissions, particularly due to the conversion of forests, grasslands, and peatlands in regions like Indonesia and Malaysia (Fulton et al., 2017). Additionally, there is evidence suggesting that subsidized rural electrification programmes have contributed to the over-extraction of groundwater in certain areas of India (Sharma, 2015). Projections indicate that the rising energy demands in the coming years could worsen water shortages, with estimates suggesting that water requirements for energy production in Asia could increase fivefold by 2050 (Mitra & Bhattacharya, 2012). The ability to draw more water can enable driving more land to agricultural production if it is not in demand for more productive purposes such as urbanization and infrastructure development.

The intensification of land use has played a significant role in climate change, primarily by encouraging the use of fossil fuel-derived fertilizers, increasing farm mechanization, and the practice of burning crop residues in certain regions of Asia that would typically decompose naturally in a more extensive farming system (Nani et al., 2011). Additionally, the expansion of agricultural activities in tropical areas has been linked to a warming effect on the atmosphere, as surface brightening leads to a decrease in net radiation that fails to offset the temperature rise caused by diminished transpiration (Duveiller et al., 2017).

Measures to strengthen land management in Asia

Addressing the transformations of land in Asia requires careful consideration of the region's food security and developmental priorities. This section explores various strategies aimed at curbing the demand for agricultural land while also highlighting the risks associated with intensification. Enhancing agricultural productivity presents a straightforward opportunity to alleviate the pressure for additional land. Despite the potential, productivity improvements in rainfed areas have been disappointingly low, and past initiatives to bolster rainfed agriculture have not yielded significant results. This situation underscores the critical need for increased investment in research and development focused on rainfed farming systems.

Sustainable agriculture in Asia is significantly represented by organic farming, which holds considerable promise for further growth. Additionally, climate-smart agriculture (CSA) offers a dual advantage by helping to mitigate climate change while also lessening its adverse effects on agricultural practices. Implementing food labelling can play a crucial role in reducing negative environmental impacts by educating consumers about the ecological footprint of their food choices, including water usage, chemical inputs, and carbon emissions. Furthermore, the current subsidy system often creates counterproductive incentives within the agricultural sector, highlighting the necessity for more targeted support for impoverished farmers rather than blanket subsidies. Reforming commodity price support is essential to eliminate market distortions and to promote the production of food that is sustainably sourced.

Addressing food–fuel conflicts is crucial, and the Government of India has taken steps through the National Biofuel Policy, which restricts biofuel production to non-food feedstock cultivated on degraded lands. To enhance the effectiveness of such policies, it is vital to engage with farmer organizations, including cooperatives and advocacy groups, that actively discourage the cultivation of biofuel crops on fertile land.

In India and East Asian nations, the adoption of renewable energy in agriculture has significantly decreased the reliance on grid electricity and coal, lessened the water needed for thermal power generation, and played a role in mitigating greenhouse gas emissions. The use of biogas has not only diminished the need for fossil fuels but has also encouraged farmers to maintain livestock, producing manure that benefits crop yields and soil health, while simultaneously bolstering the energy security of rural communities.

Implementing water-efficient technologies can alleviate the need for excessive pumping, thereby freeing up water resources for other critical sectors. These technologies can be expanded through farmer training, offering incentives, reforming electricity subsidies, establishing fair water pricing, involving water user associations, and implementing irrigation rationing.

On the demand side, tackling food waste emerges as a vital policy focus for both developed and developing nations. Strategies to minimize food wastage include enhancing storage and transportation infrastructure, enforcing food handling regulations, altering consumption habits, fostering the food processing industry, adjusting food pricing, and managing seasonal agricultural surpluses. Effective food waste policies must encompass the entire food lifecycle and require collaboration across multiple sectors and stakeholders.

Urban agriculture presents a promising solution to alleviate the strain on agricultural soils. By cultivating food within city limits, urban farming can significantly cut down on carbon emissions linked to food transportation, reduce water usage, and lessen the overall ecological footprint of urban areas. It is crucial to evaluate the potential for food production in cities, particularly in Asian regions where challenges like space constraints, electricity shortages, and water scarcity are prevalent. To maximize the benefits of urban farming, it should be integrated with renewable energy sources, water conservation techniques, organic farming practices, and effective waste management strategies, all while adapting agricultural methods to fit the unique microclimates and infrastructure of urban settings.

Effective land governance is essential to prevent environmental degradation, social tensions, and inefficient outcomes. This approach involves creating collaborative frameworks that unite various stakeholders from different sectors to reach a consensus on sustainable land use and management. Currently, land governance often suffers from fragmentation, overlooking the interconnectedness of different land types and their collective impact on the well-being of communities. By adopting integrated land governance, agencies such as the Ministry of Housing can better comprehend how their decisions affect land managed by other entities, like the Ministry of Agriculture, fostering a more holistic approach to land management.

Barriers to scaling policy solutions

A range of obstacles can be identified that hinder the capacity of Asian nations to effectively respond to land use changes and related climate change vulnerabilities. These obstacles can be categorized into scientific and technical challenges, technology-related issues, financial constraints, and governance deficiencies. One significant hurdle in addressing the impacts of land use-related climate change within the region is the insufficient financial resources dedicated to tackling the extensive challenges associated with natural resource management and other land-related sectors. Despite achieving a level of self-sufficiency in agricultural research due to years of support from international development organizations like CGIAR, IFAD, DFID, JIRCAS, and ACIAR, the actual funding allocated to maintain and build upon the initial successes in agricultural research and development has not kept pace with the region's evolving needs. The land-related focus has largely been limited. Consequently, the progress in technological advancement and its widespread adoption has not reached the necessary speed and scale to meet the sustainable land management demands of the region.

Sometimes, even though governments introduce several policy measures, they may fail to achieve the intended outcomes and impacts. For example, to curtail the depopulation of rural areas and dwindling farmers, the Government of Japan has introduced several policy measures over the years (MAFF, 2024). Despite these measures, a constant decline in agricultural land and agricultural population could be observed. This happens when the pull factors are stronger than the push factors or if the policies are made with wrong assumptions. In this case, the earlier factor may have played a major role than the latter.

Other ways of inadequate policy effectiveness could be lying within the capacity of governments to effectively implement policies. The surveys carried out by the author reveal that the capacities of governments, development agencies, and educational institutions within the region vary significantly when it comes to integrating climate change considerations into land use management, as illustrated in Figure 11.5. The data suggests that governmental bodies generally exhibit deficiencies across various capacity dimensions, including policy development and execution, strategic direction, planning, project management, external collaboration, and organizational effectiveness. In contrast, non-governmental development agencies and educational institutions demonstrate superior capabilities in most of these domains, highlighting a stark difference from governmental performance. Consequently, enhancing the government's overall capacity to tackle complex challenges such as climate change is imperative. To achieve this, governments need to engage multiple stakeholders in the formulation and execution of agricultural policies. This can be accomplished by bolstering external relationships with relevant organizations and stakeholders involved in the sector. Furthermore, governments should strive to identify mutually beneficial pathways that leverage the strengths and address the weaknesses of diverse stakeholders, thereby fostering collaborative actions. Organizational capacity encompasses not only the presence of qualified personnel but also financial resources and the readiness to engage with other stakeholders. Given the inherent characteristics of governmental structures, the effectiveness of their services is largely contingent upon their ability to collaborate with various stakeholders to fulfil community needs.

Development agencies require reliable access to data to effectively evaluate climate change risks on land associated with their projects and programmes. Much of this data is typically sourced from formal and governmental channels, making timely access to high-quality information crucial for empowering non-governmental development agencies. This access enables these organizations to offer valuable guidance and support to governments in the formulation and execution of agricultural policies. Governments need to view development organizations not as external entities but as partners in the collaborative process. Although universities and educational institutions in the region have historically faced budget constraints and a lack of focus on climate change related to land, there is a noticeable shift occurring. An increasing number of universities and colleges are now incorporating climate change courses into their curricula, leading to a rise in research on climate-related issues and heightened awareness among students. The collaboration between educational and research institutions and the private sector is vital for staying abreast of the latest advancements and fostering joint solutions. Agricultural universities should establish dedicated departments that focus exclusively on climate change topics. Presently, many agricultural universities in the region are gradually enhancing their environmental departments, with climate change often being addressed within this broader context. Given its growing significance, the subject of climate change necessitates the establishment of specialized graduate schools, laboratories, and centres of excellence within the region.

Conclusions

The significance of land in the lives, livelihoods, and economies of Asian nations cannot be overstated. However, increasing pressures from both anthropogenic and natural sources have resulted in detrimental effects on land resources across the region. These adverse consequences are subsequently undermining the resilience of communities and nations alike. Although efforts are underway to adapt to climate change, there remains a critical need to enhance the focus on land-related issues. Despite the introduction of various policies, their effectiveness is often compromised by flawed assumptions in their formulation, inadequate implementation, and a mismatch between the intensity of policy investments and the pressures faced.

A fundamental transformation in the approach to land management is essential across all sectors, extending beyond agriculture alone. It is evident that any alterations in land use within one sector, such as agriculture, can lead to significant and interconnected repercussions in other sectors. The consequences of decisions regarding agricultural land use extend beyond the agricultural sector, influencing a multitude of other sectors. The impacts of these effects are multidirectional, operating in various ways rather than being confined to a single trajectory.

The current fragmented governance of land, managed by various ministries overseeing different economic domains, underscores the necessity for a fundamental shift in land governance practices. Such a transformation should promote enhanced policy coordination among the ministries responsible for land management, ensuring that decisions made within one sector do not inadvertently harm others. Consequently, there is an urgent requirement for a system that comprehensively visualizes the repercussions of land use decisions in an integrated manner. This necessitates the advancement of integrated land use decision support systems, which remain underexplored and underdeveloped for policymaking in the Asian context.

Enhancements in food production and productivity within the region have primarily resulted from considerable intensification, which has unfortunately led to detrimental effects on agricultural ecosystems. Despite these production advancements contributing to the alleviation of poverty, hunger, and malnutrition for millions, future increases in food production and productivity must be achieved without imposing further burdens on the environment and society.

Strengthening institutional capacity is vital for the formulation of effective policies and their targeted execution. A significant disparity in capacity among various stakeholders underscores the urgent need for ongoing capacity-building initiatives, particularly through the incorporation of climate change adaptation into existing land resource development frameworks. Transitioning from ad hoc solutions to strategic interventions that prioritize the creation of sustainable adaptation benefits is critical for improving resilience against vulnerabilities in the region.

Acknowledgements

This work was supported by the Environmental Research and Technology Development Fund [JPMEERF20221C06] of the Environmental Restoration and Conservation Agency of Japan and Ministry of Foreign Affairs, Japan research project 'Analysis of the transformation of the security environment triggered by climate change and Japan's diplomatic and security approach' (2023–2025). The author extends sincere thanks to the reviewers and editors for their constructive feedback, which significantly enhanced the quality of the chapter.

References

Behera, B., Haldar, A., & Sethi, N. (2024). Agriculture, food security, and climate change in South Asia: A new perspective on sustainable development. *Environment, Development and Sustainability*, *26*, 22319–22344. https://doi.org/10.1007/s10668-023-03552-y

Blondin, S. (2023). The dual relationship between human mobility and climate change in Central Asia: Tackling the vulnerability of mobility infrastructure and transport-related environmental issues. In R. Sabyrbekov, I. Overland, & R. Vakulchuk (Eds.), *Climate change in Central Asia. Springer briefs in climate studies*. Springer. https://doi.org/10.1007/978-3-031-29831-8_9

Chandio, A. A., Jiang, Y., Amin, A., Ahmad, M., Akram, W., & Ahmad, F. (2023). Climate change and food security of South Asia: Fresh evidence from a policy perspective using novel empirical analysis. *Journal of Environmental Planning and Management*, *66*(1), 169–190. https://doi.org/10.1080/09640568.2021.1980378

Dietrich, J. P., Schmitz, C., Müller, C., Fader, M., Lotze-Campen, H., & Poppa, A. (2012). Measuring agricultural land-use intensity – a global analysis using a model-assisted approach. *Ecological Modelling*, *232*(10), 109–118.

Duveiller, G., Hooker, J., & Cescatti, A. (2017). The mark of vegetation change on Earth's surface energy balance. *Nature Communications*, *9*. https://doi.org/10.1038/s41467-017-02810-8

FAO. (2016). *State of the world's forests: Forests and agriculture. Land-use challenges and opportunities*. FAO.

FAO. (2024). *FAOSTAT*. Retrieved October 6, 2024, from http://www.fao.org/faostat

Fulton, L., Mejia, A., Arioli, M., Dematera Demeter, K., & Lah, O. (2017). Climate change mitigation pathways for Southeast Asia: CO2 emissions reduction policies for the energy and transport sectors. *Sustainability*, *9*(1160), 1–16. https://doi.org/10.3390/su9071160

Huang, S., Ghazali, S., Azadi, H., Moghaddam, S. M., Viira, A. H., Janečková, K., Sklenička, P., Lopez-Carr, D., Köhl, M., & Kurban, A. (2023). Contribution of agricultural land conversion to global GHG emissions: A meta-analysis. *Science of the Total Environment*, *876*. https://doi.org/10.1016/j.scitotenv.2023.162269

Junaid, M., & Gokce, A. (2024). Global agricultural losses and their causes. *Bulletin of Biological and Allied Sciences Research*, *2024*(1), 66. https://doi.org/10.54112/bbasr.v2024i1.66

Klasen, S., & Waibel, H. (2015). Vulnerability to poverty in South-East Asia: Drivers, measurement, responses, and policy issues. *World Development*, *71*, 1–3. https://doi.org/10.1016/j.worlddev.2014.01.007. Epub 2014 Mar 3. PMID: 32287933; PMCID: PMC7130570.

MAFF. (2024). *Census of agriculture and forestry (1960–2005), a survey of agricultural structure (2009, 2011)*. Ministry of Agriculture Forestry and Fisheries, Government of Japan. Retrieved October 20, 2024.

Mitra, B., & Bhattacharya, A. (2012). Long long-term energy scenario and water use – a case study on India (IGES Policy Brief, 8). Institute for Global Environmental Strategies.

Nani, B., Sitaula, B., & Bajracharya, R. (2011). Agricultural intensification in South Asia and its contribution to greenhouse gas emission: A review. *Asian Journal of Water, Environment and Pollution*, *8*(3), 11–17.

O'Connor, T. (2009). Water usage in biodiesel production. *Paper presented at the water energy sustainability symposium* (pp. 13–16). Groundwater Protection Council, Salt Lake City, UT.

Prabhakar, S. V. R. K. (2021). A succinct review and analysis of drivers and impacts of agricultural land transformations in Asia. *Land Use Policy*, *102*. https://doi.org/10.1016/j.landusepol.2020.105238

Prabhakar, S. V. R. K., & Elder, M. (2009). Biofuels and resource use efficiency in developing Asia: Back to basics. *Applied Energy*, *86*(1), S30–S36.

Ravanera, R., & Gorra, V. (2011). *Commercial pressures on land in Asia: An overview*. The International Land Coalition.

Sharma, S. (2015). *The political economy of the water-energy nexus in Indian irrigation*. Retrieved March 6, 2024, from http://blogs.lse.ac.uk/southasia/2015/09/09.

Shaw, R., et al. (2022). Asia. In *Climate change 2022: Impacts, adaptation and vulnerability. Contribution of working group II to the sixth assessment report of the Intergovernmental Panel on Climate Change*. Cambridge University Press (pp. 1457–1579). https://doi.org/10.1017/9781009325844.01

Tan, N. Q., Dinh, N. C., Linh, N. H. K., Hung, P. X., Kien, N. D., Phuong, T. T., & Tinh, B. D. (2023). Climate change vulnerability and poverty nexus: Evidence from coastal communities in central Vietnam. *Journal of Agriculture and Environment for International Development (JAEID)*, *117*(1), 61–84.

Thomas, K. A. (2023). Compelled to compete: Rendering climate change vulnerability investable. *Development and Change*, *54*, 223–250. https://doi.org/10.1111/dech.12756

Vlek, P., Khamzina, A., Azadi, H., Bhaduri, A., Bharati, L., Braimoh, A., & Taheri, F. (2017). Trade-offs in multi-purpose land use under land degradation. *Sustainability*, *9*(2196), 1–19.

Wasantha, A., Hoang, V., & Wilson, C. (2015). Impact of agricultural chemicals on the environment and human health in Asia: Are there solutions? In S. Managi (Ed.), *Handbook of environmental economics in Asia* (pp. 321–348). Routledge.

12 Dynamic of agriculture and sustainability of the oasis ecosystem

The case of Mansoura (Algerian Sahara)

Tayeb Addoun

Introduction

Agriculture experienced an inclusive recovery during the second half of the 20th century in the Algerian desert through the various state agricultural policies and various agricultural development programs. The launch of the law on accession to agricultural landownership (APFA) in 1983 resulted in the development of thousands of hectares across the different Saharan regions, thereby creating a new agricultural dynamic translated by the emergence of new production systems.

Mansoura is one of the 13 oases that encompass the Ghardaïa region (Algerian Sahara); it is part of the territory of the Châamba[1] territory. Over the past 30 years, this oasis has been known for various agricultural transformation phenomena in rural areas and rural development problems that most of us often see in recent years. It has been characterized for two decades by a unique production system commonly called the oasis system; this system is known in traditional palm groves in small family farms (Zenkhri et al., 2014). The oases are the basis for human survival and development in dry lands of Sahara (Chen et al., 2024).

The map of oases has barely changed and has only become more dense (BRL Engineering, 1999). Since independence to today, various agricultural land development programs in the Sahara region have experienced significant development (Addoun et al., 2022), the different agricultural policy programs that have been implemented in Algeria were characterized by the absence of continuity (Hadeid et al., 2021; Zenkhri et al., 2014).

In the 1960s, agricultural policy was marked by a partial modification of the production system, which had been based on the production of vineyards intended for local consumption. The decade of the 1970s was that of heavy institutional intervention by the state in the management of the sector and agrarian reform and agricultural development in the Sahara (Hamamouche et al., 2018) through the initiation of investments in agriculture.

In the Mansoura oasis, we can distinguish the old palm groves from the new lands of agricultural development (Bouammar & Bekhti, 2008). Through successive agricultural policies since the 1970s, Mansoura has benefited from agricultural projects within the framework of agricultural investment, which ensured

DOI: 10.4324/9781032672779-15

improvement in economic productivity and social anchoring, especially for rural landscapes (Rouabhi et al., 2016), today we are witnessing a major transformation of rural landscape dueto agricultural activity. While the city urbanization and urban sprawls are proceeding apace, increasing more and more the demand for new (Bański et al., 2023).

Indeed, the development of agricultural land began in 1983, within the framework of the law relating to APFA; 83–18 (JORA n°1373, 1983) and Executive Decree 97–483 relating to the concession of agricultural lands (JORA n°83, 1997), which focus on the availability of massive water resources and lands suitable for development (Addoun et al., 2022). This policy created a spectacular agricultural dynamic, and it restored the agricultural sector in the Mansoura region.

Moreover, the Algerian Sahara has been the subject for five decades of the implementation of agricultural development programs aimed at improving the country's food security and the living conditions of the population (Idda et al., 2017). Indeed, Saharan agriculture based on self-subsistence must ensure income, meet the food needs of a growing and urbanized population, and contribute to the country's food security (Kouzmine, 2005; Djennane, 1990), which contributes to the emergence of new forms of agriculture in the arid and hyper arid regions of southern Algeria.

Recently, some interesting studies have been carried out on Saharan agriculture, particularly on the sustainability of the oasis system and the future trajectories of modern agriculture (Hamamouche et al., 2018), professional dynamics of young people in Saharan neo-agriculture (Ouendeno et al., 2015), the issues between modes of land development (Hadeid et al., 2018), and technology adopted in the context of Saharan agricultural innovation (Naouri et al., 2020), the future of irrigated agriculture, and the issue of sustainability of groundwater supply (Bouarfa et al., 2020). All this research shows that the challenges of the two oasis agricultural systems (traditional and modern) no longer operate in isolation. The rural landscape of the oasis is undergoing transformations with the emergence of new forms of land exploitation and land use and the control of modern irrigation methods.

In the oasis of Mansoura, differences in the socio-spatial organization of agricultural activity can be observed, particularly in the socioeconomic conditions of exploitation of the old agricultural system. The latter is subject to degradation and marginalization in relation to the profound changes that oasis societies are experiencing in general, while the new agricultural system is a subject of particular interest on the part of public authorities (MADR, 2013). It seemed interesting to us that, based on a diagnosis, we could determine the causes that generated this state of agricultural investment in the Mansoura oasis and the role of rural development tools in organizing human–land relations.

In this chapter, we have aimed to identify the dynamics of agriculture and the sustainability of the oasis system in the Mansoura oasis and the surrounding extensions. In the following step, we speculate on the future diversity of the agro-system, land use and irrigation problems in the region of Mansoura.

Material and methods

The study focused on the analysis of agricultural land and irrigation problems in the oasis of Mansoura through aspects, positive and negative with its relationship of development programs. Our approach is based on interviews conducted with farmers, agricultural institutions, and local authorities.

Study area

Our work took place in the Oasis of Mansoura, which is located in the center of the northern part of the Northern Sahara. Its geographic coordinates are 2° 30′ and 03°45′ East longitude and 28° 87′ and 32°13′ North latitude (DPSB, 2017). It is located 70 km south of the capital of the wilaya of Ghardaïa on national road no.1; also called African Union Road, which passes through Ghardaïa, linking Algiers with Tamanrasset. Mansoura constitutes an important relay between the north and the south of the country. It is limited to the west by the commune of Brezina (wilaya of El Bayadh), to the east by the commune of Rouissat (wilaya of Ouargla), to the north by the commune of Sebseb, and to the south by the commune of Hassi El Lefhal (Figure 12.1).

The commune of Mansoura is chosen for its agricultural dynamism symbolizing many of the new challenges in the development of oasis agriculture land and the emergence of a new relationship between traditional agriculture and modern agriculture (Naouri et al., 2020). Also, we chose the Mansoura region as a case study because of its marked agricultural appeal in space and in terms of the presence of ancient oases and the irrigated perimeter. Mansoura is well known for its large number of dense palm groves in its northern part and its numerous agricultural farms, which provide an important source of income for local farmers.

Data and method

This research focuses on the diagnosis of agricultural development mechanisms in the commune of Mansoura and the analysis of the positive and negative aspects of these programs.

Our approach was based on interviews carried out with representatives of agricultural institutions (i.e. engineers and technicians in agriculture) and local farmers.

Interviews were carried out with 90 farmers, representing a sampling fraction of 1.9% of the communal population (4.723 habitation in 2020), agricultural institutions, and local authorities of Mansoura; in the form of a semi-open survey. It included several questions on the characteristics of the population that exploits it; for example household characteristics, age and educational status, and description of the farm such as on the method of irrigation, land status, agricultural system and on the means and production systems implemented.

Data were collected between the periods 2022–2023 about agricultural areas in the oasis of Mansoura form Directorate of agricultural services (DSA). Our survey

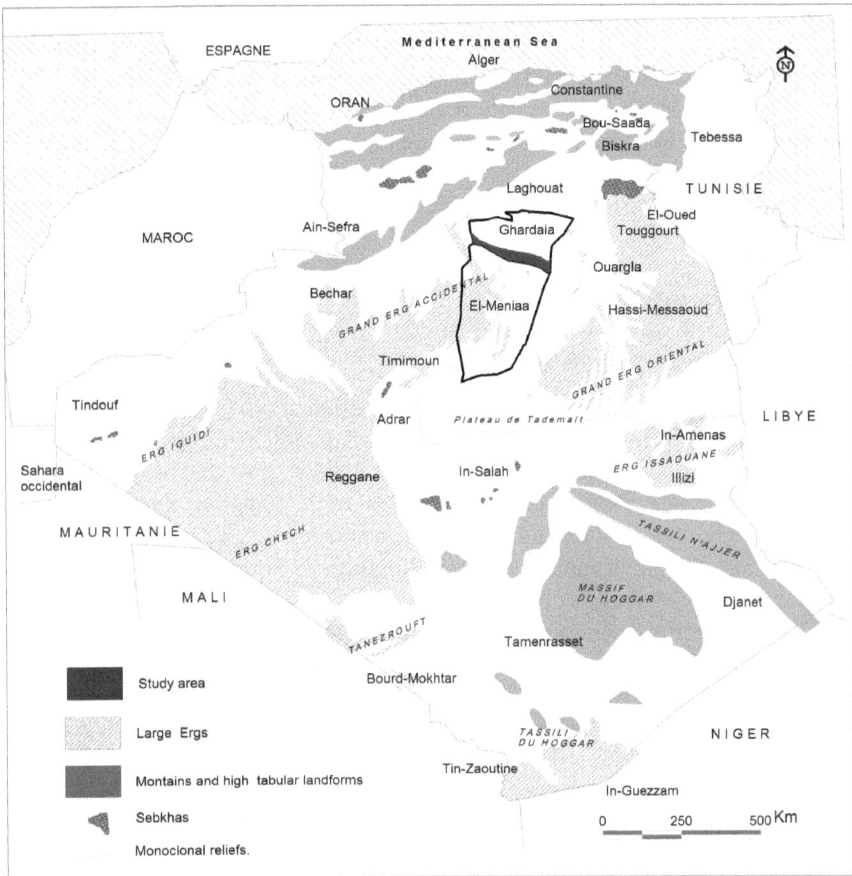

Figure 12.1 Location of the study area.

Source: Author's design from the Data of Google Earth.

was conducted between September and October 2022 by distributing a semi-open questionnaire among 90 farmers working on the old palm groves and some agricultural development farms, which were randomly selected. Like any other scientific research that focuses, above all, on concrete observations, our approach consisted of analyzing the facts and drawing relevant conclusions.

Our survey was conducted between September and October 2022 by distributing a semi-open questionnaire among 90 farmers working on the old palm groves and some agricultural development farms, which were randomly selected. Like any other scientific research that focuses, above all, on concrete observations, our approach consisted of analyzing the facts and drawing relevant conclusions.

The field survey which was carried out during the period 2022–2023 revealed that, despite the commune of Mansoura being classified as a rural area with a high proportion of farmers, 68.35% of the populations have become sedentary and live in urban areas or on the outskirts of urban areas such as Ain Lossig and Mansoura Nouvelle. In fact, this trend toward progressive sedentarization is the result of the appearance of new production agricultural systems and new needs linked to amenities in this locality, particularly economic and social structures (education, health, water, electricity, etc.).

Results and discussion

Mansoura: a predominantly rural oasis

The oasis is an isolated inhabited place in an arid or more generally hostile environment whose location is linked to the possibility of exploiting a resource, in particular water for the practice of irrigated agriculture. Also, the oasis is an important element that has contributed to the stability of the Saharan environment since antiquity, where nature factors are less difficult to resist life. The creation of human life in the Saharan environment ensures life and meets the various needs of the population (Addoun et al., 2020). Also, the oasis is not only an agricultural space but also a temporary residence, representing a summer refuge to escape the heat and where secondary houses are built (Bensalah et al., 2018) especially in the palm grove of the Mzab valley, unlike the capital of Mansoura, which is relatively less ancient compared to the Mzab region, where the practice of irrigated oasis agriculture is found.

Continuing toward the south of the Ghardaïa region and along the trans-Saharan caravan trade routes, we find the commune of Mansoura, which includes the capital Mansoura nouvelle and the secondary village of Ain Loussig; its recent origin has nomadic traditions belonging to the Châamba tribes settled around water points. Very distinct from those of the wilaya by their small size, these towns dream of a typically rural character.

Today, the rural commune of Mansoura covers an area of 4,812.55 hectares for a population of 4,723 inhabitants recorded at the end of 2020, with an average density of 0.72 inhabitants/km^2 (ONS, 2020). It is part of the Châamba territory and was created in the 1940s; it is a young commune supported by the voluntarism of the state following the administrative division of 1984.

This oasis, made of the ancient palm groves, consists of the small individual farms, often less than one hectare. The palm trees are dense and cover the entire surface, with several varieties of soft dates intended for self-consumption and the rest intended for local marketing. According to Houichiti et al. (2021), the human diet of the time was not very diversified and based mainly on the consumption of dates; for example Akerboucht, Deglet Nour and Ghars, and cereals. But the modern agriculture lands are created in the small depressions called Oued Aghzalet outside the palm groves to increase agricultural investment. As for livestock (sheep, goats, and camels), their diet mainly depended on small pastures in the Saharan rangelands.

Socioeconomic characteristics of farmers

The socioeconomic environment directly affected the development of agriculture in this region because of the interrelated relationships between the different factors that make agriculture socially acceptable and economically fruitful. Also, the socioeconomic factors in the oasis of Mansoura affected the development of the agricultural areas, which is located 5 km northeast and south of Mansoura; it extends over an area of more than 630 hectares (DAS, 2013).

Thus, after the analysis of the socioeconomic features, the results showed that the majority of the farmers were married and had a middle school education. Their ages are between 40 and 60 years. This is not a condition for practicing agricultural activity; only the category under 40 years of age is linked to study and work in the economic and industrial field. More than 80% of farmers have a family size between three and seven members, and 15% have a family size of fewer than four members. About 50% of households live in extended families (housing for grandparents).

Concerning the level of education, about 4% of farmers have a university level, 15% of the farmers have a technical education, 15% have a preparatory education, and 26% had no education level at all while 40% of them read and write. All this is a criterion that limits the practice of agriculture.

Regarding personal and socioeconomic characteristics, more than three-fourths of the farmers surveyed practiced agriculture as their main activity, which gives more hope to the agricultural sector and suggests that farmers will not give up their farms because it represents the only source of stable income throughout the year.

The majorities of farms have private legal status and represent 86% of all farms surveyed. This characterizes the old perimeters and even in those development perimeters where there is a transfer of ownership of the land from one beneficiary to another by purchase. This is confirmed by farmers from other wilayas who invest in the area by purchasing land from farmers who have benefited from the APFA law.

Evolution of land use

The useful agricultural area (UAA) increased from 3,146 hectares in 1984 to 6,151 hectares in 1988 through the application of Law 83–18 relating to accession to agricultural landownership (APFA) which provided benefits to 3,319 beneficiaries with a total area of 9,423 hectares (DAS, 2013). At the local level, the agricultural area in the commune of Mansoura almost doubled from 1,360 hectares in 2006 to 2,438 hectares in 2016 (Figure 12.2) (DPSB, 2017). This policy encouraged the privatization of public land to improve its agricultural production through the increase in UAA.

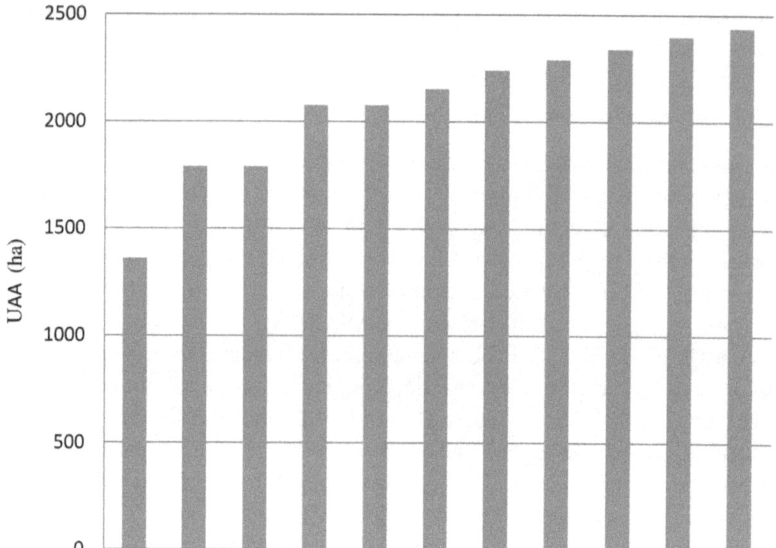

	2006	2007	2008	2009	2010	2011	2012	2013	2014	2015	2016
Sup (ha)	1360	1790	1790	2076	2076	2153	2241	2290	2340	2398	2438

Figure 12.2 Evolution of the useful agricultural area of the commune of Mansoura (2006–2016).

Source: Authors from the statistical yearbook 2006–2017, Directorate of Programming and Budget Monitoring, DPSB.

The state carried out a radical overhaul of its policy by initiating the privatization of public lands to improve its agricultural production through the increase in the UAA, implemented in the context of the 1983 law.

This policy has outlined as objectives: rural development of the region, job creation, and the expansion of rural areas near of the Oued Aghzalet in the oasis of Mansoura. All this to keep pace with population growth and reduce the rates of migration of populations from the southern regions to the north.

Until now, with the aim of making Saharan agriculture the food reserve of Algeria (Bessaoud, 2008), the policymakers have always strived to create the best land allocation system for these new agricultural lands (Alary et al., 2018). Something that agriculture in the north of the country has not been able to do. Certain events concerned agricultural land and marked the beginning of a radical change in agrarian policy. We moved from a planned socialist policy to a liberal agrarian policy, through the restructuring of socialist self-managed estates into small cooperatives of collective and individual agricultural operations in accordance with the application of Law 87–19.

However, the notable indicator during this period is that agricultural cooperatives moved from state management to free management for the benefit of modern agriculture. During the period 1990–2008, the agricultural sector experienced liberation of the sector through the regulatory consolidation of the operations of state agricultural land to the beneficiaries with a format of act of perpetual enjoyment which lasts for 99 years. This formula is similar to that of accession to landownership (Law 83–18), except that the State provides almost total financial support, particularly in terms of water mobilization, electrification, and opening rural paths and roads; what we saw in the perimeter of new Mansoura.

The environmental factor indicates that Mansoura was affected by the phenomenon of sand up, which has always existed in the Sahara, but the local peasantry tried to limit it by using locally produced materials – dry palms and windbreak. However, the lack of maintenance of the artificial dunes, the "afreg" designed to shelter the palm grove, is one of the main causes of their silting up, which is a brake on agricultural activity.

Dynamics in date palms production and their evolution

The date palm produce and its evolution and how it is structured in different oases are essential for its dynamic conservation and sustainability of land. In the Mansoura region, we notice agricultural attraction in space, in terms of the presence of the number of palm trees and the date production of palm groves; about 650,687 trees and the 175,407 quintals per year (DPSB, 2017), meaning the average yield of each palm tree is about 3.7 kg of dates per palm per year. The following table (Table 12.1) combines the date palm production of the area of Mansoura between 2001 and 2016).

This period witnesses the emergence of new agricultural plantations in the outside oasis of Mansoura and the introduction of new production techniques through the Law 18/83 on the Accession to the Agricultural Fundamental Property and the Improvement (PNDA). Based on the results (Table 12.1), we observe the growth of palm trees in production. The fourth column of the table provides changes in real crop output between consecutive years, starting from 2001 to 2016, about 15 years.

The remaining columns provide the number of palm trees in production to total date palm cultivars. The production value of the palm trees during the period 2001–2016 increased and was recorded at about 11,998 quintals. That varied between a minimum value of about 6,420 quintals in the year 2001 and a maximum of about 18,418 quintals in 2016, so the production has doubled twice in ten years. This agricultural transformation is explained by the extensive investment that has been granted in recent years outside the oasis and which has affected the yield of the ancient oasis.

Land use in the region

The land structure and use patterns in Mansoura area for different agricultural activities vary greatly based on topographic, geographical, and environmental factors. The land use structure in the area is influenced by agriculture, types of crops grown, nature of livestock, agricultural methods, and integration of different agricultural systems (such as crop–livestock systems and organic farming). Land use

Table 12.1 Evolution of production of date palms at the Oasis of Mansoura (2001–2016)

Mansoura	*Total number of palm trees*	*Number of palm trees in producing*	*Date production (Q) per year*
2001	31,250	17,210	6,420
2002	32,000	18,802	6,530
2003	32,460	18,200	6,520
2004	32,460	18,200	6,520
2005	35,343	19,492	9,639
2006	36,810	20,600	8,372
2007	40,860	22,620	9,770
2008	41,470	23,710	10,500
2009	46,160	26,239	11,375
2010	46,571	27,560	12,336
2011	46,000	29,020	12,316
2012	47,550	30,580	12,500
2013	48,071	32,510	13,236
2014	49,077	33,347	16,455
2015	42,070	33,710	14,500
2016	42,535	38,440	18,418

Source: DPSB, 2017.

patterns in Mansoura vary according to agricultural potential generally, there is traditional agriculture within the oasis; it is largely based on date production, where the overall landscape appearance is dominated by dense palm oases. In the old oasis, the land is used by the family as a common pattern within the oasis, given that the land is inherited. The land is divided into small plots and various consumer products are grown inside it and irrigated with traditional irrigation systems. Sometimes, we find modern modifications to the old irrigation systems to save and rationalize water use.

Today, this form of land occupation is witnessing a significant decline and scarcity of water resources, with scarcity of rainfall due to climate change. Add to this the new social and economic transformations that have been greatly affected by modernity, where interest in the old oasis has clearly begun to decline, and signs of neglect and reluctance of local youth to work the land have begun to appear due to their preoccupation with jobs in other economic sectors such as administration and services and their preference for investments in new agricultural expansions supported by the state. On the other hand, the structure and composition of the land outside the oasis have been different due to the development of the agricultural system and the establishment of agricultural environments far from urban centres. Successive state policies have played a major role in the development of agriculture.

The agricultural landscape outside the oasis shows a modern agricultural system and a diversity of production, with land areas varying from two hectares to more than 100 hectares. With the arrival of investors from outside the region, they have contributed significantly to the dynamism of agriculture and the activity of agricultural land. What we observed during the field visits is the concentration of agricultural investments in the Oued Aghzalet region through a specialization in large-scale consumer crops. In fact, we cannot deny the contributions of the state that were made available within the agricultural development process by equipping the new agricultural areas located along Oued Aghzalet with wells equipped with electricity, preparing rural roads etc. from facilitating access to bank loans to finance various activities accompanying agricultural activities such as drilling operations, rehabilitation of pumping infrastructure, and maintenance of agricultural water equipment (pumps, storage basins, irrigation network, and so on).

From our field observations as well, it has become clear to us that farmers are moving toward diversifying production. Indeed, agriculture appears relatively diverse, unlike what it was in the past. It includes the cultivation of fodder, grains, and market crops (potatoes, tomatoes, peppers etc.) in addition to tree farming, which the region did not know before, such as fruit trees, seasonal fruits, and widely consumed citrus fruits such as lemons and oranges. All these modern products and the agricultural gains that accompanied them were not known in the traditional system. Modern irrigation has contributed to the expansion of new agricultural activity outside the oasis in several forms through drip and sprinkler irrigation. According to the opinions of the investors investigated, the newly adopted irrigation techniques are considered among the best methods for agricultural reclamation in extremely hot desert areas known for high levels of evaporation and low rainfall, as artificial distillation reduces water consumption to a minimum while achieving greater agricultural efficiency.

Irrigation and agriculture land problems

The agricultural sector of the commune of Mansoura is experiencing problems and obstacles that negatively affect its production for different reasons: the lack of flow in the wells dedicated to irrigation, the non-allocation of wells directly addressed to the culture, and the spread of diseases due to the discharge of water near agricultural lands. These problems have caused damage to the palm trees of the commune themselves subject to old age and the slowdown of the process of their renewal which must be considered.

Among these problems encountered in irrigation in the oasis of Mansoura, interesting results were obtained by field investigations and analysis of survey questionnaire, as there are several motivations for these problems of irrigation, classified in the Table 12.2.

All of these are indicators of the lack of an adequate agricultural policy to safeguard agricultural land and irrigation demand increases. In addition to the threats of soil degradation, there is a problem of land reconstruction again, which is accompanied by environmental damage and economic losses, which takes many years.

Table 12.2 Agriculture irrigation problems in the oasis of Mansoura

	Natural problems	*Constraints related to the physical environment*	*Problems of which man is the main cause*	*Technical problems and material resources*
Agriculture irrigation problems	The lack of irrigation water which is a major element for the development of agriculture	Low rainfall Excessive temperatures in summer Significant spring frosts	Illegal irrigation by waste water constitutes a danger for the health of the population and even for the fertility of the soil	Non-compliance with technical itineraries (cultivation methods) by farmers, directly influencing production and yield
	Watersheds exposed to erosion	High frequency of sirocco	Grazing, extraction of stone, and sand; and illegal constructions (of which the forest is often threatened) especially the construction on lands of good agricultural value, have caused the loss of several hundred hectares of agricultural land	Lack of agricultural resources and equipment, the majority of which are obsolete

Source: Author's work.

The agricultural sector – advantages and constraints

Advantages

The municipality benefits from a considerable source of natural energy produced by a fairly long period of sunshine and a relatively high temperature during the year, which allows the production of a varied and high-quality range of early maturing crops.

All the land in the region is suitable for possible agricultural development thanks to the alluvium and fertilizing elements that are deposited there and the infiltration water that accumulates there.

The region contains a free aquifer that feeds the Continental Intercalaire aquifer and groundwater such as water tables (potential to be preserved); the Continental Intercalaire aquifer covers the entire territory of the municipality, and the Complexe Terminal aquifer contains significant unexploited water resources offering immense possibilities for the agricultural development of the municipality.

Constraints

Pluviometry characterized by low precipitation, particularly high temperature, very frequent and violent winds that contribute to drying out crops, and very high evaporation, especially in summer when water needs are at their maximum, mean that no agriculture can be considered without irrigation or protection.

Cultivable soils are rare and limited to areas of water accumulation and alluvial deposits.

Given the climatic conditions of the municipality of Mansoura, farms plantations are excessively expensive because they absolutely require fencing, a water supply, and permanent irrigation. The cost per hectare far exceeds the standards of the north of the country.

The population is highly concentrated in the urban area; such a population structure does not encourage agricultural development. One of the most serious consequences noted is the destruction of the surrounding palm grove; therefore, extensions can be considered only outside the municipality.

Land degradation factors and sustainability of the oasis system

The deterioration of the environmental situation of oasis systems threatens the ability of natural resources to regenerate. The pressure that manifests itself in population growth and the development of various human activities, particularly with the change in lifestyles and the emergence of new agricultural practices, is clearly beginning to affect the balance of the oasis ecosystem. The most striking manifestations of environmental deterioration are evident in the increase in pumping and the decrease in water capacity, in terms of both quantity and quality.

The impact of modern agriculture and the methods used for irrigation and agriculture have been a factor in weakening the traditional oasis system, as the desert farmer has become inclined to farm using modern methods, which cause great harm to the land, such as artificial fertilizers and focusing on more profitable crops while moving away from the cultivation of the more adaptable palm trees. The focus on vegetables, grains, and some fruit trees has also become demanding in terms of water consumption. In recent years, the area of the oasis has decreased as a result of the fragmentation it is exposed to due to urban expansion, which has begun to invade the agricultural environment of the oasis, thus taking up significant areas of its lands.

Since about 2010, agricultural production and its quality have declined. The old oasis has become characterized by a lack of manpower and the reluctance of local youth to work there because they are busy with modern jobs that are more stable and require less effort. The youth of the region prefer to work in the oil basins in particular, which has greatly affected the economic return of the oasis.

Conclusion

The oases of southern Algeria have faced major challenges in recent years due to significant investments and various agricultural development processes. The Mansoura oasis, like the rest of the desert regions, has experienced dynamic changes at several levels. As a result of adopting modern agricultural methods, using different irrigation techniques, and rationalizing the use of water resources. Also, by the intervention of various development programs developed and implemented by the state at the national level.

An analysis of the agricultural system of the commune of Mansoura clearly shows that this region, like most desert regions, is still characterized by the system of ancient oases, even if it differs in its form and origin from the rest of the ancient oases that we know in the Great South. These ancient systems force us to think about sustainability and the challenges of confronting any sudden climate change or social and economic transformation in the future.

Despite the measures taken by the authorities and stakeholders to develop the agricultural sector in the southern regions, the impact of this development policy has remained very limited at the level of the old regime, unlike the perimeters outside the oasis that reflect a positive development in the field of agricultural dynamism, and this is what we have already observed in the Mansoura region.

The state's development policy has led to the emergence of a great spirit of enterprise and entrepreneurship among small investors who aspire to improve the conditions of agricultural production and increase the productive potential of land while opening up the region geographically through the marketing of products and attracting new investors who work to promote and develop the agricultural sector.

Note

1 The Châamba tribes arrived in northern Africa with the last invasion of the Hilalis in the 11th century. This is an Arab tribe, first settled in the Metlili region, where they were famous for their cattle breeding and nomadic life.

References

Addoun, T., & Hadeid, M. (2020). Rôle des acteurs dans le développement touristique local de la ville de Zelfana au sud-est algérien: Analyse par la méthode MACTOR. *Journal International des Territoires et des Espaces Urbains – JITEU*, (5), 153–167.

Addoun, T., & Hadeid, M. (2022, January/February). Typology of agricultural farms in the South-East of Algerian Sahara: The case of Zelfana oasis. *Revista Agrária Acadêmica*, *5*(1). https://doi.org/10.32406/v5n1/2022/102-116/agrariacad

Alary, V., Aboul-Naga, A., Osman, M. A., Daoud, I., Abdelraheem, S., Salah, E., & Bonnet, P. (2018). Desert land reclamation programs and family land dynamics in the Western Desert of the Nile Delta (Egypt), 1960–2010. *World Development*, *104*, 140–153. https://doi.org/10.1016/j.worlddev.2017.11.017

Bański, J., & Mazurek, D. (2023). Identification of the spatial extent of the peri-urban area: The case of three cities in Poland. *Moravian Geographical Reports*, *31*(3), 165–173. https://doi.org/10.2478/mgr-2023-0015

Bensalah, I., Yousfi, B., Menaa, N., & Bougattoucha, Z. (2018). Urbanisation de la vallée du M'zab et mitage de la palmeraie de Ghardaïa (Algérie): un patrimoine oasien menace. *Revue belge de géographie*, *2*. https://doi.org/10.4000/belgeo.24469

Bessaoud, O. (2008). L'agriculture et la paysannerie en Algérie, les grands handicaps. In N. Benghabrit & M. Haddab (Eds.), *L'Algérie 50 ans après, état des savoirs en sciences sociales et humaines, 1954–2004* (pp. 359–384). CRASC.

Bouammar, B., & Bekhti, b. (2008). Le développement de l'économie agricole oasienne: entre la réhabilitation des anciennes oasis et l'aménagement des nouvelles palmeraies. *Le Chercheur*, *6*(1112–3613), 19–24. UKM Ouargla (Algérie).

Bouarfa, S., Brelle, F., & Coulon, C. (2020). *Quelles agricultures irriguées demain? Répondre aux enjeux de la sécurité alimentaire et du développement durable* (p. 212). Éditions Quæ.

BRL Engineering. (1999). *Plan study general development director for Saharan regions, lot I* (94 p.). Development Indicators.

Chen, Y., Fang, G.-H., Li, Z., Zhang, X., Gao, L., Elbeltagi, A., El Shaer, H., Duan, W.-L., Wassif, O. M. A., Li, Y.-P., Luo, P.-P., Selmi, A., Yu, R., Yang, J.-H., Hu, Y., Liu, C.-X., Long, Y.-X., Malik, I., Fu, A.-H., . . . Gao, Y. (2024). The crisis in oases: Research on ecological security and sustainable development in Arid regions. *The Annual Review of Environment and Resources, 49*, 1–20. https://doi.org/10.1146/annurev-environ-111522-105932

DAS. (2013). *Annuaire statistique 2013, direction des services agricoles de la wilaya de Ghardaia* (pp. 1–13). Ministère de l'Agriculture et du Développement Rural.

Djennane, A. (1990). Constat de situation dans des zones Sud des oasis algériennes. In V. Dollé, & G. Toutain (Eds.), *Les systèmes agricoles oasiens* (pp. 29–40). Options Méditerranéennes: Série A. Séminaires Méditerranéens, n° 11. CIHEAM.

DPSB. (2017). *Annuaire statistique 2017* (p. 218). Direction de la Programmation et du Suivi Budgétaire de la wilaya de Ghardaia.

Hadeid, M., Bellal, S.-A., Ghodbani, T., & Dari, O. (2018). L'agriculture au Sahara du sud-ouest algérien: entre développement agricole moderne et permanences de l'agriculture oasienne traditionnelle. *Cahiers Agricultures, 27*, 15005. https://doi.org/10.1051/cagri/2017060

Hadeid, M., Ghodbani, T., Dari, O., & Bellal, S.-A. (2021). Saharan agriculture in the Algerian oasis: Limited adaptation to environmental, social and economic change. In S. Diop, P. Scheren, & A. Niang (Eds.), *Climate change and water resources in Africa: Perspectives and solutions towards an imminent water crisis* (pp. 239–253). https://www.springerprofessional.de/en/saharan-agriculture-in-the-algerian-oasis-limited-adaptation-to-/18869980

Hamamouche, M. F., Kuper, M., Amichi, H., Lejars, C., & Ghodbani, T. (2018). New reading of Saharan agricultural transformation: Continuities of ancient oases and their extensions (Algeria). *World Development, 107*, 210–223. https://doi.org/10.1016/j.worlddev.2018.02.026

Houichiti, R., Bouammar, B., & Bissati, S. (2021). Dynamique oasienne et évolution des modes d'exploitation agricole au sud d'Algérie (cas de la region de Ghardaïa). *Management & Gouvernance. Entreprises – Territoires – Sociétés, 25*(2), 105–112.

Idda, S., Bonté, B., Mansour, H., Bellal, S.-A., & Marcel, K. (2017). Monument historique ou système bien vivant? Les foggaras des oasis du Touat (Algérie) et leur réalimentation en eau par pompage. *Cah Agric, 26*, 55007. https://doi.org/10.1051/cagri/2017049.

JORA. (1983, août 16). La loi d'accession la propriété foncière agricole n° 83–18 du 13 août 1983. *Journal Officiel de la République Algérienne, 1373.*

JORA. (1997, décembre 15). Décret exécutif n°497–483 du 15 décembre 1997, fixant les modalités, charges et conditions de la cession des parcelles de terre du Domaine privé de l'Etat dans les périmètres de mise en valeur agricole. *Journal Officiel de la République Algérienne, 83.*

Kouzmine, Y. (2005). Les villes sahariennes algériennes et le développement urbain durable, ville réelle et ville normative. *Bulletin de la Société Neuchâteloise de Géographie*, n° 49, 85–103. http://www.cairn.info/revue-annales-de-geographie-2009-6-page-659.htm

MADR. (2013). *Ministère de l'Agriculture et du Développement Rural, les principales mutations du secteur de l'agriculture et du développement rural depuis 1962 dans la wilaya de Ghardaia.* Rapport de synthèse. Version première. Ministère de l'agriculture et du développement rural.

Naouri, M., Kuper, M., & Hartani, T. (2020). The power of translation: Innovation dialogues in the context of farmer-led innovation in the Algerian Sahara. *Agricultural Systems, 180.* https://doi.org/10.1016/j.agsy.2020.102793

ONS. (2020). *Office National des Statistiques*. annexe Ouargla.

Ouendeno, M.-L., Daoudi, A., & Colin, J.-P. (2015). Les trajectoires professionnelles des jeunes dans la néo-agriculture saharienne (Biskra, Algérie) revisitées par la théorie de l'agricultural ladder. *Cah Agric*, *24*, 396–403. https://doi.org/10.1684/agr.2015.0793.

Rouabhi, A., Mekhlouf, A.-H., Mokhneche, S., & Elkolli., N. (2016). Farming transitions under Socio-economic and climatic constraints in the southern part of Sétif, Algeria. *Journal of Agriculture and Environment for International Development – JAEID*, *110*(1), 139–153, https://doi.org/10.12895/jaeid.20161.429. https://www.jaeid.it/index.php/jaeid/article/view/11110

Zenkhri, S., Kouri, L., Karabi, M., Kemassi, A., & Ould El Hadj, M.-D. (2014). Impact Du Plan National De Développement Agricole (pnda) Sur Les Systèmes De Production Oasiens Dans La Région D'ouargla (sahara Septentrional Sud Est Algérien). *Revue ElWahat pour les Recherches et les Etudes*, *7*(2), 95–100. https://www.asjp.cerist.dz/en/downArticle/2/7/2/1189

Part III

Farmland in planning and policy

13 Targeting agricultural land expansion across Africa

Pathways through effective nature protection, sustainable intensification, and strengthening land governance

Chinwe Ifejika Speranza, Chima Iheaturu, Frank Mintah, Ingrid Kjelsen, Robin Hartmann, Juri Fitz, Paule Pamela Tabi Eckebil, Georges A. Agonvonon, Giulia F. Curatola Fernández, Vladimir R. Wingate, Samuel Hepner, Phydias Agossou, and Felicia O. Akinyemi

Introduction

Africa, spanning approximately 30.4 million km^2 and comprising about 20% of the Earth's terrestrial surface, encompasses diverse agroecological zones ranging from humid rainforests, and sub-humid areas to semi-arid and arid regions, each characterised by distinct agricultural practices. The agricultural sector employs about 65% of Africa's active labour population (African Union Commission and African Union Development Agency – NEPAD; AUC & AUDA-NEPAD, 2022) and accounts for 25% or more of gross domestic product (GDP) in many African countries (Alliance for a Green Revolution in Africa (AGRA), 2016), making it a major land use and key economic sector across the continent. The relatively high agricultural GDP is attributed to progress towards achieving the goals of the African Union's Comprehensive Africa Agriculture Development Programme (CAADP) (AGRA, 2016; African Union Development Agency AUDA-NEPAD, 2024). However, consistently achieving the 2003 Maputo declaration on CAADP and the subsequent CAADP-2014 Malabo target of allocating at least 10% of national expenditure to agriculture and rural development remains difficult (AUDA-NEPAD, 2024). The 2023 CAADP biennial review report showed that only three member countries (Burundi, Ethiopia, and Mauritania) achieved this target in that period, indicating that the continent is not on track to achieve the CAADP-Malabo target by 2025 (AUDA-NEPAD, 2024).

Growth in agricultural production across many African countries remains below the global average. Between 1980 and 2018, cereal yields in sub-Saharan Africa grew by just 38%, roughly one-third of the growth achieved in South and Southeast Asia during the same period (Fuglie, 2019). While land productivity in sub-Saharan Africa increased from USD185 in 1990 to USD330 in 2016, it remains

DOI: 10.4324/9781032672779-17

lower than in North Africa and other regions, as well as below SSA's annual population growth rate during this period (International Food Policy Research Institute (IFPRI), 2020). The 2021 Africa Agriculture Status Report revealed that 75% of agricultural production growth since 2000 resulted from cropland expansion, while only 25% stemmed from yield improvements (AGRA, 2021). Kapuya et al. (2023) show that cereal yields across Africa are 60% below the global average and confirm that areal expansion and extensification have been the main pathways to increase food production.

Additionally, the rising rural populations in many African countries have been linked to increasing land scarcity (AGRA, 2022). During 1961–2020, land use associated with current per capita daily food supply in Africa ranged from about 12 m^2 in Central Africa to 45 m^2 in West Africa. However, it is expected to remain stable or slightly decline for all African regions to between 12 m^2 and about 38 m^2 in 2050 (Abe-Inge & Kwofie, 2023). This slight decline suggests that a similar amount of land will be required for food production in the future.

Further, secure land tenure is often not given in many African countries. Insecure land tenure can constrain farmers' investment in improving land, adopting sustainable land management practices, and accessing credit, which in turn limits agricultural productivity and food security (Deininger & Byerlee, 2012; von Braun et al., 2023). Insecure land tenure is often attributed to the post-colonial parallel land governance systems of statutory land tenure and customary land tenure and their grey-in-betweens. Considering that most land in Africa, especially sub-Saharan Africa is under customary tenure, it is critical to identify pathways to improve land tenure security for men, women, and youth farmers. Many of Africa's largely smallholder farmers manage lands of 2–5 hectares and do not have formal titles, and women, who play a key role in African agriculture, have even lesser land rights than men (Doss et al., 2013). Land reforms, although they might have mixed effects (Boone, 2018), have been recommended to improve tenure rights of smallholder men and women farmers and enhance agricultural production (Ali et al., 2014; von Braun et al., 2023).

Given the challenges of weak land tenure, potential land scarcity, and reliance on cropland expansion to boost agricultural production, it is crucial to explore the mechanisms driving this expansion and potential solution-options. This chapter thus aims to (1) identify the types of land use and land cover (LULC) targeted by agricultural land expansion and (2) examine response options to agricultural land expansion and dynamics, while considering implications for natural habitat conservation, biodiversity protection, deforestation reduction, greenhouse gas (GHG) emission mitigation (noting that Africa is not a net emitter), and agricultural productivity.

This chapter makes a critical contribution to the discourse on sustainable land management and agricultural production in a context of human population pressure, food insecurity, and the need for nature and biodiversity conservation by examining potential response strategies.

Methods

To achieve the study's objectives, we conducted a spatial analysis of LULC changes and reviewed existing literature to contextualise observed transitions, dynamics, their broader implications, and potential response strategies.

The spatial analysis focused on LULC transitions in Africa from 2000 to 2020, using remote sensing data products from the Global Land Analysis and Discovery (GLAD) data set (Potapov et al., 2022a). The GLAD data set offers global land cover classifications at a 30-metre spatial resolution. It features high thematic detail and validation accuracies above 85% for most land cover classes, making it suitable for analysing LULC changes (Potapov et al., 2022a). Specifically, the analysis prioritised agricultural land (cropland) conversions, while other LULC transitions were examined for contextual understanding.

LULC data for 2000 and 2020 were sourced directly from the GLAD data set repository. These data were processed in ArcGIS Pro® and the R statistical environment (R Core Team, 2021) to standardise the LULC classes, ensuring consistency across the study period. The data sets for both years were clipped to the African boundary in ArcGIS Pro®, and their attribute tables were harmonised by mapping their numeric class values to corresponding LULC types.

A transition matrix approach (Pontius et al., 2004) was employed in R to quantify changes between LULC classes. For each cell in the LULC rasters, the LULC class in 2000 was compared to the class in 2020, and the transitions were tallied in the transition matrix. This matrix allowed us to identify significant land use change pathways, including the conversions of LULC types from and to agricultural land as well as other conversions. Transition areas were calculated by multiplying the number of cells in each transition type by the cell area (0.0009 km^2). The resulting areas were then summarised as percentages of the total study area to illustrate the relative magnitude of each transition.

After the spatial analysis, the review focused on extracting insights related to identified key dynamics, providing a nuanced perspective that complemented the spatial analysis and deepened our understanding of the complex issues surrounding agricultural land expansion across Africa.

Results and discussion

Agricultural land expansion as a dominant land use change

The analysis of LULC transitions in Africa between 2000 and 2020 revealed significant changes in cropland distribution and extent. Figure 13.1 illustrates the complex patterns of LULC transitions across the continent during this period, with cropland dynamics playing a central role in landscape transformation.

The most significant cropland expansion occurred through the conversion of dense short vegetation (shrubland) to cropland, accounting for 21% of the total cropland area in 2020. This is represented by a prominent flow in Figure 13.1

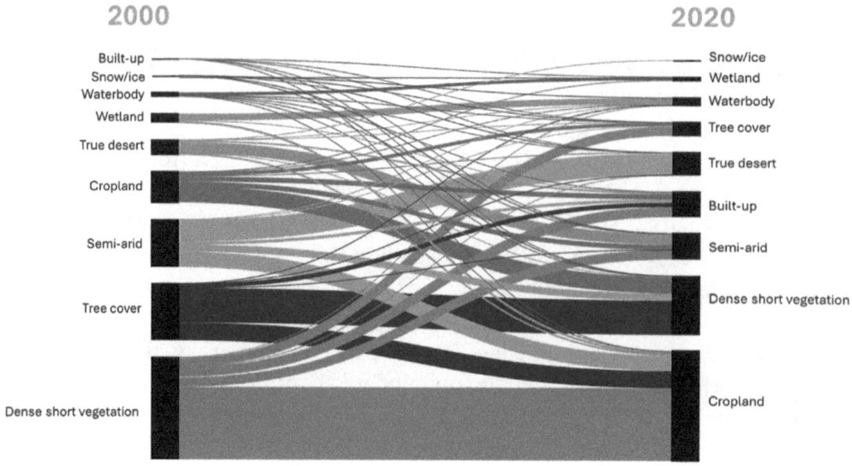

Figure 13.1 Land Use and Land Cover (LULC) transitions across Africa from 2000 to 2020 based on remote sensing data products from the Global Land Analysis and Discovery (GLAD) data set (Potapov et al., 2022a). The flows represent conversion pathways (gains and losses), with their widths proportional to the percentage of total land area converted. Distinct shades represent the different LULC classes. Persisting LULC areas are not shown.

Source: Authors

from the dense short vegetation category to cropland. The second largest source of new cropland was from tree cover, contributing 5.2% to the 2020 cropland area. Semi-arid areas added 4.5% to cropland expansion, while smaller but notable conversions occurred from true desert (0.5%), wetland (0.6%), and water body (0.1%) to cropland.

While cropland expanded in many areas, it also experienced losses. The largest conversion from cropland was to dense short vegetation, with about 6% of the 2000 cropland area transitioning, shown as a significant reverse flow in Figure 13.1. Other notable losses of cropland included conversions to semi-arid areas (2%), tree cover (1%), and built-up areas (2%). Smaller percentages of cropland were converted to wetland (0.2%), true desert (0.1%), and water body (0.1%).

The overall change in cropland area from 2000 to 2020 was a net increase of 33% relative to the 2000 cropland area. This net gain is reflected in Figure 13.1 by the larger cumulative flows entering the cropland category compared to those leaving it.

Implications of agricultural land expansion and dynamics across Africa

Cropland transitions and their dynamics have significant implications for maintaining natural land cover, safeguarding biodiversity, and mitigating LULC change-associated GHG emissions. Although Africa remains a net carbon sink, high deforestation rates linked to agricultural expansion threaten the continent's role in global climate regulation. Most cropland expansion in sub-Saharan Africa occurs at the expense of forests and areas of dense short vegetation, underscoring the need for robust measures to protect these ecosystems.

The expansion of cropland across Africa represents a significant component of global land use change. Potapov et al. (2022b) report that between 2000 and 2019, cropland in Africa increased by 34%, amounting to 53.2 ± 39.4 Mha. However, this expansion has coincided with a decline in per capita cropland area, reflecting the dual pressures of rapid population growth and limited land availability. Agricultural expansion is a primary driver of deforestation and biodiversity loss, emphasising the urgent need to enhance productivity on existing croplands to meet rising food demands without further degrading ecosystems (Jayne et al., 2014; Jayne & Sanchez, 2021).

In sub-Saharan Africa, key drivers of cropland expansion include rapid population growth and state-led development policies. Proximate drivers such as agricultural mechanisation, the expansion of cash crop production, declining soil fertility, and climate change further exacerbate this trend (Jellason et al., 2021). Addressing these challenges requires a dual approach: protecting natural ecosystems from encroachment while enhancing the productivity of existing croplands through sustainable intensification and regenerative practices.

However, the relationship between agricultural intensification and land expansion is nuanced. Adolph et al. (2023) highlight that in contexts with weak forest governance and limited alternative livelihoods, increased agricultural productivity can paradoxically drive further land conversion, as higher profits incentivise additional expansion. These findings underscore the necessity of pairing intensification efforts with robust forest governance, livelihood diversification, and conservation incentives to ensure productivity gains do not aggravate environmental degradation.

To mitigate the environmental and social implications of cropland transitions, a multifaceted approach is essential. Here, we explore strategies that include forest protection, leveraging protected areas and community-managed landscapes, sustainable and agroecological intensification, and policy integration to address the interconnected challenges of agricultural land expansion and loss of biodiversity and natural ecosystems.

Addressing agricultural land expansion through effective protected areas and community-managed areas

Improving the effectiveness of protected areas and community-managed areas can constrain agricultural land expansion. Strong land institutions and governance were found to constrain agricultural expansion (Jellason et al., 2021), while linking law enforcement with community-based regulations can increase biodiversity conservation compliance (Atuo et al., 2020).

However, nature protection policies remain insufficient, with implementation challenges often exacerbated by weak institutional frameworks and competing land use priorities. While protected areas are expected to conserve natural habitats, including forests, ineffective governance and enforcement of protected areas have enabled encroachment by other land uses including agriculture (Sassen et al., 2013; Bryant & Mitchell, 2021; Kouassi et al., 2023). Protected areas in Africa, in particular along the Gulf of Guinea, have been less effective in preventing encroachment by croplands (Li et al., 2024). Various studies emphasise enforcement of land use as a critical challenge, especially in protected areas, where ineffective enforcement led to encroachment and conversion of forest reserves to croplands, for example cocoa in Cote d' Ivoire (Kouassi et al., 2023), in the Mo River landscape of Togo (Diwediga et al., 2017), in Cameroon (Ngouhouo-Poufoun et al., 2024), and in Kenya (Mwangi et al., 2017).

Enforcing protected status of land can forestall agricultural expansion into forests and natural ecosystems (Sassen et al., 2013). In analysing the effectiveness of protected areas in East Africa, Riggio et al. (2019) report that since gazettement, only 6.8% of East African protected areas were transformed to agriculture or other human use from 2001 to 2016 and that only 1.6% of the strict protected areas became converted thereby showing that strict protection might be the most effective approach to safeguard natural habitats in that region's context. However, strict protection contrasts the approaches of Indigenous Peoples and local communities who already sustainably manage natural ecosystems (Sze et al., 2021) and, in such contexts, constrain their human rights, right to land and livelihoods, hence the need for inclusive and equitable governance arrangements (Tauli-Corpuz et al., 2020; Kashwan et al., 2021).

Further, community-managed natural ecosystems and sacred natural sites have been effective in maintaining biodiversity and in some cases preventing agricultural expansion (Mintah et al., 2024). Local initiatives through collective action have compensated the inadequate state capacity to enforce laws and regulations (Mintah et al., 2024). Therefore, to improve protection of forests and other natural ecosystems, there is a need to recognise and support Indigenous Peoples and community-based natural resources management including state recognition of traditional governance of natural sacred sites (Mintah et al., 2024). Such strategies align with the Kunming-Montreal Global Biodiversity Framework.

Addressing agricultural land expansion through sustainable and agroecological intensification

Increasing the productivity of Africa's agricultural lands such as through sustainable intensification and agroecological regenerative practices is critical to reducing agricultural land expansion pressures. With arable land availability finite, improving the efficiency of existing agricultural systems becomes imperative. Farmers must be equipped to adopt informed practices, such as integrated soil fertility management, the informed use of organic and inorganic fertilisers, and water-efficient irrigation techniques (Vanlauwe et al., 2023). Enhanced access to agricultural inputs and extension services is crucial, as is improving farmer education to support the adoption of advanced land management techniques. For instance, informed fertiliser use has been linked to increased yields and improved soil health, although it requires careful application to avoid or reduce adverse environmental effects such as nutrient leaching and nitrous oxide emissions (Kihara et al., 2020).

Sustainable intensification (SI) involves enhancing crop yields per unit area in a cost-effective, environmentally friendly way that improves profitability, livelihoods, biodiversity, and agroecological services, ensuring the future viability of staple crops or cash crops like cocoa (Kongor et al., 2024). SI strategies include improved land management, closing yield gaps via irrigation (Mueller et al., 2012; Senthilkumar, 2022), integrated pest and disease management, integrated soil fertility management, mixed cropping, agroecological agroforestry (Wezel et al., 2015; Garbach et al., 2016; Pretty, 2018; Ollinaho & Kröger, 2021), and farmer training with incentives to adopt sustainable practices.

Diverse forms of agricultural intensification such as "ecological intensification", "sustainable intensification", and "agroecological intensification" have similar principles and practices, but the last two forms have been found in Europe, to target differing land-sparing or land-sharing outcomes (Levidow, 2018). Wezel et al. (2015, p. 1283, 1292) argue that sustainable intensification is more broadly about producing more yields from the same land area while conserving resources, minimising "negative environmental impacts and enhancing natural capital" and ecosystem services, whereas ecological intensification gives more weight to "understanding and intensification of biological and ecological processes and functions in agroecosystems". Agroecological intensification goes further and highlights the systems approach and encompasses more cultural and social perspectives (Wezel et al., 2015). By referring to sustainable, ecological, and agroecological intensification in this chapter, we acknowledge their similarities but also their differences as well as the contexts that condition their adoption.

While some farmers may choose certain intensification practices, agroecological intensification (AEI) can reduce the adverse effects associated with the use of agrochemicals such as pesticide residues while achieving comparable yields and supporting a multifunctional agriculture that provides additional ecosystem services. Garbach et al. (2016, p. 11) define AEI as agricultural practices that "integrate ecological principles and biodiversity management into farming systems with the aims of increasing farm productivity, reducing dependency on external inputs, and sustaining or enhancing ecosystem services". Examples of AEI cropping systems are organic agriculture, conservation agriculture (without synthetic chemicals), precision agriculture, or system of rice intensification, including land management practices such as cover cropping, intercropping with nitrogen-fixing legumes, incorporating crop residues, and agroforestry (Garbach et al., 2016).

The achievement of land-sharing and land-sparing goals, as set forth in the CAADP-Malabo Declaration, hinges on the implementation of effective irrigation strategies. Currently, Northern Africa has the greatest extent of irrigated cropland in Africa, at 4.8 million hectares while other regions lag behind (AUDA-NEPAD, 2024). Regional cereal yields remain below the global average: Northern Africa (2.43 t/ha), Central Africa (1.20 t/ha), Southern Africa (1.76 t/ha), Eastern Africa

(1.34 t/ha), and Western Africa (1.40 t/ha) (Kapuya et al., 2023). It is unlikely that the CAADP-Malabo target of doubling irrigated areas by 2025 will be met, even though annual growth has reached 3.9% (AUDA-NEPAD, 2024). Despite the existence of considerable irrigation potential, the practice of smallholder irrigation remains underexplored.

Increased mechanisation is also central to meeting productivity goals under the CAADP-Malabo Declaration but remains underutilised in sub-Saharan Africa, where only 10% of farm power is mechanised, compared to higher global averages (AUDA-NEPAD, 2024). Sub-Saharan Africa had only 221,000 tractors in 2000 compared to Asia's seven million and Europe's ten million (FAO, 2016). Most farm power in sub-Saharan Africa relies on human labour (65%) and draught animals (25%) and mechanisation uptake faces barriers such as machine availability, high costs, and skill gaps (AUDA-NEPAD, 2024). Tractor rental services are emerging in countries like Nigeria (Takeshima & Lawal, 2020), and policy measures such as reduced tariffs on machinery parts could further encourage adoption (Kapuya et al., 2023). Expanding access to mechanisation for women, who perform 60–80% of food production labour in sub-Saharan Africa, could significantly reduce work burdens (FAO, 2016).

Soil fertility challenges persist in Africa, with average fertiliser use (41.82 kg/ha) falling short of the CAADP-Malabo target of 50 kg/ha. Only four countries – Ethiopia, Egypt, Malawi, and Morocco – have met this target (AUDA-NEPAD, 2024). Fertiliser availability and affordability are major challenges, compounded by supply chain disruptions from the Russia–Ukraine conflict (AUDA-NEPAD, 2024). Integrated soil fertility management and rehabilitation of degraded soils can improve fertiliser efficiency, while incentivising use through extension services is crucial (Vanlauwe et al., 2023).

The effective use of plant protection products is vital for mitigating crop pests like the fall armyworm (Obour et al., 2022). However, agrochemical overuse risks soil salinisation and environmental degradation (Omuto et al., 2024). Biological control methods, such as push-pull technology and botanical pesticides like neem oil, offer sustainable alternatives (Kammo et al., 2019). Integrated pest and disease management has potential, but adoption remains low, calling for agroecological crop protection approaches (Deguine et al., 2021).

High-quality seeds are essential for productivity gains. The progress towards the CAADP-Malabo target of a 100% increase in quality seed availability by 2025 is uneven, with 19 countries on track and 20 making notable progress (AUDA-NEPAD, 2024). Enhancing formal and informal seed systems, supporting Farmer Managed Seed Systems, and fostering public-private collaborations are key to improving seed access and adoption (Ayenan et al., 2021; Breen, 2024).

Yet social-ecological contexts often influence farmer choices of land management practices (Akrofi-Atitianti et al., 2018; Takele et al., 2023; Oduniyi, 2021; Waaswa et al., 2021; Volken & Bottazzi, 2024). Thus, relevant, effective, and sustained policies that enable informed choices of various agricultural and agroecological intensification practices are critical.

Addressing agricultural land expansion through strengthening land governance and policy integration

Strengthening land governance is a critical step, particularly through the enforcement and reform of existing land use plans. Given that most land in sub-Saharan Africa remains under customary tenure despite post-colonial statutory land regimes, mechanisms to bridge customary and modern statutory land tenure systems are essential. Supporting collective landownership registration can help preserve communal lands, while individual land registration for both men and women has been shown to enhance perceptions of tenure security (Msangi et al., 2023), although sometimes formalising land tenure leads to mixed outcomes (Asaaga et al., 2020). Whether land tenure security enables the adoption of sustainable land management practices depends also on other factors, including access to extension services, credit, and modern agricultural inputs (Asaaga et al., 2020). Generally, land tenure security often incentivizes investments in land quality improvements, which in turn support agricultural productivity (AUDA-NEPAD, 2024).

Maintaining better protection of forests and natural ecosystems while promoting sustainable or agroecological intensification could benefit from policy integration. Rather than treating agricultural interventions disparately from forestry and nature conservation, as usually is the norm, policy strategies could be mutually reinforcing and could offer maximum benefit for livelihoods as well as the environment.

In the forestry sector, responses towards addressing deforestation have focused on reforestation, through afforestation programmes to counteract desertification (Nwilo et al., 2020) or launched campaigns to promote reforestation (Manzoor et al., 2022; Ocloo et al., 2024). Other responses include the enactment or promulgation of acts or laws; for example the Forest Plantation Development Fund Act 2000 in Ghana, which was implemented to increase tree cover by allowing farmers the rights to trees and timber on their land (Manzoor et al., 2022). The "modified Taungya system" in northern Ghana has also been implemented and aimed at restoring forest through growing tree by farmers (Manzoor et al., 2022). While states historically have adopted the establishment of protected areas as an approach to conserve forests and natural ecosystems, existing multi-party initiatives in Africa include the African Union Great Green Wall Initiative, intended to check the advance of the Sahara desert (Goffner et al., 2019; Ellison & Ifejika Speranza, 2020), and the African Forest Landscape Restoration Initiative (AFR100) (Mansourian & Berrahmouni, 2021). The emphasis on growing new trees has obscured addressing the drivers of forest loss, as deforestation pressures continue to increase.

On the other hand, the implementation of policies in the agriculture sector had a negative spillover effect on forests. Government policies to increase national income from agricultural commodities have driven conversions from natural vegetation to oil palm, rubber, cocoa, or coffee plantations (Adzigbli et al., 2024). Various policies, including the Structural Adjustment Policies (1983–1990), drove an expansion of agricultural lands, mechanisation, input support for farmer organisations, and commercialisation of agroforestry in the quest to industrialise agriculture and raise its national economic contributions (Adzigbli et al., 2024). Such related policies liberalised the agricultural sector for foreign investments, allowing for large-scale land acquisition and green grabbing (Bourgoin et al., 2023).

While recognising these policy failures to effectively halt or reverse land conversion from agriculture due to weak or no enforcement of laws (Diwediga et al., 2017), in many instances, there has been weak coordination of the sectors involved. Addressing the agriculture expansion-deforestation challenge through an integrated approach of sustainable intensification or agroecological intensification, nature protection, and sustainable use, in a coordinated manner, could be an effective way to tackle agriculture-related deforestation.

While this contribution has focused mainly on land-based responses to agricultural land expansion, other strategies that can equally contribute to addressing cropland expansion include reducing the proportion of African population that is dependent on agriculture by training them for employment in the tertiary sectors. Equally, forest and landscape restoration can be initiated to restore areas already degraded. Further, protecting agricultural lands from conversion to non-agricultural uses is equally important to maintaining food security and minimising additional pressure on forests and other natural ecosystems. While this strategy has been adopted in other world regions (cf. Oliveira et al., 2019), it is still to be explored in many African countries. These aspects can be explored in future studies of agricultural land expansion across Africa.

Conclusions

This chapter has shed light on the transitions and dynamics of agricultural land expansion across Africa, using cropland expansion as the main proxy, highlighting its implications for food security, forest conservation, and sustainable land use. The results of this research thus contribute to the efforts to understand and find solutions to the complex problems of sustainable land management across Africa.

The analysis identified low food production in Africa as a key driver of decisions by African countries to enhance agricultural productivity. Reducing cropland expansion is viewed as a secondary, albeit important, benefit of these strategies. This chapter highlights strategies to address cropland expansion, emphasising the importance of sustainable intensification approaches, including agroecological and organic practices, improved land governance, and reinforcing inclusive Indigenous Peoples and community-based nature conservation strategies and sustainable use of biodiversity.

Crucially, the chapter reveals that while African countries are striving to improve agricultural productivity, enhance food security, and alleviate pressure on forests and ecosystems, the set targets are unlikely to be achieved by 2025. This underscores the urgent need for coordinated efforts in research, policy development, and practical implementation. Disparities in agricultural intensification are evident, with countries in northern and southern Africa advancing in mechanisation, irrigation, fertiliser use, and improved seeds while other regions lag. The CAADP-Malabo Declaration has spurred various initiatives across Africa, yet there remains a critical need to strengthen science-policy dialogues. This ensures that strategies and monitoring frameworks are consistently informed by scientific evidence.

Finally, an integrative approach that extends beyond a land management and natural ecosystems approach to agricultural expansion, to include socio-economic drivers, gender, and intersectionality as well as wider policy coherence factors, is required. However, this is beyond the scope of the present chapter. Further aspects include the contributions of technological measures such as leveraging information and communication technologies to enable precision nutrient application and improve land management practices. Examining the political ecology and economy of agricultural intensification can uncover constraints, trade-offs, and strategies to achieve more equitable outcomes for the different stakeholders and beneficial outcomes for nature.

Funding and acknowledgements

This research is part of the SUSTAINFORESTS research project funded by the European Research Council (ERC) under the European Union's Horizon 2020 research and innovation programme (grant agreement No 101001200). This publication contributes to the Initiative Afrique at the University of Bern, to the Programme on Ecosystem Change and Society (https://pecs-science.org/), and to the Global Land Programme (https://www.glp.earth/).

References

Abe-Inge, V., & Kwofie, E. M. (2023). Accounting for environmental-nutrition trade-offs in African food systems. In *Africa agriculture status report* (pp. 143–162). AGRA. Retrieved December 15, 2024, from https://agra.org/wp-content/uploads/2024/08/Africa-Agriculture-Status-Report-2023-Empowering-Africas-Food-Systems.pdf-_compressed.pdf

Adolph, B., Jellason, N. P., Kwenye, J. M., Davies, J., Dray, A. G., Waeber, P. O., Jeary, K., & Franks, P. (2023). Exploring farmers' decisions on agricultural intensification and cropland expansion in Ethiopia, Ghana, and Zambia through serious gaming. *Land*, *12*(3), Article 3. https://doi.org/10.3390/land12030556

Adzigbli, W. K., Duku, E., Atampugre, G., Fürst, C., & Nyarko, B. K. (2024). Agricultural land use policies and landscape dynamics: Evidence from rainforest agroecological zone. *Land Use Policy*, *142*, 1–13. https://doi.org/10.1016/j.landusepol.2024.107184

African Union Commission and African Union Development Agency – NEPAD. (2022). *AUC & AUDA-NEPAD second continental report on the implementation of agenda 2063* (AUC & AUDA-NEPAD, pp. 1–154). https://au.int/en/documents/20220210/second-continental-report-implementation-agenda-2063

AGRA. (2016). *Progress towards agriculture transformation in Sub-Saharan Africa* (Africa Agriculture Status Report, Issue 4, pp. 1–296). AGRA. Retrieved December 15, 2024, from https://agra.org/wp-content/uploads/2017/09/aasr-report-2016-press0409201601.pdf

AGRA. (2021). *A decade of action: Building sustainable and resilient food systems in Africa: Building sustainable and resilient food systems in Africa* (No. 9; Africa Agriculture Status Report, Issue 9, pp. 1–256). Retrieved December 15, 2024, https://agra.org/wp-content/uploads/2021/09/AASR-2021-A-Decade-of-Action-_Building-Sustainable-and-Resilient-Food-Systems-in-Africa.pdf

AGRA. (2022). *Accelerating African food systems transformation* (No. 10; Africa Agriculture Status Report, Issue 10, pp. 1–119). Retrieved December 15, 2024, https://agra.org/wp-content/uploads/2022/09/AASR-2022.pdf

Akrofi-Atitianti, F., Ifejika Speranza, C., Bockel, L., & Asare, R. (2018). Assessing climate smart agriculture and its determinants of practice in Ghana: A case of the cocoa production system. *Land*, *7*(1), 30. https://doi.org/10.3390/land7010030

Ali, D. A., Deininger, K., & Goldstein, M. (2014). Environmental and gender impacts of land tenure regularization in Africa: Pilot evidence from Rwanda. *Journal of Development Economics*, *110*, 262–275. https://doi.org/10.1016/j.jdeveco.2013.12.009

Asaaga, F. A., Hirons, M. A., & Malhi, Y. (2020). Questioning the link between tenure security and sustainable land management in cocoa landscapes in Ghana. *World Development*, *130*, 104913. https://doi.org/10.1016/j.worlddev.2020.104913

Atuo, F. A., Fu, J., O'Connell, T. J., Agida, J. A., & Agaldo, J. A. (2020). Coupling law enforcement and community-based regulations in support of compliance with biodiversity conservation regulations. *Environmental Conservation*, *47*(2), Article 2. https://doi.org/10.1017/s0376892920000107

AUDA-NEPAD. (2024). *Fourth biennial review report of the African Union commission on the implementation of the Malabo declaration* (No. 4, pp. 1–169). Retrieved December 15, 2024, from https://au.int/sites/default/files/documents/43556-doc-EN_4th_CAADP_Biennial_Review_Report-COMPLETE.pdf

Ayenan, M. A. T., Aglinglo, L. A., Zohoungbogbo, H. P. F., N'Danikou, S., Honfoga, J., Dinssa, F. F., Hanson, P., & Afari-Sefa, V. (2021). Seed systems of traditional African vegetables in Eastern Africa: A systematic review. *Frontiers in Sustainable Food Systems*, *5*, 1–12. https://doi.org/10.3389/fsufs.2021.689909

Boone, C. (2018). Legal empowerment of the poor through property rights reform: Tensions and trade-offs of land registration and titling in Sub-Saharan Africa. *The Journal of Development Studies*, *55*(3), Article 3. https://doi.org/10.1080/00220388.2018.1451633

Bourgoin, J., Diop, D., Jahel, C., Interdonato, R., & Grislain, Q. (2023). Beyond land grabbing narratives, acknowledging patterns and regimes of land control in Senegal. *Land Use Policy*, *131*, 1–11. https://doi.org/10.1016/j.landusepol.2023.106753

Breen, C., Ndlovu, N., McKeown, P. C., & Spillane, C. (2024). Legume seed system performance in sub-Saharan Africa: Barriers, opportunities, and scaling options. A review. *Agronomy for Sustainable Development*, *44*(20), Article 2. https://doi.org/10.1007/s13593-024-00956-6

Bryant, C., & Mitchell, M. I. (2021). The political ecology of cocoa in Ghana: Past, present and future challenges. *Natural Resources Forum*, *45*(4), Article 4. https://doi.org/10.1111/1477-8947.12232

Deguine, J.-P., Aubertot, J.-N., Flor, R. J., Lescourret, F., Wyckhuys, K. A. G., & Ratnadass, A. (2021). Integrated pest management: Good intentions, hard realities. A review. *Agronomy for Sustainable Development*, *41*(38), Article 3. https://doi.org/10.1007/s13593-021-00689-w

Deininger, K., & Byerlee, D. (2012). The rise of large farms in land abundant countries: Do they have a future? *World Development*, *40*(4), Article 4. https://doi.org/10.1016/j.worlddev.2011.04.030

Diwediga, B., Agodzo, S., Wala, K., & Le, Q. B. (2017). Assessment of multifunctional landscapes dynamics in the mountainous basin of the Mo River (Togo, West Africa). *Journal of Geographical Sciences*, *27*(5), Article 5. https://doi.org/10.1007/s11442-017-1394-4

Doss, C., Meinzen-Dick, R., & Bomuhangi, A. (2013). Who owns the land? Perspectives from rural ugandans and implications for large-scale land acquisitions. *Feminist Economics*, *20*(1), Article 1. https://doi.org/10.1080/13545701.2013.855320

Ellison, D., & Ifejika Speranza, C. (2020). From blue to green water and back again: Promoting tree, shrub and forest-based landscape resilience in the Sahel. *Science of the Total Environment*, *739*, 1–14. https://doi.org/10.1016/j.scitotenv.2020.140002

FAO. (2016). *The state of food and agriculture* (Climate Change, Agriculture and Food Security, pp. 1–194). FAO. https://openknowledge.fao.org/server/api/core/bitstreams/07bc7c6e-72e5-488d-b2f7-3c1499d098fb/content

Fuglie, K., Gautam, M., Goyal, A., & Maloney, W. F. (2019). *Harvesting prosperity: Technology and productivity growth in agriculture*. World Bank. https://doi.org/10.1596/978-1-4648-1393-1

Garbach, K., Milder, J. C., DeClerck, F. A. J., Montenegro de Wit, M., Driscoll, L., & Gemmill-Herren, B. (2016). Examining multi-functionality for crop yield and ecosystem services in five systems of agroecological intensification. *International Journal of Agricultural Sustainability*, *15*(1), Article 1. https://doi.org/10.1080/14735903.2016.1174810

Goffner, D., Sinare, H., & Gordon, L. J. (2019). The Great Green Wall for the Sahara and the Sahel Initiative as an opportunity to enhance resilience in Sahelian landscapes and livelihoods. *Regional Environmental Change*, *19*(5), Article 5. https://doi.org/10.1007/s10113-019-01481-z

International Food Policy Research Institute (IFPRI). (2020). *Agricultural Total Factor Productivity (TFP), 2000-2016* (Version V1) [Dataset]. Harvard Dataverse. https://doi.org/10.7910/DVN/PJDGTJ

Jayne, T. S., Chamberlin, J., & Headey, D. D. (2014). Land pressures, the evolution of farming systems, and development strategies in Africa: A synthesis. *Food Policy*, *48*, 1–17. https://doi.org/10.1016/j.foodpol.2014.05.014

Jayne, T. S., & Sanchez, P. A. (2021). Agricultural productivity must improve in sub-Saharan Africa. *Science*, *372*(6546), Article 6546. https://doi.org/10.1126/science.abf5413

Jellason, N. P., Robinson, E. J. Z., Chapman, A. S. A., Neina, D., Devenish, A. J. M., Po, J. Y. T., & Adolph, B. (2021). A systematic review of drivers and constraints on agricultural expansion in Sub-Saharan Africa. *Land*, *10*(3), Article 3. https://doi.org/10.3390/land10030332

Kammo, E. Q., Suh, C., Mbong, G. A., Djomo, S. H., Chimi, N. L. L., Mbeungang, D. L., Mafouasson, H., Meseka, S., & Menkir, A. (2019). Biological versus chemical control of fall armyworm and Lepidoptera stem borers of maize (Zea mays). *Agronomie Africaine*, *31*(2), Article 2.

Kapuya, T., Mutyasira, V., Haddad, L., & Keizire, B. (2023). A stocktake of Africa's food systems. In *Empowering Africa's food systems for the future* (Issue 11, pp. 52–80). AGRA. Retrieved December 15, 2024, from https://agra.org/wp-content/uploads/2023/09/AASR-2023.pdf

Kashwan, P., Duffy, R., Massé, F., Asiyanbi, A. P., & Marijnen, E. (2021). From racialized neocolonial global conservation to an inclusive and regenerative conservation. *Environment: Science and Policy for Sustainable Development*, *63*(4), 4–19. https://doi.org/10.1080/00139157.2021.1924574

Kihara, J., Bolo, P., Kinyua, M., Nyawira, S. S., & Sommer, R. (2020). Soil health and ecosystem services: Lessons from sub-Sahara Africa (SSA). *Geoderma*, *370*, 1–12. https://doi.org/10.1016/j.geoderma.2020.114342

Kongor, J. E., Owusu, M., & Oduro-Yeboah, C. (2024). Cocoa production in the 2020s: Challenges and solutions. *CABI Agriculture and Bioscience*, *5*(102), Article 1. https://doi.org/10.1186/s43170-024-00310-6

Kouassi, A. K., Zo-Bi, I. C., Aussenac, R., Kouamé, I. K., Dago, M. R., N'guessan, A. E., Jagoret, P., & Hérault, B. (2023). The great mistake of plantation programs in cocoa agroforests – let's bet on natural regeneration to sustainably provide timber wood. *Trees, Forests and People*, *12*, 1–9. https://doi.org/10.1016/j.tfp.2023.100386

Levidow, L. (2018). Sustainable intensification: Agroecological appropriation or contestation? In D. H. Constance, J. T. Konefal, & M. Hatanaka (Eds.), *Contested sustainability discourses in the agrifood system* (1st ed., pp. 19–41). Routledge/Earthscan. https://doi.org/10.4324/9781315161297

Li, G., Fang, C., Watson, J. E. M., Sun, S., Qi, W., Wang, Z., & Liu, J. (2024). Mixed effectiveness of global protected areas in resisting habitat loss. *Nature Communications*, *15*, Article 1. https://doi.org/10.1038/s41467-024-52693-9

Mansourian, S., & Berrahmouni, N. (2021). *Review of forest and landscape restoration in Africa 2021* (1st ed.). FAO and AUDA-NEPAD. https://openknowledge.fao.org/handle/20.500.14283/cb6111en

Manzoor, S. A., Griffiths, G. H., Robinson, E., Shoyama, K., & Lukac, M. (2022). Linking pattern to process: Intensity analysis of land-change dynamics in Ghana as correlated to past socioeconomic and policy contexts. *Land*, *11*(7), Article 7. https://doi.org/10.3390/land11071070

Mintah, F., Pamela, T. E. P., Oberlack, C., & Speranza, C. I. (2024). Why do forests persist and re-emerge amidst tropical deforestation pressures? Archetypes of governance and impact pathways. *Forest Policy and Economics*, *169*, 1–18. https://doi.org/10.1016/j.forpol.2024.103352

Msangi, H. A., Waized, B., Löhr, K., Sieber, S., & Ndyetabula, D. W. (2023). Development outcomes of land tenure formalization under customary and statutory land tenure systems in Tanzania: A multinomial endogenous switching regression approach. *Agriculture & Food Security*, *11*(1). https://doi.org/10.1186/s40066-022-00403-3

Mueller, N. D., Gerber, J. S., Johnston, M., Ray, D. K., Ramankutty, N., & Foley, J. A. (2012). Closing yield gaps through nutrient and water management. *Nature*, *490*(7419), Article 7419. https://doi.org/10.1038/nature11420

Mwangi, H., Lariu, P., Julich, S., Patil, S., McDonald, M., & Feger, K.-H. (2017). Characterizing the intensity and dynamics of land-use change in the Mara River Basin, East Africa. *Forests*, *9*(1), Article 1. https://doi.org/10.3390/f9010008

Ngouhouo-Poufoun, J., Chaupain-Guillot, S., Ndiaye, Y., Sonwa, D. J., Yana Njabo, K., & Delacote, P. (2024). Cocoa, livelihoods, and deforestation within the Tridom landscape in the Congo Basin: A spatial analysis. *PLOS ONE*, *19*(6), Article 6. https://doi.org/10.1371/journal.pone.0302598

Nwilo, P. C., Olayinka, D. N., Okolie, C. J., Emmanuel, E. I., Orji, M. J., & Daramola, O. E. (2020). Impacts of land cover changes on desertification in northern Nigeria and implications on the Lake Chad Basin. *Journal of Arid Environments*, *181*, 1–10. https://doi.org/10.1016/j.jaridenv.2020.104190

Obour, P. B., Arthur, I. K., & Owusu, K. (2022). The 2020 Maize production failure in Ghana: A case study of Ejura-Sekyedumase municipality. *Sustainability*, *14*(6), Article 6. https://doi.org/10.3390/su14063514

Ocloo, M. D., Huang, X., Fan, M., & Ou, W. (2024). Study on the spatial changes in land use and landscape patterns and their effects on ecosystem services in Ghana, West Africa. *Environmental Development*, *49*, 1–20. https://doi.org/10.1016/j.envdev.2023.100947

Oduniyi, O. S. (2021). Factors driving the adoption and use extent of sustainable land management practices in South Africa. *Circular Economy and Sustainability*. https://doi.org/10.1007/s43615-021-00119-9

Oliveira, E., Leuthard, J., & Tobias, S. (2019). Spatial planning instruments for cropland protection in Western European countries. *Land Use Policy*, *87*, 1–12. https://doi.org/10.1016/j.landusepol.2019.104031

Ollinaho, O. I., & Kröger, M. (2021). Agroforestry transitions: The good, the bad and the ugly. *Journal of Rural Studies*, *82*, 210–221. https://doi.org/10.1016/j.jrurstud.2021.01.016

Omuto, C. T., Kome, G. K., Ramakhanna, S. J., Muzira, N. M., Ruley, J. A., Jayeoba, O. J., Raharimanana, V., Owusu Ansah, A., Khamis, N. A., Mathafeng, K. K., Elmobarak, A. A., Vargas, R. R., Koetlisi, A. K., Dembele, D., Diawara, M., Mbaikoubou, M., Maria, R. M., Adam Boukary, I., Malatji, A., . . . Nyamai, M. (2024). Trend of soil salinization in Africa and implications for agro-chemical use in semi-arid croplands. *Science of the Total Environment*, *951*, 1–9. https://doi.org/10.1016/j.scitotenv.2024.175503

Pontius, R. G., Jr., Shusas, E., & McEachern, M. (2004). Detecting important categorical land changes while accounting for persistence. *Agriculture, Ecosystems & Environment*, *101*(2–3), 251–268. https://doi.org/10.1016/j.agee.2003.09.008

Potapov, P., Hansen, M. C., Pickens, A., Hernandez-Serna, A., Tyukavina, A., Turubanova, S., Zalles, V., Li, X., Khan, A., Stolle, F., Harris, N., Song, X.-P., Baggett, A., Kommareddy, I., & Kommareddy, A. (2022a). The global 2000–2020 land cover and land use change dataset derived from the landsat archive: First results. *Frontiers in Remote Sensing*, *3*. https://doi.org/10.3389/frsen.2022.856903

Potapov, P., Turubanova, S., Hansen, M. C., Tyukavina, A., Zalles, V., Khan, A., Song, X.-P., Pickens, A., Shen, Q., & Cortez, J. (2022b). Global maps of cropland extent and change show accelerated cropland expansion in the twenty-first century. *Nature Food*, *3*(1), Article 1. https://doi.org/10.1038/s43016-021-00429-z

Pretty, J. (2018). Intensification for redesigned and sustainable agricultural systems. *Science*, *362*(6417), eaav0294. https://doi.org/10.1126/science.aav0294

R Core Team. (2021). *The R Project for statistical computing* [Computer software]. https://www.R-project.org

Riggio, J., Jacobson, A. P., Hijmans, R. J., & Caro, T. (2019). How effective are the protected areas of East Africa? *Global Ecology and Conservation*, *17*, 1–12. https://doi.org/10.1016/j.gecco.2019.e00573

Sassen, M., Sheil, D., Giller, K. E., & ter Braak, C. J. F. (2013). Complex contexts and dynamic drivers: Understanding four decades of forest loss and recovery in an East African protected area. *Biological Conservation*, *159*, 257–268. https://doi.org/10.1016/j.biocon.2012.12.003

Senthilkumar, K. (2022). Closing rice yield gaps in Africa requires integration of good agricultural practices. *Field Crops Research*, *285*, 1–19. https://doi.org/10.1016/j.fcr.2022.108591

Sze, J. S., Carrasco, L. R., Childs, D., & Edwards, D. P. (2021). Reduced deforestation and degradation in Indigenous Lands pan-tropically. *Nature Sustainability*, 5. https://doi.org/10.1038/s41893-021-00815-2

Takele, A., Abelieneh, A., & Wondimagegnhu, B. A. (2023). Determinants of adoption of land management practices among farmers in Western Lake Tana and Beles River watersheds (Ethiopia) as a climate change adaptation strategy. *Cogent Food & Agriculture*, *9*(1). https://doi.org/10.1080/23311932.2023.2170951

Takeshima, H., & Lawal, A. (2020). Evolution of agricultural mechanization in Nigeria. In X. Diao, H. Takeshima, & X. Zhang (Eds.), *An evolving paradigm of agricultural mechanization development: How much can Africa learn from Asia?* (pp. 423–456). International Food Policy Research Institute. https://doi.org/10.2499/9780896293809_13

Tauli-Corpuz, V., Alcorn, J., Molnar, A., Healy, C., & Barrow, E. (2020). Cornered by PAs: Adopting rights-based approaches to enable cost-effective conservation and climate action. *World Development*, *130*, 104923. https://doi.org/10.1016/j.worlddev.2020.104923

Vanlauwe, B., Amede, T., Bationo, A., Bindraban, P., Breman, H., Cardinae, R., Couede, A., Chivenge, P., Corbeels, M., Dobermann, A., Falconnier, G., Fatunbi, W., Giller, K., Harawa, R., Kamau, M., Merckx, R., Palm, C., Powlson, D. S., Rusinamhodzi, L., . . . Groot, R. (2023). *Fertilizer and soil health in Africa the role of fertilizer in building soil health to sustain farming and address climate change*. International Fertilizer Development Center – IFDC. https://ifdc.org/resources/fertilizer-and-soil-health-in-africa-the-role-of-fertilizer-in-building-soil-health-to-sustain-farming-and-address-climate-change/

Volken, S., & Bottazzi, P. (2024). Sustainable farm work in agroecology: How do systemic factors matter? *Agriculture and Human Values*. https://doi.org/10.1007/s10460-024-10539-6

von Braun, J., Ulimwengu, J. M., Nwafor, A., & Nhlengethwa, S. (2023). *The overview chapter* (No. 11; Empowering Africa's food systems for the future, pp. 14–26). AGRA. Retrieved December 15, 2024, from https://agra.org/wp-content/uploads/2024/08/Africa-Agriculture-Status-Report-2023-Empowering-Africas-Food-Systems.pdf-_compressed.pdf

Waaswa, A., Oywaya Nkurumwa, A., Mwangi Kibe, A., & Ngeno Kipkemoi, J. (2021). Climate-smart agriculture and potato production in Kenya: Review of the determinants of practice. *Climate and Development*, *14*(1), 1–16. https://doi.org/10.1080/17565529.2021.1885336

Wezel, A., Soboksa, G., McClelland, S., Delespesse, F., & Boissau, A. (2015). The blurred boundaries of ecological, sustainable, and agroecological intensification: A review. *Agronomy for Sustainable Development*, *35*(4), Article 4. https://doi.org/10.1007/s13593-015-0333-y

14 Development policy and its influence on land use

A study of institutional markets in Maranhão, Brazil

Laudelina Alves Ribeiro, Uelson Serra Garcia, Jefferson Andronio Ramundo Staduto, and Ana Cecília Kreter

Introduction

The "High Human Development Index Plan (HDI+)" of Maranhão state, launched in 2016, is a comprehensive project comprising various action lines implemented by the government of Maranhão, a state located in the northeast – one of the poorest regions – of Brazil (Figure 13.1). Maranhão has 217 municipalities, and HDI+ aims to combat extreme poverty in the 30 municipalities with the lowest Human Development Index (HDI) in the state (SAF, 2020). The HDI+ has been executing development projects in both urban and rural areas of these municipalities. However, the selected 30 are small and rural, with many family farms. Maranhão is ecologically diverse, covered by the Cerrado and Amazon biomes, with a small fragment of the Caatinga biome.

In rural areas, government procurement of agricultural products (institutional markets) and other agricultural policy instruments is implemented by the public sector's technical and rural extension service (TRES). These markets have been promoting more sustainable production on small farms and enhancing the well-being of the rural population (Sambuichi et al., 2014; Kroth et al., 2020). The strategies employed in these projects encompass a sustainable territorial development approach and public policies aimed at valuing social, cultural, economic, political, institutional, and environmental diversity (SAF, 2021b).

HDI+ integrates federal programs, such as the Food Acquisition Program and the National School Feeding Program (NSFP), which have extensive coverage across Brazil. Additionally, the Family Farming Procurement Program of the Maranhão state government strengthens institutional purchases. However, NSFP has the potential for greater financial support (Sambuich et al., 2014).

NSFP is considered one of the largest and most comprehensive school feeding programs in the world. Providing basic meals to students integrates federal, state, and municipal government efforts, encourages participation and social control, and fosters local development by valuing and facilitating access to family farming (BRASIL, 2017). According to Kroth et al. (2020), NSFP can be considered a public health policy, as it promotes healthy eating habits.

DOI: 10.4324/9781032672779-18

Also, NSFP aims to support sustainable production, increase collective actions among family farmers, and boost income (Cunha et al., 2017; Cecconello et al., 2023; Moreira et al., 2023), with a focus on reducing food and nutritional insecurity (Vogt & Kaiser, 2008). This program is strongly associated with the efficiency and reach of TRES, ensuring that producers deliver quality food using highly sustainable agricultural practices (Camargo et al., 2013; Santana & Lima, 2018; Moreira et al., 2023).

Given this context, this chapter focuses on analyzing the set of HDI+ municipalities from a sustainable rural development perspective, particularly evaluating institutional markets supporting land use change and promoting sustainable agriculture. The institutional market programs, which comply with various regulations, drive land use change and promote sustainable development. This chapter also contributes to filling a gap in the literature by evaluating the HDI+, particularly the potential of institutional purchases, articulated with TRES, which in turn fosters sustainable land use on family farms in the Cerrado and Amazon biomes.

National School Feeding Program (NSFP)

Farmers' decisions to adopt specific agricultural practices and their capacity to advance toward more sustainable farming systems are influenced by external factors such as knowledge, skills, markets, public policies, resources, and landownership structures. Therefore, promoting the adoption of different practices and agricultural systems to enhance sustainability requires understanding how variability in individual, family, farm, and regional characteristics affects farmers' responses to incentives and disincentives (NRC, 2010).

Creating, operating, and evolving alternative food supply chains have garnered significant attention from scholars, professionals, and agricultural agencies (Heiss et al., 2015). Regional institutional markets support sustainable agriculture, including schools and other large-scale food establishments (Vogt & Kaiser, 2008). Agricultural production programs directed at institutions represent new alternative ways to organize food supply chains (Heiss et al., 2015). Integrating the practical operational aspects of agricultural policy with food and nutritional security policy is essential for developing new ways of producing and consuming (Schmitt, 2005).

Institutional markets have a strong adherence to sustainable regional food systems, which aim to maximize the potential for regional food self-sufficiency, achieve a minimum quantity and quality of food for regional residents, and sustain and develop the resources of regional food systems (Lawless et al., 1996). Starr et al. (2003) and Vogt and Kaiser (2008) consider regional food as part of a sustainable development model that addresses the interrelated concerns of environmental sustainability, agricultural sustainability, food quality, and safety, as well as providing healthy and nutritious food to community members.

The Brazilian institutional market of NSFP was established on June 17, 2013, with guidelines set by the National Education Development Fund – NEDF (FNDE, 2016). NSFP emphasizes the development of alternative agri-food models such as organic agriculture, agroecology, and extractivism. Sustainability is incorporated into its social, economic, environmental, political, cultural, and ethical focus.

NEDF mandates social control by the community to ensure healthy and adequate food, teaching and learning in food and nutrition education, care of enrolled students, support for sustainable development, and right-to-school feeding with food and nutritional security (BRASIL, 2013). Since 2009,[1] at least 30% of food purchases were required to come from family farms, with a preference for products from agrarian reform settlements, traditional Indigenous communities, and quilombola communities[2] (BRASIL, 2009).

NSFP resources are managed by municipal, state, and federal governments, which are responsible for their use and supplementation. On the one hand, NSFP has contributed to providing healthy food to students in public schools; on the other hand, it has encouraged family farm production and sustainability of the local agri-food sector through more conservation-oriented farming practices and production of organic and/or agroecological products (Lopes et al., 2021; Amaral & de Carvalho Guimarães, 2023; Teixeira, 2024). In Latin America, this program is considered the largest universal food supplement program, judged by the number of students served and the volume of investment (Libermann & Bertolini, 2015).

Institutional purchases through various procurement methods serve as an educational tool for both public management and a segment of family farming (Rigon & Bezerra, 2014). The organization of producers and a TRES policy could potentially contribute to expanding production and promoting a sustainable land use agricultural model (Kroth et al., 2019; Sheriff et al., 2020; Sahu et al., 2023). Moreover, these public policies can encourage local economies, reduce social vulnerability, promote public health, and particularly contribute to sustainable agricultural production (Cunha et al., 2017; Cecconello et al., 2023; de Souza et al., 2023).

HDI+ Plan of the State of Maranhão

The HDI+ was promoted by the State of Maranhão (IMESC, 2015), and began implementation in January 2016, with the intention of reducing extreme poverty and social inequalities in both urban and rural areas through a sustainable territorial development approach. According to SAF (2021a), in rural areas, its strategies are conducted by the State Secretariat of Family Agriculture (SAF), which includes the State Agency for Agricultural Research and Rural Extension of Maranhão and the Institute of Colonization and Lands of Maranhão, aiming to promote income generation through agricultural production activities and the dissemination of technologies targeted at family farmers. Through TRES, the HDI+ plan aims to promote production by applying conservation and preservation practices to 9,000 family farms (SAF, 2021a).

Most of the municipalities (17) participating in the HDI+ in Maranhão are in the Cerrado biome (Figure 14.1), where forest legislation requires farms to preserve 20% of their areas with native vegetation cover. The remaining 13 municipalities are within the Amazon biome, where preservation areas cover 80% of the farms, thus affecting land use (Oliveira et al., 2017).

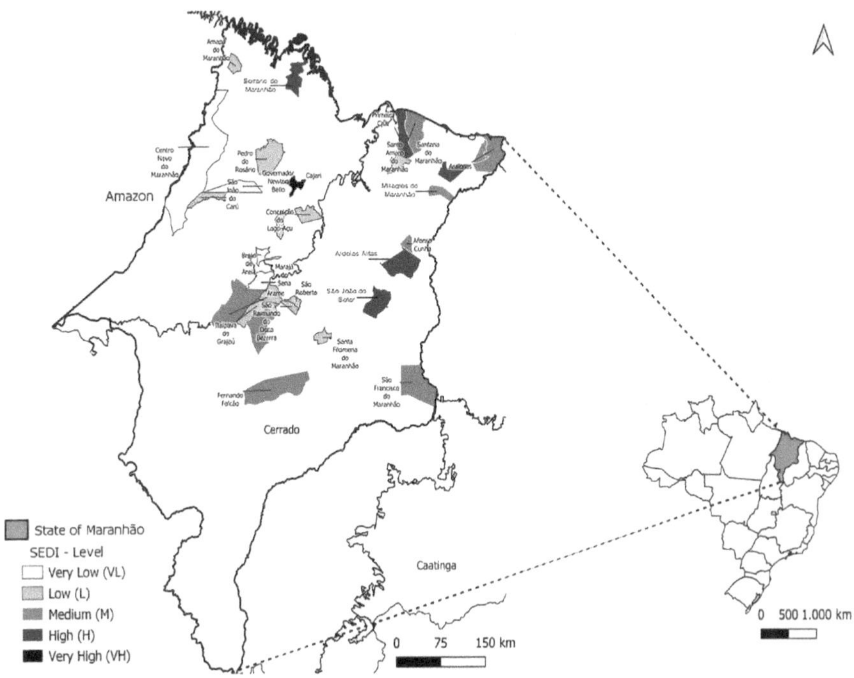

Figure 14.1 Map of the 30 municipalities belonging to the "High HDI Plan".

Source: IBGE (2021b). Elaborated by the authors.

Note: SEDI – Socioeconomic and Environmental Development Index.

NSFP provides for both centralized and decentralized procurement. The HDI+ adopts the decentralized model, in which the Maranhão government transfers the funds it receives from NEDF directly to public schools, which are responsible for implementing the program (SEDUC, 2015). HDI+ promotes both the supply and consumption of food in public schools, thus forming a strategy for socioeconomic, territorial, and sustainable development aimed at reducing extreme poverty in the municipalities with the lowest HDI in Maranhão.

Methodological Procedures

In this chapter, factor analysis was applied, a method aimed at synthesizing information, known as factors, by grouping variables with minimal loss of information. This technique is recommended when the analyzed variables have a high correlation coefficient among them and seeks to identify the performance of a set of interdependent elements in a small number of factors (Hair Júnior et al., 2009; Fávero & Belfiore, 2017). Factorial analysis groups social, economic, and environmental parameters for municipalities in the HDI+ index, thereby allowing for the analysis of the characteristics of the most significant factors for this group of municipalities.

The model is composed of a theoretical correlation matrix P_{pxp}, which is linearly linked with the standardized variables and the common factors, as presented in Equation 14.1 (Mingoti, 2005).

$$Z_p = l_{p1}F_1 + l_{p2}F_2 + \cdots + l_{pm}F_m + \varepsilon_p \tag{14.1}$$

Where:

Z_p are variables that are linearly linked with the new random variables ($F_j = 1, 2, \ldots, m$) that need to be identified; and ε_p is the vector of random errors, corresponding to measurement errors and the variation of Z_p.

The method for extracting factors through principal components followed the adequacy of the factor analysis model using the Kaiser-Meyer-Olkin (KMO) statistic and Bartlett's test of sphericity (Bartlett, 1954; Mingoti, 2005; Hair Júnior et al., 2009; Fávero & Belfiore, 2017).

To assess the levels of rural development, the Socioeconomic and Environmental Development Index (SEDI) was used, based on the studies by Melo and Parré (2007) and Stege and Parré (2013), as shown in Equation 14.2.

$$GI = \frac{\Sigma_{i=1}^{N}\left(W_i F_i\right)}{\Sigma_{i=1}^{N}\left(W_i\right)} \tag{14.2}$$

Where:

GI is the gross index–the weighted average of the factor scores;

W_i is the percentage of variance explained by each factor;

F_i are the factor scores.

The SEDI values are calculated through interpolation between 100 and 0. The criteria presented by Melo and Parré (2007) and Stege and Parré (2013) were applied. The SEDI was stratified according to the mean and the number of standard deviations in the sample: Very High (VH), High (H), Medium (M), Low (L), and Very Low (VL).

The variables used in estimations were based on Kageyama (2004), Stege and Parré (2013), and Staduto et al. (2018). In the end, 12 variables were used in the model (Table 14.1) out of the 31 variables initially considered.

Determination and Characterization of Factors

Bartlett's test was statistically significant at 1%, and the KMO value was 0.542, which is higher than the critical value of 0.5. These results indicate that the factor analysis technique is appropriate for application to the dataset in Table 14.2 (Hair Júnior et al., 2009). The principal components method was applied using Varimax orthogonal rotation, from which five factors were extracted. Subsequently, factors with eigenvalues greater than 1 were selected, resulting in a cumulative variance of 82.61% (Table 14.2), with an average variance of 16.52%.

Table 14.1 Description of the variables, data sources, and extra information used

Variable	*Source*	*Unit*
(ID4) – Farm sizes	2017 Census of Agriculture (IBGE, 2021a)	Average area (hectare)
(ID7) – Population density	2022 Demographic Census (IBGE, 2023a)	Inhabitants per square kilometer
(ID8) – Cattle herd	2017 Municipal Livestock Research (IBGE, 2023b)	Cattle per area of farms
(ID9) – Goat herd	2017 Municipal Livestock Research (IBGE, 2023b)	Goat per area of farms
(ID11) – NSFP funds	2017 FNDE (2017)	Percentage of the Budget Executed
(ID13) – Forest and agro-forestry preserved areas	2017 Census of Agriculture (IBGE, 2021a)	Percentage of the area
(ID15) – Farms with production involving vegetal extraction	2017 Census of Agriculture (IBGE, 2021a)	Percentage of the farm
(ID18) – Farm with production aimed at self-consumption	2017 Census of Agriculture (IBGE, 2021a)	Percentage of the farm
(ID24) – Percentage of the buffalo population	2017 Municipal Livestock Research (IBGE, 2023b)	Buffalo per area (square root of the area)
(ID26) – Deforested area	Assis et al. (2019)	Hectare (ha)
(ID28) – Literate farmer	2017 Census of Agriculture (IBGE, 2021a)	Percentage of farmers
(ID30) – Gini index	2010 SUS (2023)	The index between 0 and 1

Source: Research data. Elaborated by the authors.

Table 14.2 Eigenvalue, variance, and accumulated variance

Factor	*Eigenvalue*	*Variance explained by the factor (%)*	*Accumulated variance (%)*
F1	3.03	20.15	20.15
F2	2.12	16.36	36.50
F3	1.97	15.93	52.43
F4	1.59	15.58	68.01
F5	1.20	14.59	82.61

Source: Research data. Elaborated by the authors.

From the identification of the five extracted factors, factor scores and commonalities associated with each variable were obtained (Table 14.3). Communality represents the percentage of each variable's variance explained by the factors. Values greater than 0.5 are considered satisfactory for a variable's contribution to the factor's structure (Hair Júnior et al., 2009).

Table 14.3 Variables, factor scores, and commonalities of the 30 municipalities in Maranhão with the lowest HDI

Variable	*Factor*					*Community*
	F1	*F2*	*F3*	*F4*	*F5*	
ID4	−.381	−.063	**−.782**	−.350	−.006	0.883
ID7	−.129	.042	**.877**	−.197	.021	0.826
ID8	−.151	.153	−.106	**.890**	.020	0.850
ID9	.105	−.360	.176	**.727**		0.746
ID11	−.122	.164	−.183	−.125	**−.846**	0.807
ID13	**.770**	−.061	.190	−.463	.179	0.879
ID15	.058	**.927**	.070	.034	−.177	0.900
ID18	**.818**	−.346	.179	.196	.098	0.869
ID24	−.222	**.620**	.436	.045	.406	0.790
ID26	**.791**	.174	−.349	−.087		0.799
ID28	−.505	**.608**	−.232	−.331	.218	0.835
ID30	.102	.106	−.132	.011	**.831**	0.730

Source: Research data. Elaborated by the authors.

Factor 1 concentrates on most of the analyzed variables. It comprises three variables – ID13, ID18, and ID26 – with a variance of 20.15%, all of which have positive factor scores. These variables are related to deforestation and preservation areas, agroforestry systems, and farms with production aimed at self-consumption, respectively. Despite the positive correlation between deforestation and preservation variables, such a scenario reflects that even poorer areas with larger forested regions are highly susceptible to native vegetation replacement.

The production aimed at self-consumption reflects small-scale farms that are unable to participate in local markets or global food chains. These farms collectively contribute to the process of replacing native vegetation cover. It is important to note that these municipalities also have large farms that harm the environment at large scale.

Public policies that prioritize and promote conservation and preservation of natural resources through agricultural practices and sustainable land use in less-developed areas may slow down deforestation. One such mechanism is ensuring the generation and commercialization of surpluses from family farmers through institutional purchases, such as those facilitated by NSFP (Triches & Grisa, 2015). Additionally, NSFP is strongly associated with TRES, which promotes farm production within a region by adopting conservationist agricultural techniques and practices (Camargo et al., 2013; Santana & Lima, 2018; Moreira et al., 2023).

Factor 2 comprises variables ID15, ID24, and ID28, with a variance of 16.36%. These variables represent farms whose two main economic activities are related to plant product extraction and buffalo breeding, and literate farmers, respectively. The economic activities in ID15 and ID24 highlight their importance in the poorer municipalities. These activities are correlated because buffalo breeding is typically practiced within the same areas where babassu palm is grown, with its nuts being harvested by local women (de Oliveira et al., 2022). According to Shiraishi Neto et al. (2021), since buffalo breeding was introduced into Maranhão, the activity has often been practiced extensively with free-ranging animals in natural floodplains. These authors highlight the negative environmental impacts of this activity in floodplains – characterized by inundated plains – as it is a sensitive ecosystem.

Factor 3, composed of variables ID4 and ID7, with a variance of 15.93%, relates to average farm areas and population density, respectively, with the first variable having a negative sign. This factor indicates that in the municipalities participating in HDI+, where population density is higher and shows economic dynamics (Raiher et al., 2020), the average farm area tends to be smaller. Therefore, in municipalities where larger areas are used for agricultural production, the population is smaller and, consequently, tends to have lower economic dynamics.

Factor 4, composed of variables ID8 and ID9 – cattle and goat herds per area, respectively – has a variance of 15.58%. This factor highlights the importance and prevalence of livestock farming in municipalities with the lowest socioeconomic indicators. In addition to large-scale livestock farming, small-scale livestock farming, such as goat farming, is observed, ensuring the productive activity of small-family farmers. Raising animals in municipalities with low social and economic development guarantees these farmers access to low-cost protein and serves as a source of economic resources in times of financial need (Costa et al., 2022).

Factor 5 consists of variables ID11 and ID30, with a variance of 14.59%. This factor shows the importance of an inverse correlation between NSFP values and Gini indexes. Therefore, in municipalities where the efficiency of institutional procurement funds, through NSFP, is higher, income concentration is lower. In municipalities with low human development, public policies related to institutional markets ensure food security and fight social inequality, which is reflected in the Gini index, strongly associated with poverty in Brazil (Barros et al., 2000).

Human capital, as captured by the variable ID28, represents the most fundamental form of education, namely literacy. The selection of this criterion was due to the extremely low level of educational attainment and its quality in rural areas of Brazil (Pereira & Castro, 2019). Improving education can enhance productivity levels in agricultural labor (Behrman & Birdsall, 1983) and is central to the social development (Sen, 1997) of rural communities (Bebbington, 1999). However, public policy on rural education targeting this population is marginally significant in Brazil (Maia & Buainain, 2015). In this context, TRES has the potential to build the capacity of farmers to adopt new technologies (Norton & Alwang, 2020), thereby increasing their skills and human capital.

Additionally, TRES has been crucial for the effectiveness of NSFP's institutional market because it promotes collective actions that help achieve the production scale required to ensure a regular supply (Santana & Lima, 2018; Camargo et al., 2013). Moreover, TRES is strongly linked to training farmers for sustainable land use by adopting conservationist agricultural practices and preserving natural resources (Rivera & Qamar, 2003; Sheriff et al., 2020). Thus, the institution guides rural producers toward a more sustainable, equitable, and economically viable agricultural scenario (Sahu et al., 2023).

TRES is a public good, and since it plays a role in spreading conservationist agricultural techniques for free in rural areas, it positively affects all assisted producers (Rivera & Qamar, 2003; Lambin et al., 2014; Moreira et al., 2023). However, in Brazil, even in São Paulo, the richest state in the country, TRES's structure is very deficient. For this reason, part of it is supplemented by nongovernmental and private sectors (Camargo et al., 2013); however, in the poorer states of the country, farmers rely on the public sector. In Maranhão, only 5.20% of farms are served by TRES. This coverage drops to 2.92% when considering the HDI+ farms. The absence of this service reduces access to NSFP (Santana & Lima, 2018).

Socioeconomic and Environmental Development Index

The Socioeconomic and Environmental Development Index (SEDI), calculated to capture the multidimensional development of HDI+ municipalities, allows for ranking them and analyzing the channels that impact them. SEDI average and standard deviation were 45.29 and 23.58, respectively, indicating a heterogeneous distribution (Table 14.4). Most of these municipalities fell into the Low development category. Only 20% of the municipalities achieved the best results in the Very High and High classes – 3.33% and 16.67%, respectively. Figure 14.1 shows the map of SEDI levels and Appendix 1 presents the results of the SEDI, Gross Index (GI), and the ranking and category for the analyzed municipalities in Maranhão state.

The municipality of Cajari topped the SEDI ranking and had the best performance in two factors, F2 and F3. It was also notable for having the highest percentages of buffalo herds and farms with literate producers (74.92%). As expected, human capital contributed to socioeconomic and environmental development. The municipality holds the second position in babassu extraction, surpassed only by Serrano do Maranhão, which has 52.20% and ranks second in the SEDI.

Although extractivism is highly conservationist, extractive products are not typically directed to school feeding programs within the region. This justifies Cajari's

Table 14.4 Socioeconomic and environmental development categories, SEDI intervals, and the total number of municipalities

Category	*Upper limit*	*Lower limit*	*Number of municipalities*	*Percentage of municipalities (%)*
VH	100.00	90.47	1	3.33
H	90.46	66.88	5	16.67
M	66.87	43.29	9	30.00
L	19.70	43.28	11	36.67
VL	0.00	19.69	4	13.33

Source: Research data. Elaborated by the authors.

high SEDI ranking despite not using NSFP funds. Primeira Cruz, which is also located in an extractivism territory (carnauba palm), characterized by fishing and livestock, has the largest goat herd per area (ID9) in the state and does not direct these products to school feeding. The same is true for Santana do Maranhão, which has the largest cattle herd per area (ID8) but whose product is not on the NSFP procurement list, only the processed product (mesocarp flour), but this technology is inaccessible to most producers. In this municipality, it becomes evident how economic factors can boost SEDI ranking without utilizing NSFP resources.

Factor 1-grouped environmental variables were the most expressive for Aldeias Altas, classified as High development. This municipality is located within the babassu-producing region and stands out for having farms that produce for self-consumption. However, it has the second-largest deforested area, with São João do Soter holding the first place, followed by Aldeias Altas, Fernando Falcão, Araioses, and Arame. All these municipalities are within the Cerrado biome, except Arame, which is in a Cerrado-Amazon ecotone (IMESC, 2015). Despite the predominance of extractivism in Aldeias Altas, agriculture and livestock farming prevail, as in the other HDI+ municipalities (IMESC, 2015). Once again, it is worth noting that the number of farmers served by TRES is negligible, although its potential to contribute to sustainable land use would be enormous for these municipalities.

TRES should support both family farmers in adopting conservation practices and the specific needs of quilombola and Indigenous communities. These groups hold distinct land-based worldviews that often diverge from the market-driven perspectives prevalent in society (Bonin, 2015). Notably, seven HDI+ municipalities encompass Indigenous territories, as follows:

- Arame: Timbira ethnicity;
- Centro Novo do Maranhão: Urubu Kabor and Guajajara ethnicities;
- Fernando Falcão: Canela ethnicity;
- Itaipava do Grajaú: Timbira and Guajajara ethnicities; and
- Governador Newton Bello, Jenipapo dos Vieiras, and São João do Carú, Guajajara ethnicity.

Meanwhile, the quilombola population is distributed across four municipalities: Aldeias Altas, Pedro do Rosário, Serrano do Maranhão, and Cajari (IMESC, 2015).

In 2017, only eight of the 30 HDI+ municipalities (26.67%) achieved the minimum requirement of 30% of PNEA funds for the purchase of foods from family farms (FNDE, 2017). Even those municipalities that managed to meet the NSFP minimum were classified as Medium, Low, or Very Low development. NSFP has the potential to expand its coverage and thus contribute to increasing the HDI and sustainability of these municipalities.

The municipalities of Belágua and Conceição do Lago Açu executed 29.64% and 29.93% of the required funds, respectively – close to the minimum requirement. However, eight municipalities did not allocate any resources to family farming, as expected (FNDE, 2017). Kroth et al. (2020) highlight that only 26.7% and

42.1% of Brazilian municipalities in 2011 and 2016, respectively, met the 30% minimum for family farming. This performance indicates that there are barriers to the effective implementation of NSFP. In addition to the absence of TRES, there is also a lack of motivated and trained managers to access NEDF resources for institutional purchases.

Aldeias Altas, São João do Soter, and Serrano do Maranhão were classified as High development and allocated up to 20% of NSFP funds. This suggests that institutional procurement programs can contribute to increasing SEDI in municipalities with low HDI in Maranhão. Moreover, they align with the HDI+ goal of ensuring that family farmers have access to markets and, consequently, to income and quality of life (SAF, 2015).

In Maranhão in 2017, 39.17% of municipalities met the minimum requirement of 30% NEDF for NSFP in purchasing products from family farms. Among the HDI+ municipalities, this percentage was only 26.7%. Therefore, there is still significant room for growth in this institutional market in the 30 least socioeconomically developed municipalities in Maranhão.

Final Remarks

This chapter analyzes Maranhão state municipalities from the "High HDI Plan" (HDI+), with a focus on the Brazilian School Feeding Program (NSFP). NSFP's institutional market supports sustainable land use and provides nutritious food to public schools. The 30 lowest-HDI municipalities showed a heterogeneous economic structure, especially in land use, influenced by biome-dictated natural cover preservation requirements (20% Cerrado, 80% Amazon).

Deforestation is a significant challenge in this region. Municipalities with extensive forests and self-consumption small-scale farmers without access to local markets or global food chains are associated with deforestation. Adopting conservationist practices can reduce deforestation for these producers.

Municipalities allocating more NEDF resources to NSFP for small producers have lower income concentration, highlighting NSFP's potential to improve quality of life. Larger farms correlate with smaller populations and less economic dynamism, emphasizing the importance of small-scale agriculture for wealth generation, income distribution, and environmental quality.

The SEDI shows that for municipalities of HDI+, only 20% achieved Very High or High development categories. Some HDI+ municipalities without NSFP access attained high sustainable development indices due to the prevalence of extractive and livestock activities, which are conservationist despite animal production. However, these products are not on the NSFP procurement list, only the processed product (mesocarp flour), but this technology is inaccessible to most producers. Small-scale farmers prevail in all municipalities, including those focused on extractivism. Therefore, NSFP can promote sustainable production, in addition to being an opportunity to improve the quality of life of the rural population.

We conclude that NSFP has the potential to directly promote the expansion of sustainable production in family farming and the production of organic and agroecological products with reduced environmental impact and indirectly the production of babassu by purchasing mesocarp flour, thereby altering land use. Potentially, it can also slow the rate of illegal deforestation, especially in municipalities located in the Amazon biome. Furthermore, we conclude that several factors contribute to NSFP budgets falling below the minimum 30% threshold or even reaching zero, with two factors particularly significant: a) the willingness of local governments to access available funds; and b) the extent to which TRES can encourage participation in the NSFP and sustainable land use through the implementation of conservation agriculture practices.

HDI+ municipalities have the potential to expand extractivism through the production of babassu and consolidate a bioeconomy; however, buffalo production poses a threat to its expansion. Moreover, strengthening small-scale organic and agroecological production could provide a more optimistic outlook for these municipalities. Institutional markets can contribute to a set of actions that promote more sustainable production and lead to changes in land use. However, future setbacks on environmental issues can be expected due to human actions and climate change. Nevertheless, a better scenario might occur if there is positive action from the public sector, especially through projects led by TRES.

Notes

1 Law n° 11,947 of June 16.

2 Quilombolas are individuals residing in rural territories established by slaves who escaped from captivity during and after the era of the slave system, occupying free and remote lands. These territories also include lands acquired through inheritance, donations, payment for services rendered to the state, and purchases. Such territories are referred to as Quilombos (Schmitt et al., 2002).

References

Amaral, A. B., & de Carvalho Guimarães, J. (2023). O panorama atual das conexões entre alimentação escolar, PNAE e agricultura familiar. *Contribuciones a Las Ciencias Sociales*, *16*(10), 24459–24485. https://doi.org/10.55905/revconv.16n.10-337

Assis, L. F. F. G., Ferreira, K. R., Vinhas, L., Maurano, L., Almeida, C., Carvalho, A., Rodrigues, J., Maciel, A., & Camargo, C. (2019). TerraBrasilis: A spatial data analytics infrastructure for large-scale thematic mapping. *ISPRS International Journal of Geo-Information*, *8*, 513. https://doi.org/10.3390/ijgi8110513

Barros, R. P. D., Henriques, R., & Mendonça, R. (2000). Desigualdade e pobreza no Brasil: retrato de uma estabilidade inaceitável. *Revista Brasileira de Ciências Sociais*, *15*, 123–142. https://doi.org/10.1590/S0102-69092000000100009

Bartlett, M. S. (1954). A note on the multiplying factors for various c^2 approximations. *Journal of the Royal Statistical Society*, *16*(2), 296–298.

Bebbington, A. (1999). Capitals and capabilities: A framework for analyzing peasant viability, rural livelihoods and poverty. *World Development*, *27*(12), 2021–2044. https://doi.org/10.1016/S0305-750X(99)00104-7

Behrman, J. R., & Birdsall, N. (1983). The quality of schooling: Quantity alone is misleading. *The American Economic Review*, *73*(5), 928–946. https://www.jstor.org/stable/1814663

Bonin, I. (2015). Cosmovisão indígena e modelo de desenvolvimento. *Jornal Porantim, Encarte Pedagógico*, *5*. https://cimi.org.br/cosmovisao-indigena-e-modelo-de-desenvolvimento/

Camargo, R. D., Baccarin, J. G., & Silva, D. B. P. D. (2013). O papel do Programa de Aquisição de Alimentos (PAA) e do Programa Nacional de Alimentação Escolar (PNAE) no fortalecimento da agricultura familiar e promoção da segurança alimentar. *Temas de Administração Pública*, *8*(2), 1–21.

Cecconello, E. R. F., Moro, L. D., Foguesatto, C. R., Breichtenbach, R., Neckel, A., Spanhol, C. P., Vieira-Filho, J. E. R., & Mores, G. D. V. (2023). Challenges and potentialities of sustainability in the institutional food market of family farming. *Sustainability*, *15*(22), 15796. https://doi.org/10.3390/su152215796

Costa, D. A., Souza, C. L., Sivieiro, A., Aragao, W. A., Oliveira, A. V. D., Beber, P. M., Silva, J. M. A., & Nascimento, J. S. C. (2022). A produção animal da agricultura familiar na reserva extrativista Cazumbá – Iracema, Acre, Brasil. *Revista Brasileira de Agropecuária Sustentável*, *12*(1), 98–106.

Cunha, W. A. D., Freitas, A. F. D., & Salgado, R. J. D. S. F. (2017). Efeitos dos programas governamentais de aquisição de alimentos para a agricultura familiar em Espera Feliz, MG. *Revista de Economia e Sociologia Rural*, *55*(3), 427–444. https://doi.org/10.1590/1234-56781806-94790550301

de Oliveira, P. J. A., Mukhi, U., Quental, C., & de Oliveira Cerqueira Fortes, P. J. (2022). Connecting businesses and biodiversity conservation through community organizing: The case of babassu breaker women in Brazil. *Business Strategy and the Environment*, *31*(5), 2618–2634. https://doi.org/10.1002/bse.3134

de Souza, S. R. G., Vale, D., do Nascimento, H. I. F., Nagy, J. C., da Silva Junior, A. H. M., Rolim, P. M., & Seabra, L. M. J. (2023). Food purchase from family farming in public institutions in the northeast of Brazil: A tool to reach sustainable development goals. *Sustainability*, *15*, 2220. https://doi.org/10.3390/su15032220

Fávero, L. P., & Belfiore, P. (2017). *Manual de análise de dados*. Elsevier.

Fundação Nacional de Desenvolvimento da Educação (FNDE). (2016). *Aquisição de produtos da agricultura familiar para a alimentação escolar*. https://www.fnde.gov.br/programas/pnae/pnae-area-gestores/pnae-manuais-cartilhas/item/8595-manual-de-aquisição-de-produtos-da-agricultura-familiar-para-a-alimentação-escolar

Fundação Nacional de Desenvolvimento da Educação (FNDE). (2017). *Dados da agricultura familiar: aquisições agricultura familiar 2017*. Retrieved May 25, 2021, from https://www.fnde.gov.br/index.php/programas/pnae/pnae-consultas/pnae-dados-da-agricultura-familiar

Hair Júnior, J. F., Black, W. C., Babin, B. J., Anderson, R. E., & Tatham, R. L. (2009). *Análise multivariada de dados* (6th ed.). Bookman.

Heiss, S. N., Sevoian, N. K., Conner, D. S., & Berlin, L. (2015). Farm to institution programs: Organizing practices that enable and constrain Vermont's alternative food supply chains. *Agriculture and Human Values*, *32*, 87–97. https://doi.org/10.1007/s10460-014-9527-1

Instituto Brasileiro de Geografia e Estatística (IBGE). (2021a). *Censo Agropecuário 2017: Resultados definitivos*. Sistema IBGE de Recuperação Automática (SIDRA). Retrieved June 30, 2021, from https://sidra.ibge.gov.br/pesquisa/censo-agropecuario/censo-agropecuario-2017Instituto Brasileiro de Geografia e Estatística (IBGE). (2021b). *Malha municipal*. Retrieved October 20, 2021, from https://www.ibge.gov.br/geociencias/organizacao-do-territorio/15774-malhas.html?=&t=o-que-e

Instituto Brasileiro de Geografia e Estatística (IBGE). (2023a). *Censo Demográfico de 2022*. Sistema IBGE de Recuperação Automática (SIDRA). Retrieved October 20, 2023, from https://sidra.ibge.gov.br/Tabela/4714

Instituto Brasileiro de Geografia e Estatística (IBGE) (2023b) Pesquisa Municipal de Pecuária 2017. https://sidra.ibge.gov.br/tabela/3939 Retrieved October 20, 2023.

Instituto Maranhense de Estudos Socioeconômicos e Cartográficos (IMESC). (2015). *Plano de ação mais IDH: diagnóstico preliminar*. Retrieved May 21, 2021, from http://imesc.ma.gov.br/atlas/Home/diagnostico

Kageyama, A. (2004). Desenvolvimento rural: Conceito e medida. *Cadernos de Ciência & Tecnologia, Brasília*, *21*(3), 379–408.

Kroth, D. C., Geremia, D. S., & Mussio, B. R. (2020). Programa Nacional de Alimentação Escolar: Uma política pública saudável. *Ciencia & Saude Coletiva*, *25*(10), 4065–4076.

Kroth, D. C., Zonin, V. J., Coletti, T., Simões, W., & Von Dentz, E. (2019). Análise da operacionalização da política de aquisição de alimentos da agricultura familiar por parte

dos municípios: o caso do PNAE em três microrregiões do Oeste Catarinense. *Redes. Revista do Desenvolvimento Regional*, *24*(1), 138–162. https://doi.org/10.17058/redes.v24i1.9697
Lambin, E. F., Meyfroidt, P., Rueda, X., Blackman, A., Börner, J., Cerutti, P. O., Dietsch, T., Jungmann, L., Lamarque, P., Lister, J., Walker, N. F., & Wunder, S. (2014). Effectiveness and synergies of policy instruments for land use governance in tropical regions. *Global Environmental Change*, *28*, 129–140. https://doi.org/10.1016/j.gloenvcha.2014.06.007
Lawless, G., Stevenson, G. W., Hendrickson, J., & Cropp, R. (1996). *The farmer-food buyer dialogue project: Toward a more regional food system*. University of Wisconsin-Madison Center for Cooperatives, Madison, WI.
Libermann, A. P., & Bertolini, G. R. F. (2015). Tendências de pesquisa em políticas públicas: uma avaliação do Programa Nacional de Alimentação Escolar-PNAE. *Ciência & Saúde Coletiva*, *20*, 3533–3546. https://doi.org/10.1590/1413-812320182311.28012016
Lopes, I. D., Basso, D., & Pauli, R. I. P. (2021). Limits and potentialities of the national school food program in the perception of regional socialactors. *Gestão & Regionalidade*, *37*(110). https//doi.org/10.13037/gr.vol37n110.6144
Maia, A. G., & Buainain, A. M. (2015). O novo mapa da população rural brasileira. *Confins. Revue franco-brésilienne de géographie*, (25), https://doi.org/10.4000/confins.10548
Melo, C. O., & Parré, J. L. (2007). Índice de desenvolvimento rural dos municípios paranaenses: determinantes e hierarquização. *Revista de Economia e Sociologia Rural, Rio de Janeiro*, *45*(2), 329–365.
Mingoti, S. A. (2005). *Análise de dados através de métodos de estatística multivariada: uma abordagem aplicada*. Editora UFMG.
Moreira, I. R. J., Freitas, A. F. D., Alves Júnior, A., de Freitas, A. F., Bernardo, J. S., & da Silva, S. M. (2023). Family farming cooperatives and associations and the institutional market created by the National School Feeding Program (PNAE) in Minas Gerais, Brazil. *Sustainability*, *15*, 5202. https://doi.org/10.3390/su15065202
National Research Council (NRC). (2010). *Toward sustainable agricultural systems in the 21st century*. National Academies Press.
Norton, G. W., & Alwang, J. (2020). Changes in agricultural extension and implications for farmer adoption of new practices. *Applied Economic Perspectives and Policy*, *42*(1), 8–20. https://doi.org/10.1002/aepp.13008
Oliveira, T. S., Filho Andrade, V. S., Rocha, V. M., & Menezes, J. S. (2017). Os impactos do desmatamento e queimadas de origem antrópica sobre o clima da Amazônia brasileira: um estudo de revisão. *Revista Geográfica Acadêmica*, *11*(2), 157–181.
Pereira, C. N., & Castro, C. N. D. (2019). Educação: contraste entre o meio urbano e o meio rural no Brasil. *Boletim Regional, Urbano e Ambiental*, *21*. https://repositorio.ipea.gov.br/bitstream/11058/9661/1/BRUA21_Ensaio5.pdf
Raiher, A. P., Gobbo, I. S. F., & Sene, T. S. (2020). Fundo de participação dos municípios e o desenvolvimento: uma análise dos municípios de pequeno porte do Brasil. *Emancipação*, *20*(especial), 1–15. https://doi.org/10.5212/Emancipacao.v.20.14361.004
Rigon, A. S., & Bezerra, I. (2014). Food security and nutrition, family farming and institutional purchases: Challenges and potentialities. *Demetra: Food, Nutrition & Health*, *9*(2), 435–444. http://dx.doi.org/10.12957/demetra.2014.11628
Rivera, W. M., & Qamar, M. K. (2003). *Agricultural extension, rural development and the food security challenge*. Food and Agriculture Organization of the United Nations.
Sahu, K. K., Bardhan, R., Chouhan, N. S., Dixit, D., Tripathi, S., Pandey, A., & Ahmed, R. (2023). A comprehensive review on role of agricultural extension services in the sustainable development of global agriculture. *International Journal of Environment and Climate Change*, *13*(10), 3514–3525. https://doi.org/10.9734/ijecc/2023/v13i103021
Sambuichi, R. H. R., Galindo, E. P., Oliveira, M. D., & Moura, A. D. (2014). Compras públicas sustentáveis e agricultura familiar: a experiência do Programa de Aquisição de Alimentos (PAA) e do Programa Nacional de Alimentação Escolar (PNAE). In R. H. R. O. Sambuichi, A. P. M. D. O. Silva, M. A. C. D. O. Oliveira, & M. Savian (Eds.), *Políticas agroambientais e sustentabilidade: desafios, oportunidades e lições aprendidas* (Vol. 1, pp. 75–104). IPEA.

Santana, L. S. L., & Lima, F. A. X. (2018). Os programas PAA e PNAE e seus efeitos sobre um grupo de mulheres do Assentamento Lagoa do Serrote II. *Revista De Extensão E Estudos Rurais*, *7*(1), 311–336. https://doi.org/10.36363/rever712018311-336

Schmitt, C. J. (2005). Aquisição de alimentos da agricultura familiar Integração entre política agrícola e segurança alimentar e nutricional. *Revista de política agrícola*, *14*(2), 78–88. https://seer.sede.embrapa.br/index.php/RPA/article/view/539

Secretaria de Educação do Estado do Maranhão (SEDUC). (2015). *Alimentação escolar*. Retrieved May 25, 2021, from https://www.educacao.ma.gov.br/alimentacao-escolar

Secretaria de Estado de Agricultura Familiar (SAF). (2015). *Governo organiza municípios do "Mais IDH" para comercialização na agricultura familiar*. Retrieved May 21, 2021, from http://saf.ma.gov.br/governo-organiza-municipios-do-mais-idh-para-comercializacao-na-agricultura-familiar/

Secretaria de Estado de Agricultura Familiar (SAF). (2020). *Sistema da agricultura familiar – informativo*. Retrieved September 28, 2021, from https://saf.ma.gov.br/files/2021/01/Informativo_SAF_2020.pdf

Secretaria de Estado de Agricultura Familiar (SAF). (2021a). *Mais IDH valorizando a unidade de produção familiar*. Retrieved May 21, 2021, from http://saf.ma.gov.br/download-saf-2020/

Secretaria de Estado de Agricultura Familiar (SAF). (2021b). *Mais IDH*. Retrieved September 2021, from https://saf.ma.gov.br/mais-idh/

Sen, A. (1997). Editorial: Human capital and human capability. *World Development*, *25*(12), 1959±1961.

Sheriff, S. Y., Sule, H., Abubakar, B. Z., Maikasuwa, M. A., & Agbomakha, F. I. (2020). Challenges militating against the effectiveness of knowledge management in sustainable land use and agricultural production among extension workers in Sokoto state, Nigeria. Agricultural Sciences, *11*, 912–920. https://doi.org/10.4236/as.2020.1110059

Shiraishi Neto, J., Lima, R. M., & Brito, C. S. (2021). Conflitos socioambientais em sítio Ramsar: Brutalidade e expulsões na região da Baixada Ocidental Maranhense. *Novos Cadernos NAEA*, *24*(2), 63–82.

Staduto, J. A. R., Orlandi, M., & Chioveto, A. T. (2018). Desenvolvimento rural do Estado do Mato Grosso por bioma: uma análise do "vazio." *Revista Política e Planejamento Regiona*, *5*(3), 260–283.

Starr, A., Card, A., Benepe, C., Auld, G., Lamm, D., Smith, K., & Wilken, K. (2003). Sustaining local agriculture: Barriers and opportunities to direct marketing between farms and restaurants in Colorado. *Agriculture and Human Values*, *20*, 301–321.

Stege, A. L., & Parré, J. L. (2013). Fatores que determinam o desenvolvimento rural nas microrregiões do Brasil. *Confins*, *19*.

Teixeira, K. D. C. (2024). *Programa Nacional de Alimentação Escolar (PNAE): o papel dos agentes implementadores e as dinâmicas locais na aquisição de alimentos orgânicos e agroecológicos* (Thesis of Master Science). UFSCar. https://repositorio.ufscar.br/handle/ufscar/20135

Triches, R. M., & Grisa, C. (2015). Entre mudanças e conservadorismos: uma análise dos programas de aquisição de alimentos (PAA e PNAE) a partir da retórica da intransigência. *Revista Nera*, v. esp, *26*, 10–27.

Vogt, R. A., & Kaiser, L. L. (2008). Still a time to act: A review of institutional marketing of regionally-grown food. *Agriculture and Human Values*, *25*, 241–255. https://doi.org/10.1007/s10460-007-9106-9

Appendix

Appendix 14.1 Socioeconomic and Environmental Development Index (SEDI), gross index (GI), and the ranking and category for the analyzed municipalities in Maranhão State, Brazil

Municipality	*SEDI*	*GI*	*Ranking*	*Category*
Cajari	100.00	1.03	1	VH
Serrano do Maranhão	87.60	0.79	2	H
Santana do Maranhão	83.30	0.71	3	H
São João do Soter	73.61	0.53	4	H
Primeira Cruz	73.42	0.53	5	H
Aldeias Altas	68.07	0.43	6	H
Araioses	66.16	0.39	7	M
Água Doce do Maranhão	56.19	0.20	8	M
Fernand o Falcão	56.05	0.20	9	M
Arame	52.17	0.13	10	M
São Francisco do Maranhão	50.73	0.10	11	M
Santo Amaro do Maranhão	50.46	0.10	12	M
Jenipapo dos Vieiras	48.35	0.06	13	M
Afonso Cunha	47.68	0.04	14	M
Milagres do Maranhão	45.88	0.01	15	M
Santa Filomena do Maranhão (MA	39.19	−0.11	16	L
Belágua	38.45	−0.13	17	L
Itaipava do Grajaú	37.66	−0.14	18	L
Satubinha	37.29	−0.15	19	L
Conceição do Lago-Açu	37.25	−0.15	20	L
São Roberto	35.93	−0.18	21	L
Amapá do Maranhão	32.39	−0.24	22	L
Pedro do Rosário	29.49	−0.30	23	L
São João do Carú	28.32	−0.32	24	L
São Raimundo do Doca Bezerra	27.16	−0.34	25	L
Lagoa Grande do Maranhão	25.08	−0.38	26	L
Centro Novo do Maranhão	12.66	−0.61	27	VL
Marajá do Sena	12.00	−0.62	28	VL
Brejo de Areia	6.21	−0.73	29	VL
Governador Newton Bello	0.00	−0.85	30	VL

Source: Research data. Elaborated by the authors.

15 Rapid increases in arable land values and long-term implications to the farm sector and rural economy of the United States

2000–2023

Gregory Veeck

Introduction

Increases in prices for commodity crops combined with several decades of low interest rates, bipartisan expansions/extensions of government subsidy programs, and the increasing commodification of arable land by investors within and beyond the farming sector have pushed arable land values in the United States to highs not seen for decades (Qiu, 2022). Average US cropland values increased significantly in 2023, rising to an average of $5,460 per acre (8.1% over 2022), representing a mean annual increase of $410.00 per acre, the third highest absolute year-on-year increase since the 1970s behind only 2022 and 2013 (Munch, 2023; NASS-USDA, 2023; Reyman, 2022). Tracking the rise in arable land prices, cash rents have also jumped, with 2022 to 2023 increases from 4.4% to 7.1% depending on land quality and access to irrigation (NASS-USDA, 2023). For farm families who own all or most of their land, the rapid rise in arable land value is generally good news, representing an increase in household wealth and credit access. However, for farm families who rent all or part of their land or for younger farmers without legacy land seeking to purchase land to start or expand operations, the recent rapid increases in arable land prices present a major structural problem that varies in intensity across all farming regions in the nation (Buys et al., 2023a; Carlisle et al., 2019; Halvorson, 2023; Katchova & Dinterman, 2018).

High sales prices and, as significantly, the rapid increase in these prices especially after 2019 constitute long-term threats for several reasons, most significantly because high prices pose a major entry barrier for younger farmers. As of 2017, the year of the most recently completed US Agricultural Census taken every five years, the average age of a "farm producer" (defined as the main decision-maker on each farm) was 57.5 years, up 1.2 years from 2012 (NASS-USDA, 2023). Exploding land costs represent investment barriers to entry by tech-savvy younger famers and those of other underrepresented groups (Carlisle et al., 2019). In a comprehensive study of relative technical efficiency based on total factor productivity (TFP), a standard assessment tool of economic efficiency using data derived from the 2012 US Agricultural Census, Fried and Tauer (2016) report farms operated by farmers aged 65 and above were found to be 13% less productive than farms

DOI: 10.4324/9781032672779-19

operated by producers aged 35–44 but even 10% less efficient than farms operated by the youngest cohort in the study (<25 years of age). Scholars widely acknowledge that high land prices and rents that limit entry by new farmers contribute directly to lower TFP, and the broader adoption of new and green technologies favored by younger farmers (Buys et al., 2023b; Carlisle et al., 2019; ERS-USDA, 2023a; Fried & Tauer, 2016).

With virtually all inputs (fertilizer, fuel, chemicals, seed, equipment) also increasing in price, fast-rising land values present a major hurdle for young farmers or farmers of any age renting all or most of their land. For those just starting out or reliant on the acres they rent to make ends meet, rising rents can present as an insurmountable barrier to entry (Fried & Tauer, 2016). Margins for many farmers, especially young and underrepresented farmers farming mostly on rented land, are immediately affected economically and psychologically by even minimal changes in land prices and rental agreements (Katchova & Dinterman, 2018). "Those who lack equity from landownership have reduced access to operating lines of credit needed to fund annual equipment and input purchases; and rising land prices and rents often take away the benefits of high crop prices" (Munch, 2023). A 2022 survey by the National Young Farmers Coalition indicated that finding affordable land for purchase or rent was considered by all respondents to be the top challenge they face [in 2022] (Qiu, 2022). Barriers to entry impacting young/new farmers have many implications for the adoption of technical innovations in farming such as "green" agriculture practices or "low input sustainable agriculture" (LISA), as well as potential impacts on the nation's long-term food supply and the continuation of the sector as a highly profitable source of exports, global influence, and downstream employment (Carlisle et al., 2019).

Experts recognize the many benefits of new younger farmers and farm laborers, but "younger farmers often do not survive as farmers" due to high production costs, debt, and a lack of access to loans due to a lack of collateral [land] (Fried & Tauer, 2016). Both the rapid exit of young farmers and the aging of those who remain increase the mean age of principal producers and combine to assure sector TFP will decline over time. The rate of that decline correlates to both time and location. To illustrate, in 1982, farmers below 45 years old owned 25% of farmland in the midwestern US state of Iowa, while those over 65 owned just 29%. In 2022, however, less than 5% of the state's cropland was owned by farmers under 45 years old while 66% of the state's land is now owned by farmers over 65 (Reyman, 2023).

Demographics are also linked to location and farming practices. New York county (urban Manhattan) in New York state specializes in organic, hydroponic, and "green" agricultural products for the restaurant trade and has the lowest mean age of farmers in the United States at 31.4 years, while a county in the nearby state of New Jersey within a historical truck-farming region records the oldest mean age for all US counties at 72.5 years old (Buys et al., 2023a). On average, farmers are older in southern states and younger in the Midwest and in peri-urban and suburban areas throughout the country (NASS-USDA, 2021). Nationwide, just over 40% of cropland and pasture is owned by farmers over 65 (United States Senate Special Committee on Aging, 2023; Zulauf, 2013). This "greying" of the farm sector coming in the last half-century mirrors trends found throughout in the EU, Japan, and other advanced nations (United States Senate Special Committee on Aging, 2023).

This chapter introduces recent changes in arable land costs and the implications of these changes on the future of the agricultural sector of the United States. The first section, excluding this introduction, will highlight current research related to why arable land prices in the United States have risen so quickly, especially after 2019. The second section presents information related to changes to the price of arable land for the period from 2000 to 2023 from a historical, geographical, or regional perspective. The third section briefly assesses the long-term implications of recent historically high prices for the American farming sector and the nation's rural economies.

A "perfect storm"

As in all nations, there are many interrelated macroeconomic factors influencing the rise in cropland and pasture prices in the United States, exclusive of issues including land quality, drainage, and the anticipated productivity of any given plot of land. Macroeconomic factors include historically high prices for major commodity crops, including corn, soybeans, cotton, and wheat; a robust residential housing market encroaching on proximate cropland; and diverse government subsidy programs extended and expanded during the COVID-19 pandemic if contingent on the crops grown. In addition, lobbyists concerned with farm labor shortages and the "greying" of the American farm workforce noted previously have pushed through legislation that quickly expanded the H-2A (farm worker visa) program, moderating the downward pressure that regional and seasonal labor shortages might have on cropland and pasture values. Collectively, these economic issues have converged to create what Jason Henderson, a prior dean of the College of Agriculture at Purdue University and former official at the Federal Reserve Bank of Kansas City, recently called a "perfect storm" stimulating fast-rising farmland values while raising concerns from government and academic researchers (Carlisle et al., 2019; Katchova & Dinterman, 2018; Qiu, 2022).

***Commodity price increases and cropland prices*:** Since 2000, demand for globally traded agricultural commodities has generally increased and consequently wholesale prices in real terms have also increased. China's entry into the WTO in 2001 helped move prices upwards in the early years of the millennium. Since, climate change causing both regional flooding and drought has shifted grain stocks downward in many parts of the world, especially the Middle East, Sahelian Africa, and South Asia. War in Ukraine, but as importantly political and social unrest in many additional parts of the world including Southeast Asia and East Africa, has further reduced gross production and increased dependence on global markets (Kroeger, 2023; Malakhail et al., 2023). Because of these shortfalls and numerous other regional issues that further restricted global grain supply chains during the COVID-19 era, both US domestic and international prices have surged (Figure 15.1). High prices, with short-term (3-year, 5-year) projections of further

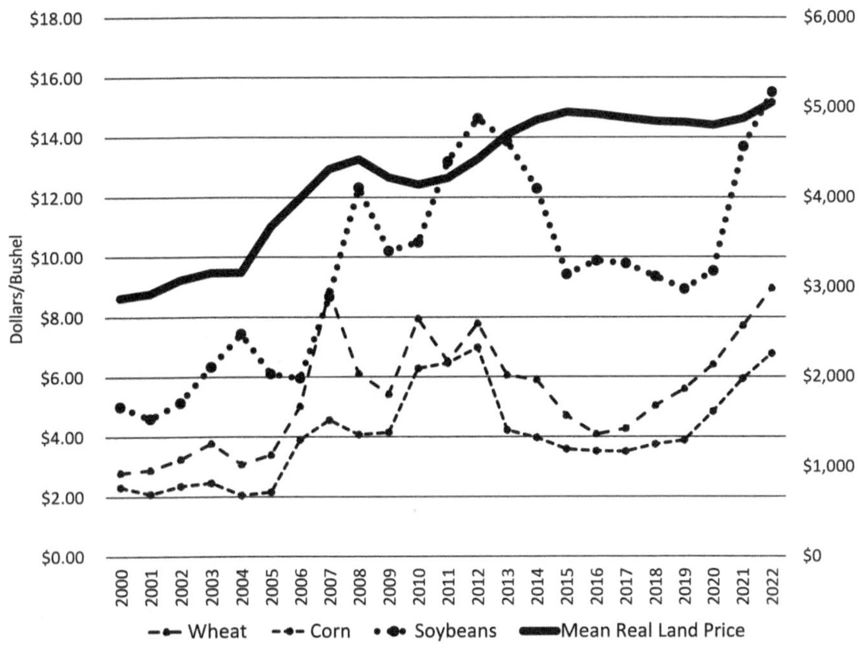

Figure 15.1 Mean annual prices for selected commodities ($/bushel) and mean value for cropland ($/acre) for USA: 2000–2022. Pearson product moment correlations for annual grain prices and annual sales price for cropland in real prices are 0.612 for corn, 0.592 for wheat, and 0.737 for soybeans (p<.05 for all).

Source: Compiled by author with data kindly provided by Macrotrends, https://www.macrotrends.net/2532/ (select commodity, wheat, corn, soybeans) and NASS-USDA, 2022.

strong demand, helped push cropland prices past previous highs recorded in the 2011–2012 agricultural year. Figure 15.1 also provides the mean value of cropland aggregated to the entire United States (right Y axis) to allow for a visual comparison of crop and land prices from 2000 to 2022. Pearson's product moment correlation analysis indicates strong positive correlations between mean annual US cropland prices and prices per bushel for major commodities such as wheat (r=0.592, p<.05), corn (r=0.612, p<.05), and soybeans (r=0.737, p<.05) for the period from 2000 to 2022. As noted previously, and to be shown directly, there are many factors that will impact sales prices for cropland and pasture. However, high commodity prices and projections for further increases post-2018 undoubtedly factor into the bidding process when cropland is bought and sold (Agamerica, 2023, p. 9).

***Growing demand for housing and cropland issues*:** As land and input prices rise, and the mean age of the farm owner/operators increases, many farm families make the often-difficult decision to sell their land. Much farmland is transferred from farm to farm, often increasing the scale of remaining farms depending on the crops grown and the location of the farm. However, given that the "next generation" of many farm families may decide not to farm or rent out their land for agriculture when zoning permits, a growing amount of farmland has been placed for sale for non-farm uses in recent years (Hunter et al., 2022; Qiu, 2022). The loss of farmland in the United States is really staggering. A study by the Farmland Information Center of the American Farmland Trust estimates that as much as 371 million acres of US farmland, approximately 41% of the national total, could change hands within the next 20 years (American Farmland Trust, Farmland Information Center 2014 updated to 2018, the United States Senate Special Committee on Aging, 2023). In a related study, Hunter et al. (2022) found that from 2001 to 2016, the nation lost or compromised 2,000 acres of farmland and ranch land every single day during that period.

The absolute number of farms is also declining. According to the 2017 Agricultural Census, there are currently 2.04 million farms and ranches in the United States, down 3.2% from just 2012 (LPGas, 2019). This 3% decline is typical for every five years of the last quarter century. Agamerica (2023) estimates that if US farmland is a $3.42 trillion market as research suggests, this means nearly $1.5 trillion in farmland could exchange ownership in the next two decades. Investment in arable land as an asset category in investment banking and retirement/pension plans has been steadily increasing.

> The prospect of land serving as a potential future development site is an attractive form of optionality to some investors. The National Association of Realtors markets farmland as a good inflation hedge with consistent returns and lower volatility than non-farm asset classes.
>
> (Lip, 2023)

In cases where cropland sales transfer arable land into other sectors, the most common uses are highly developed commercial/industrial expansions in peri-urban areas or low-density rural residential housing (Hunter et al., 2022). Hunter et al. (2022) cite very high urban housing prices and more opportunities for remote work as factors that will continue to reduce arable land stocks through rural housing expansions. As a consequence, more small farms (think potentially affordable entry farms) will be lost rather than large-acreage farms in more homogeneous farming areas with declining population (Hunter et al., 2022). Despite recent supply chain issues related to COVID-19, new construction of single homes has surged during the past decade. In 2011, 431,000 new single homes were built. This almost tripled to 1,127,000 by 2021, if down slightly to 1,008,000 in 2022 (Statista, 2023). The takeaway is that over the past quarter century, an ever-increasing amount of land for rural-sited single-family homes has been sourced from arable land stocks. With improved zoning and the greater application of comprehensive regional development planning, some of these losses may be moderated, but the current trend is readily apparent. Declining arable land stocks stimulate higher sales prices for remaining arable land and as a consequence result in further exclusion of younger farmers and other underrepresented groups (American Farmland Trust 2014 updated to 2018).

***Labor issues and cropland prices*:** While lagging behind those of a number of EU nations, American farms still rank among the most efficient when measured by relative TFP (ERS-USDA, 2023a). The labor demands of the US agricultural sector have long been maintained with mixture of two groups of workers: (1) self-employed farm operators and their family members and (2) hired workers (Moriarity, 2022). One aspect of US success in agriculture has been the ongoing substitution of technologies and innovations that steadily reduced demand for farm labor of both types since the 1940s (ERS-USDA, 2023a). Still, given the low and variable incomes associated with many farming activities, labor shortages continue to represent a major challenge to the US farm sector (Buys et al., 2023b). Family farms comprise 96% of all US farms, account for 87% of land in farms, and 82% of the value of all agricultural products sold, but as these owners age, more hired farm labor is required by the sector (ERS-USDA, 2023b; NASS-USDA, 2021).

An overlooked factor related to both labor and rising prices for cropland and pasture has been the rapid expansion of the H-2A Temporary Agricultural Workers Program, commonly called the H-2A visa program. This program, modified most recently in Fall of 2023, is designed to allow licensed/vetted farms participating in the program to hire temporary agricultural workers (up to 10 months/year) from other nations (ERS-USDA, 2023b). While not receiving a lot of international attention, this program has quietly become vital to the nation's farm economy. Crop farmers use this program to meet seasonal labor needs, but most livestock producers, such as ranches, dairies, and hog and poultry operations, are not *legally* allowed to use the program (Castillo et al., 2021). One exception is for rangeland pasture operations, as shepherds herding sheep and goats often in remote areas are allowed, reflecting the curious and complicated nature of the existing regulations of the H-2A program (ERS-USDA, 2023b).

The H-2A program specifically designed for the importation of foreign agricultural workers was started during the Reagan administration in 1987. The original legislation allowing foreign temporary workers to come and work in any sector began in 1952 as a portion of the Immigration and Nationality Act promulgated in that year. Only recently, however, has the program seen significant growth (Figure 15.2). Admitted H-2A visa recipients grew from 33,392 in 2000 to 298,000 in 2022, down from 317,000 in 2021, despite the border controls instituted during the COVID-19 pandemic (Martin, 2023). According to Philip Martin of the nonprofit Wilson Center (United States), approximately12,000 US farm employers typically hire H-2A workers each year. H-2A program participants are concentrated among relatively few large employers. The 100 largest H-2A employers account for 40% of all H-2A job certifications, and the 600 largest accounts for 70% (Martin, 2023). It is true that many H-2A program participants work on fruit and vegetable operations, but it is also clear that larger operations of all types tend to use the H-2A program most frequently and concertedly lobbied for its expansion in 2023 (Martin, 2023). Consider that just 105,453 farms (out of 2.04 million completing 2017 United States Agricultural Census) produce 75% of total off-farm sales, and 85,127 farms constitute 58% of all farmlands (NASS-USDA, 2022).

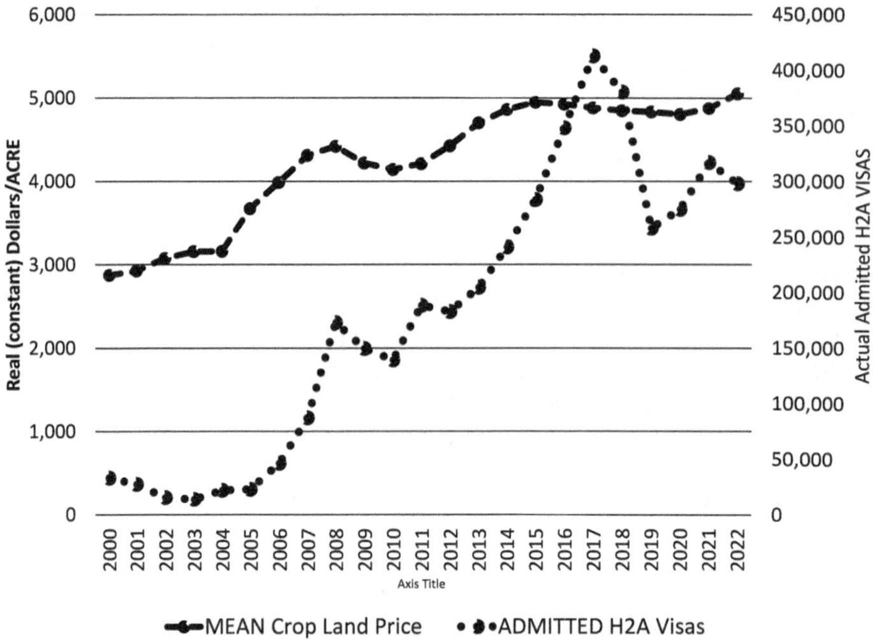

Figure 15.2 Increase in mean cropland price in real prices ($/ACRE) for the United States and the absolute number of persons admitted through the H-2A agricultural work visa program from 2000 to 2023. Pearsons Product Moment correlation for US mean cropland sales price and persons admitted through the H-2A visa program is 0.893, p=<.05).

Source: Calculated by author using data on arable land prices from the *Land Values* annual surveys provided by the National Agriculture Statistics Service (NASS) of the United States Department of Agriculture (USDA) and H-2A visa date compiled from USDA, https://www.ers.usda.gov/webdocs/publications/102015/eib–226.d

As domestic farm labor becomes increasingly difficult to secure, especially in major agricultural states such as California, Florida, Georgia, North Carolina, and Washington, support for the expanded H-2A program increased. Currently, 52% of all H-2A visas are awarded to these five states (Martin, 2023). In Fall of 2023, the program was significantly expanded reflecting ongoing labor shortages due to support from many large farming operations, especially those in California and the southeastern States. Pearson's product moment correlation analysis indicates that the number of admitted H-2A temporary farm worker visas is positively and strongly correlated with rising cropland prices (r=0.893, p<.05) (Figure 15.2). Local or regional shortages of labor clearly influence cropland prices (Agamerica, 2023). Labor shortages might potentially depress agricultural land prices, but given the 2023 expansion of the H-2A program and limited migration control, foreign labor, legal and illegal, generally meets demand for the sector, although regional shortages remain common and probably will for years to come.

A remarkable study by the Economic Research Service of the U.S. Department of Agriculture (2023b) found that approximately 70% of agricultural labor in the United States in 2020 was foreign born. Issues, economic and otherwise, associated with illegal farm labor are simply beyond the scope of this chapter, but foreign-born labor is clearly essential to the sector. The ERS-USDA (2023b) estimates that the percentage of field crop farmworkers not legally authorized to work increased from 14% in 1991 to 41% in 2020. A survey conducted that year found 30% of *hired* agricultural laborers were US born, 6% were immigrants who had obtained US citizenship, 23% were other authorized immigrants (primarily permanent residents, green-card holders, or H-2A program participants), and the remaining 41% had no legal work authorization; that is roughly half of hired crop farmworkers lack legal immigration status (ERS-USDA, 2023b). Recognition of such a high degree of dependency on undocumented labor for the agriculture sector should inform the current debate on effective reform of antiquated immigration law (ERS-USDA, 2023b; Latzke, 2023). Without doubt, localized labor shortages threaten the US agricultural sector, but it is possible that land prices remain higher in particular regions in part because of the rapid expansion of the H-2A program in these locations. Of course, while well beyond the scope of this chapter, it is clear that illegal labor is, in fact, essential for the agriculture sector of the United States at the present time.

In summary, there are many reasons that cropland and pasture prices have increased so dramatically in the last five to seven years. High wholesale commodity prices, consumption of arable land for nonfarming purposes, and programs and conditions that have allowed adequate supplies of labor are simply some of the more significant aspects. Others that could have been introduced, such as the expansion of agricultural subsidies, including ethanol credits, export credits, and COVID-19-era programs, have also played a role, but these are often commodity-specific, and regionally variable and therefore far more difficult to summarize and assess at the national level in a single chapter.

Temporal and spatial variability in US crop and pasture from 2000 to 2023

This section summarizes regional and national annual arable land sales data collected by the National Agricultural Statistics Service of the United States Department of Agriculture (NASS-USDA) as reported in the monthly *Land Values Summary* publication for the years from 2000 to 2023. Given the long time period, annual regional and national prices will be reported as real prices, adjusted annually for inflation using the Consumer Price Index (1982–1984 = 100) provided by the US Bureau of Labor (United States Bureau of Labor Statistics, 2023). For simplicity, given the international focus of this volume, state data on arable land prices are aggregated to the ten official USDA (United States Department of Agriculture) Farm Production Regions (NASS-USDA, 2022).

As might be expected, the most expensive agricultural land is found in densely populated coastal areas of regions such as the Pacific region, including the states of California, Oregon, and Washington ($8,800.00/acre; 2.47 acres=1 hectare). These states represent some of the most agriculturally diverse states in the country with extensive acreage in fruits, nuts, and vegetables as well as being major grain producers of wheat and rice (California). The American breadbasket, the Midwest region, follows close behind, at $8,540.00/acre. Both regions have extensive agricultural export supply chains which help mitigate domestic price fluctuations and quite probably fluctuations in cropland values as well. Also, higher than average prices are found within the crop regions adjacent to the Atlantic Ocean: the northeast ($7,380/acre), Appalachia ($5,190/acre), and the southeast ($5,060.00/acre). At the other end of the spectrum, low arable land prices are found in interior, drier regions with significantly lower population density, such as the Northern Plains ($4,200.00/acre) and Southern Plains ($2,480.00/acre) and the Delta region ($3,260.00/acre).

Turning to the rate of growth over time, that is the percentage increase in cropland value/acre from 2000 to 2023, for the ten USDA agricultural production regions, a significantly different pattern may be observed (Figure 15.3). While the Pacific region and those three regions bordering the Atlantic recorded the highest 2023 cropland sale prices, the 24-year percentage increase in cropland prices from 2000 to 2023 is lowest for all coastal regions and highest for the interior regions. In fact, some of the regions with the lowest land prices in 2023 still recorded the fastest percentage increases in value from 2000 to 2023, including the Northern Plains (72.5%), the Southern Plains (50.6%), and the Delta region (44.2%).

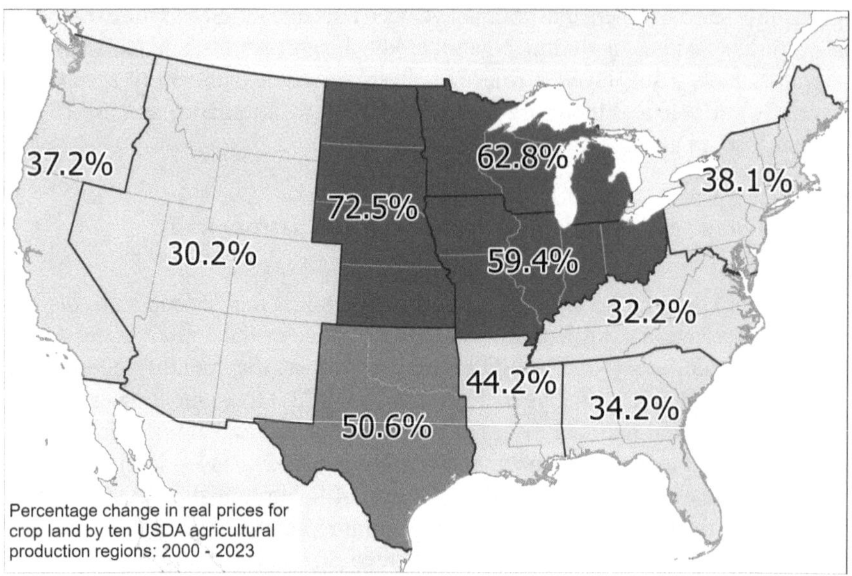

Figure 15.3 USDA farm production regions.

Source: USDA.

The coefficient of variation (CV) is a simple but effective statistical technique used to measure value-free spatial or temporal variability in a ratio-scale dataset through the creation of a ratio whereby the numerator is the standard deviation of the sample population, and the denominator is the sample mean ($CV=s/\bar{x}$). Sample means and standard deviations are absolute measures, as their values are contingent on the size and magnitude of the units of the data from which they are calculated. In contrast, the CV is a *relative* measure of dispersion, independent of the units of measure, and so ideal for analysis of spatial or temporal datasets (Lembo & McGrew, 2024).

Using the spatial-temporal cropland (USD/acre) dataset compiled for the 48 coterminous US states for the time period from 2000 to 2023, CV analyses were conducted to analyze both variations in cropland prices (1) over time for the period from 2000 to 2023 and also (2) geographically to assess the degree to which prices have varied for each particular state during the same time period. The national mean price per acre of cropland in the United States in real prices (1980–1982=100) increased from $2,875.73 in 2000 to $5,394.00 in 2023, or by 87.57% over the 24 years of the analysis. Prices for cropland in the early years of the study, that is the period from 2000 to 2014 fluctuated from state to state far more than for the latter period after 2014. To rephrase the findings, sales prices for cropland in the United States fluctuated far more across all transactions from place to place prior to 2014 than for the period from 2015 to 2023, with the greatest variability coming from sales from 2005 to 2007.

There is also geographic variability in US cropland prices over time. Results from state-scale analysis assessing temporal variability of cropland prices by state are presented as Table 15.1. These results are particularly interesting. While results for all 48 states are not provided, a selected number of states (n=12) reflecting the greatest and least variability as measured by CV values are provided. The greatest variability in cropland prices over the 24 years of the study are found for states within the Northern Plains, Midwest, and Lake States. That is, actual sales prices year to year varied more in states located within these regions – reflecting greater instability in sales from year to year and state to state in the regions. The Northern Plains region recorded the third lowest absolute sales price/acre for cropland in 2023. However, the four states incorporated in the region also reflected the greatest variability in cropland prices as measured by coefficient of variation analysis (Table 15.1) for the past 24 years. In practical terms, this makes valuation of assets and access to capital via loans based on such valuations much more fluid from year to year, potentially creating financial instability or uncertainty as crop prices, or land rents, fluctuate.

Variable prices for land year-to-year make banks nervous and invite speculation, while further restricting young and new farmers from traditional forms of credit, thus further limiting access to arable land. While unexpected, it is certain that states with the lowest 2023 mean sales price/acre, the Northern Plains, Southern Plains, and Delta Regions, all have CV values above the mean of 31.07 USD/acre. In short, those regions with the lowest mean sales prices in 2023, and typically most other years since 2000, consistently record the greatest variability in mean annual

Table 15.1 Coefficients of variation (CV) for reported cropland values calculated using years from 2000 to 2023 for selected US states

State marker	*RANKED coefficient of variation (CV) of cropland sale value (USD/acre) for selected states: 2000–2021*	*NASS-USDA survey estimates of mean cropland price by state (USD/acre) for 2023*
South Dakota	58.59	4,550
North Dakota	56.62	2,660
Nebraska	53.87	6,830
Kansas	51.02	3,440
Iowa	47.71	10,100
Minnesota	47.06	6,820
Georgia	24.39	4,200
Florida	23.61	8,760
Virginia	22.69	5,700
Tennessee	22.53	5,000
South Carolina	21.26	3,300
North Carolina	17.35	5,000

Source: Calculated by author using data on arable land prices from the *Land Values* annual surveys provided by the National Agriculture Statistics Service (NASS) of the United States Department of Agriculture (USDA).

sales price/acre. As noted earlier, there are exceptions whereby some states with the most productive and expensive farmland in the nation such as Iowa, Minnesota, Illinois, and Indiana also have high CV values, but generally regions with lower absolute values for cropland and pasture have the highest variation in annual sales prices, at least for the last 24 years.

Implications of rising agricultural land prices

For most nations represented in the chapters in this book, regional differences in arable land prices based on land quality, location, and competing uses are the norm. High agricultural land prices typically reflect higher prices in other land use/land cover categories as well as the changing conditions facing the regional rural economy. Rural places in many of these nations are equally under threat as well. The potential effects of high and unstable prices on the farm sector will be similar in many nations as well. However, the sheer size of the United States and the diversity of the nation's agricultural regions and environments mean that some aspects of cropland and pasture sales markets are unique. At the present time (2023), prices for an acre (0.404 hectare) of cropland in the United States can vary from place to place *by a factor of ten* based not only on land quality, access to irrigation, and location, but also on the potential demand for that land, both within agriculture or for non-agricultural uses, or as an asset class for portfolio investment.

In the United States, cropland and pasture prices per acre have risen steadily since 2000, spiking after 2019 to unprecedented levels. Most experts expect the trend to continue into the next decade. Cropland prices also vary significantly by region and state, with typically higher prices in those coastal regions with higher population densities and greater competition for land resources. Mean annual cropland sales prices in the Pacific, Midwest, and Northeast regions are typically two or sometimes even three times higher than those within the lowest values including the Southern Plains, the Northern Plains, and the Delta regions. Those regions with the highest arable land sales prices in 2000, the Northeast and the Pacific, continue to have the highest values in 2023, but year-on-year price increases over the same period for these regions were actually lower than for all other regions excepting Appalachia. That is, annual price fluctuations for land within these "expensive" crop regions are proportionally quite modest and more predictable vis-à-vis the Northern and Southern Plains or the Corn Belt, where land costs less but varies more from year to year.

The greater annual variability of cropland and pasture sales prices over the past quarter century as reflected in higher values for the coefficients of variation (CVs) in the 48-state analysis potentially reflects a more significant problem to US agriculture. Simply put, it is not just high sales prices for cropland in the United States that should necessarily be of greatest concern but rather high annual and regional variability in these prices. Greater variability in mean annual sales prices from year-to-year potentially creates greater destabilizing effects on the financial structures and institutions supporting the agriculture sector, directly impacting who is doing the farming and how farming is done.

For non-legacy farms and farmers renting land, annual rents typically track sales prices. As land prices increase, rents increase. When land is purchased, most farms use the land as collateral for loans for inputs and to cover operation costs. Variation in prices then impacts not only rents but also the absolute value and structure of loans.

The greater the uncertainty in land markets, potentially, the smaller the loan, the greater the collateral requirements for the same loan, and/or the higher the interest rate. High absolute prices represent barriers to young farmers. As demand for 5–10-acre rural residences increases in suburban and peri-urban areas, cropland prices will rise, further limiting who can afford to farm. There are many additional impacts on the farm sector as well. Studies have well documented the decline in TFP resulting from a "greying" farm sector. Fewer younger farmers mean slower adoption of new technologies, limited development of new products and new sales niches, less "green" farming, and fewer energy-saving cultivation practices such as minimum tillage and no-till or LISA. In short, without efforts to create a new generation of farmers, there will be less innovation of all types, and fewer acres devoted to organic agriculture, "green" farming, and renewable energy crops. These are all important adoptions that most experts feel represent the future of the most effective, efficient, and profitable agricultural sector.

But consequences extend beyond the sector. With farm consolidation, caused in part by the barriers to entry of new younger farmers due to high and unstable land prices, rural areas have steadily lost population and the amenities that could draw more people to rural places. Depopulation directly impacts the economies of rural areas in many ways, reducing tax bases, limiting consumers for small businesses, and while providing many challenges for effective support for local schools, transit, and social services as the number of working-age adults declines. Skyrocketing arable land prices especially after 2019 in the more agro-centric regions such as the Northern and Southern plains and the rural portions of the Midwest present new challenges to the farmers and local economies of these places, not to mention loss of rural culture and lifeways.

In conclusion, the nation's agricultural planners need to promote opportunities and conditions that attract and retain younger farmers to promote greater efficiency, adopt new technologies, and retain comparative advantage in internationally traded commodities, such as soybeans, corn, wheat, and cotton. Given current conditions, creative new programs, potentially paid for with reductions in legacy subsidy programs, are needed that will allow younger farmers to enter the sector and prosper. These might include new financial products and programs needed to help support entry and land purchase by younger farmers and those of other underrepresented groups but also new training programs about how to employ government programs to buy land, produce new products, and access new opportunities that will harness their enthusiasms as well. An infusion of young farmers can transform the sector and allow the United States to maintain its position among the leading agricultural nations, but such an influx will help revitalize small town and rural economies and cultures of America as well, potentially in the most cost-effective manner. Deciding how to address the effects of high land prices is a requisite first step for any of these efforts.

Acknowledgements

I wish to thank staff at the National Agricultural Statistics Services of the United States Department of Agriculture (NASS-USDA) for providing the data at both the state level and also aggregated to the ten agricultural regions. Jason Glatz of Waldo Library of Western Michigan University drafted maps presented in the chapter.

References

Agamerica. (2023). *The 2023 U.S. farmland report*. Retrieved December 13, 2023, from https://agamerica.com/farm-publications/us-farmland-report-2023/

American Farmland Trust, Farmland Information Center. (2014). *Tenure, ownership, and transition of agricultural land survey*. Last updated April 2018. Retrieved August 15, 2022 from https://farmlandinfo. org/wp-content/uploads/sites/2/2019/09/TOTAL_TalkingPoints_4-26-18.pdf

Buys, D., Green, J. J., & Robertson, M. N. (2023a, October). The average American farmer is so old they're practically a boomer. *FORTUNE*. Retrieved December 15, 2023, from https://fortune.com/2023/10/12/how-old-american-farmer-average-age-boomer-gen-x/

Buys, D. R., Robertson, M. N., & Green, J. J. (2023b). The aging agricultural workforce: Opportunities for policy development and intervention. *The Public Policy and Aging Report*, *33*(3), 105–110. https://doi.org/10.1093/ppar/prad015. Retrieved January 4, 2024.

Carlisle, L., De Wit, M. M., DeLonge, M. S., Calo, A., Getz, C., Ory, J., Munden-Dixon, K., Galt, R., Melone, B., Knox, R., Iles, A., & Press, D. (2019). Securing the future of US agriculture: The case for investing in new entry sustainable farmers. *Elementa: Science of the Anthropocene*, *7*(17), 20. https://doi.org/10.1525/elementa.356. Retrieved January 4, 2024.

Castillo, M., Simnitt, S., Astill, G., & Minor, T. (2021). *Examining the growth in seasonal agricultural H-2A labor* (EIB-226). U.S. Department of Agriculture, Economic Research Service.

Economic Research Service, United States Department of Agriculture (ERS-USDA). (2023a, September). *International agricultural productivity*. Online Report. Retrieved December 8, 2023, from https://www.ers.usda.gov/data-products/international-agricultural-productivity/

Economic Research Service, United States Department of Agriculture (ERS-USDA). (2023b, December). *Farm labor*. Online Report. Retrieved December 7, 2023, from https://www.ers.usda.gov/topics/farm-economy/farm-labor/#:~:text=In%202018%E2%80%9320%2C%2030%20percent,percent%20held%20no%20work%20authorization

Fried, H. O., & Tauer, L. W. (2016). Chapter 16: The aging U.S. farmer: Should we worry? In J. Aparicio, C. A. Knox-Lovell, & J. T. Pastor (Eds.), *Advances in efficiency and productivity* (Vol. 249, pp. 391–407). Springer International Publishing. Retrieved January 4, 2024, from https://doi.org/10.1007/978-3-319-48461-7_16

Halvorson, J. (2023). 2022 *Census of agriculture impacts the next generations of farmers*. USDA NASS Public Affairs Specialist. Retrieved January 5, 2024, from https://www.usda.gov/media/blog/2023/02/22/2022-census-agriculture-impacts-next-generations-farmers

Hunter, M., Sorensen, A., Nogeire-McRae, T., Beck, S., Shutts, S., & Murphy, R. (2022). *Farms under threat 2040: Choosing an abundant future*. American Farmland Trust. Retrieved December 3, 2023, from https://farmlandinfo.org/publications/farms-under-threat-2040/

Katchova, A. L., & Dinterman, R. (2018). Evaluating financial stress and performance of beginning farmers during the agricultural downturn. *Agricultural Finance Review*, *78*(4), 457–469. Retrieved January 5, 2024, from https://doi.org/10.1108/AFR-08-2017-0074

Kroeger, T. (2023, April). High grain prices rippled throughout the economy. *Beyond the Numbers: Prices & Spending*, *12*(7). U.S. Bureau of Labor Statistics. Retrieved December 15, 2023, from https://www.bls.gov/opub/btn/volume-12/high-grain-prices-rippled-throughout-the-economy.htm

Latzke, J. M. (2023). Ag employers offer their take on labor shortage. *Farm Progress*. Retrieved December 15, 2023, from https://www.farmprogress.com/farm-operations/ag-employers-offer-their-take-on-labor-shortage

Lembo, A. J., Jr., & McGrew, J. C., Jr. (2024). *Statistical problem solving in geography* (4th ed.). Waveland Press.

Lip, G. (2023). Farmland. *Corporate Finance Institute (CFI) Newsletter*. Retrieved January 3, 2024, from https://corporatefinanceinstitute.com/resources/commercial-real-estate/farmland/

LPGas. (2019, July). Farmers aging as fewer operations counted in latest US ag census. *Industry Update: Industry News and Numbers*, *LPGas Magazine.com*, pp. 8, 26.

Malakhail, F., Debnath, D., & Westhoff, P. (2023). Causes of food inflation in North America: COVID-19 and the Russia-Ukraine war. *Studies in Agricultural Economics*, *125*, 98–107. Retrieved December 7, 2023, from https://studies.hu/wp-content/uploads/2023/08/2574_Debnath.pdf

Martin, P. (2023, August 25). H-2A program expands in 2023. *Farm Labor and Rural Migration News*. Retrieved November 15, 2023, from https://www.wilsoncenter.org/article/h-2a-program-expands-2023

Moriarity, A. (2022, September 14). Immigrant farmworkers and America's food production: 5 Things to know. *FDC.us Blog*. Retrieved January 5, 2024, from https://www.fwd.us/news/immigrant-farmworkers-and-americas-food-production-5-things-to-know/#:~:text=1%7C%20Farmworkers%20 are%20essential% 20workers,workers%20in%20the%20United%20States

Munch, D. (2023). U.S. agricultural land values and cropland cash rents reach new highs. *Farm Bureau Market Intel*. Retrieved November 20, 2023, from https://www.fb.org/market-intel/u-s-agricultural-land-values-and-cropland-cash-rents-reach-new-highs#:~:text=This%20increase%20came%20in%20as, highest%20behind %202022%20and%202013

National Agricultural Statistics Service, United States Department of Agriculture (NASS-USDA). (2021). *Family-owned farms account for 96% of U.S. farms, according to the Census of Agriculture*. Typology report: Small family farms make up 88% of all U.S. farms. Retrieved December 12, 2023, from https://www.nass.usda.gov/Newsroom/archive/2021/01-22-2021.php

National Agricultural Statistics Service, United States Department of Agriculture (NASS-USDA). (2022, August). *Land values 2022 summary*. ISSN: 1949-1867. Retrieved January 4, 2021, from https://www.nass.usda.gov/Publications/Todays_Reports/reports/land0822.pdf

National Agricultural Statistics Service, United States Department of Agriculture (NASS-USDA). (2023). 2023 *Agricultural land: Land values and cash rents*. NASS Highlights, October No. 2023-6. Retrieved December 3, 2023, from https://www.nass.usda.gov/Publications/Highlights/2023/2023LandValues CashRents_FINAL.pdf

Qiu, L. (2022, November 14). Farmland values hit record highs, pricing out farmers. *New York Times*, Section A, p. 1. Retrieved November 29, 2022, from https://www.nytimes.com/2022/11/13/us/politics/farmland-values-prices.html#:~:text=A%20series%20of%20economic%20forces,at%20the% 20College%20of%2Agriculture

Reyman, D. (2022, Spring). Today's land market. *Stalcup AgService Today's Land Owner*, *XLII*(1). www.stalcup.com

Reyman, D. (2023, Fall). Iowa farmland ownership and tenure survey. *Stalcup AgService Today's Land Owner*, *XLVI*(3). www.stalcup.com

Statista. (2023). *Volume of new single family home construction starts in the United States from 1980 to 2022, with forecasts until 2024*. Retrieved December 4, 2023, from https://www.statista.com/statistics/184842/single-family-house-starts-in-the-united-states-since-2000/

United States Bureau of Labor Statistics. (2023). *CPI inflation calculator*. Retrieved January 3, 2023, from https://www.bls.gov/data/inflation_calculator.htm

United States Senate Special Committee on Aging. (2023). *Feeding the future* (Senate report). Ranking Member Mike Braun. Retrieved January 3, 2024, from https://www.aging.senate.gov/imo/media/doc/senate_aging_farmers_report.pdf

Zulauf, C. (2013, October 23). Putting the age of U.S. farmers in perspective. *Farmdoc-Daily*, (3), 202. Department of Agricultural and Consumer Economics, University of Illinois at Urbana-Champaign. Retrieved January 4, 2024, from https://farmdocdaily.illinois.edu/wp-content/uploads/2018/05/fdd231013.pdf

16 Urbanization and agricultural land use transformation in China

A perspective of off-farm Employment

Zhao Qianyu and Liu Hao

Introduction

With the development of urbanization in China, a significant shift of rural labor to off-farm employment sectors in urban areas has been observed. This transfer of rural labor to off-farm employment essentially represents a reallocation of production factors such as labor and capital.[1] Given that labor is a critical input in agricultural land use, the reallocation of labor raises important questions about its impact on agricultural land use and food security. Concerns have been expressed by many scholars regarding the extensive utilization or even abandonment of farmland due to rural labor loss.[2] However, statistical data suggest that these concerns might be somewhat overstated. Despite the transfer of rural labor, there has been a continuous increase in grain output and agricultural value in China. This raises the question: Is this phenomenon due to the optimization of the human–land relationship facilitated by the transfer of rural labor to off-farm employment in the urbanization process in China or is it because the inputs of machinery, fertilizers, and other factors have compensated for the reduction in labor inputs?

Agriculture is the foundation of the national economy in China. As the key input factor in agricultural land use, the rural labor transfer and its impact on land use have garnered significant attention from scholars. The out-migration of labor for off-farm employment and the inflow of remittances can alter the input or output combinations, thereby leading to changes in agricultural land use.[3] However, the exact nature of these changes remains inconclusive. Some studies argue that the large-scale transfer of rural labor to off-farm employment significantly affects the supply of agricultural labor, altering labor constraints.[4] In the context of China's small-scale farming operations and imperfect labor markets, the transfer of "non-zero-value labor" leads to aging and feminization of the agricultural labor force, negatively impacting agricultural land use.[5] As the opportunity cost of engaging in agricultural land use increases, farmers may shift to less labor-intensive crops, resulting in extensive land use or even land abandonment.[6] Other studies, however, argue that improvements in agricultural machinery services, professionalization of production management, and the existence of self-exploitation mechanisms among farmers nearly offset the negative impacts of rural labor loss.[7]

DOI: 10.4324/9781032672779-20

Additionally, the remittance effect of off-farm employment can alleviate financial constraints on agricultural land use,[8] allowing farmers to mitigate labor reduction constraints by purchasing outsourcing services and transferring land.[9] Therefore, the key to the direction of impact lies in whether the beneficial effects of rural labor transfer and farmers' self-adjustment mechanisms can offset the negative effects of labor loss. However, there is still no consensus on this issue.

These discussions on the impact of rural labor transfer to off-farm employment on agricultural land use mainly focus on land inputs and outputs. Theoretically, farmers can adjust agricultural land use scales, alter labor inputs per unit of land, increase inputs of substitute production factors, or shift to low-labor-input agricultural land use to adapt to changes in labor factors. This implies that the transfer of rural labor to off-farm employment not only affects the input–output relationship in agriculture but also leads to adjustments in the agricultural land use structure. Specifically, for grain production, which has stable income expectations, is conducive to mechanization, and is low in labor intensity, off-farm employment and grain production might complement each other, promoting the specialization of agriculture.[10] However, some studies argue that in the absence of improved labor productivity, the outflow of rural labor can reduce grain planting areas, lower the cropping index, and decrease the proportion of grain planting in the agricultural land use structure.[11]

The transfer of rural labor to off-farm employment triggers chain reactions, causing potential changes in agricultural land use efficiency and structure. Existing research provides significant insights for analyzing the transformation of agricultural land use induced by off-farm employment of rural labor. However, attention to this issue must be paid to China's unique agricultural land production relations. Most existing studies are based on cross-sectional data from questionnaire surveys, which makes it difficult to analyze the agricultural land use changes induced by off-farm employment over a long time series. Moreover, the impact of rural labor transfer to off-farm employment on agricultural land use may be influenced by whether the marginal productivity of the transferred labor is zero, indicating a potentially nonlinear relationship.[12] However, empirical analysis of this nonlinear relationship based on the Chinese context is still lacking. Additionally, while grain production is part of agricultural land use, the impact of labor transfer on grain production and other agricultural land uses may be different, and existing studies have not directly focused on this disparity or only analyzed it indirectly through planting structure adjustments. Therefore, it is necessary to further analyze the impact of rural labor transfer to off-farm employment on grain production and its role in transforming agricultural land use structure and efficiency.

The marginal contribution of this study lies in the innovative analysis of the nonlinear impact of rural labor transfer to off-farm employment on agricultural land use efficiency based on county-level data in China. It explores the changes in grain production, agricultural land use structure, and efficiency during China's rapid urbanization and industrialization process. The content is arranged as follows: the first section is the introduction, which analyzes the research background and significance, as well as the review of existing literature and the research gap;

the second section is theoretical analysis, constructing an impact framework of rural labor transfer to off-farm employment on agricultural land use transformation based on dual economy theory of urban and rural areas and proposing research hypotheses; the third section sets up the econometric regression model and provides descriptive analysis of variables; the fourth section presents and analyzes the model estimation results; and the fifth section is conclusions and discussions.

Theoretical Analysis and Research Hypotheses

Historical experience indicates that countries undergoing urbanization and industrialization accumulate a large surplus of rural labor, leading to the formation of a dual economic structure of urban and rural areas. In 1978, China reformed the rural land system by implementing the Household Contract Responsibility System (HCRS), which facilitated the combination of labor and production materials and provided an institutional foundation for the accumulation of surplus rural labor. By the late 1990s, the agricultural sector had accumulated a significant surplus of labor and began a large-scale transfer to urban and off-farm sectors, driven by the attraction of urban industrial sectors.[13] According to dual economy theory of urban and rural areas, developing countries have an excess population in the traditional agricultural sector, while land is limited, and there are often no breakthrough advancements in production technology within a certain period. As a result, after reaching a certain level of output, the yield of agricultural products can hardly increase substantially. Thus, the marginal productivity of each additional laborer in agriculture is nearly zero, sometimes even negative, and this surplus labor is referred to as "zero-value labor." The existence of "zero-value labor" causes income disparities between urban and rural areas, as well as differences in marginal productivity and wages between agricultural and industrial sectors.[14] Therefore, as long as the wage offered by the industrial sector is higher than that in the agricultural sector, there is an incentive for surplus agricultural labor to transfer to off-farm employment.

According to the research by Fei and Rains (1964), the transfer of labor to off-farm employment can be divided into three stages.[15] The first stage is the transfer of "zero-value labor" to off-farm employment, where the marginal productivity is zero, and thus this transfer does not affect the total agricultural output level. On the contrary, it optimizes the input–output relationship in agriculture due to the adjustment of the human–land relationship. The second stage is the transfer of "non-zero-value labor" to off-farm employment, where the marginal productivity is not zero. Under constant other conditions, this transfer negatively impacts agricultural output. Therefore, this stage must rely on improving agricultural labor productivity to compensate for the impact of the outflow of not entirely "surplus" farmers from the agricultural sector. The third stage is the post-Lewis turning point phase of mutual support between agricultural and industrial sectors. With a large transfer of non-surplus rural labor, agricultural product prices and rural labor costs rise significantly, reaching the Lewis turning point. At this point, labor allocation between agricultural and industrial sectors will be determined by competitive wage levels, with the agricultural sector continuing to supply surplus labor to the industrial sector and the industrial sector supporting agricultural development.

The relationship between rural labor transfer and agricultural land use in China is more complex than described by dual economy theory of urban and rural areas. Off-farm employment transfer directly affects agricultural land use by influencing changes in labor inputs. This change includes both the quantity and quality of labor inputs.[16] Additionally, the impact direction could be positive, negative, or neutral. The transfer of rural labor to off-farm employment in China has been accompanied by a significant increase in agricultural mechanization, partially substituting the labor loss.[17] Moreover, under the small-scale farming model based on the HCRS, the impact of this transfer on agricultural land use transformation requires further empirical analysis using Chinese data. However, the nonlinear relationship reflected in dual economy theory of urban and rural areas between rural labor transfer and agricultural output is crucial for constructing the regression analysis model in this study, providing significant insights and guidance.

It should be further noted that the adjustment of China's agricultural land use structure occurs under the constraints of the national food security strategy. Grain production not only has a different strategic position from other agricultural land use (which creates inertia in farmers' grain production decisions) but also differs in labor input demand. Grain production has advantages in mechanization difficulty and expected yield stability.[18] This characteristic of grain production means that with a low labor transfer rate, the proportion of grain production in crop planting increases. However, as the labor transfer increases, agricultural income is no longer the main source of household income, farmers' ability to bear agricultural risks improves, expected returns on agricultural land use investments increase, and changes in outsourcing of productive services and market demand for agricultural products lead to a decrease in the proportion of grain production in crop planting.

Based on this analysis, this study proposes the following three hypotheses for further testing:

H1: The impact of rural labor transfer to off-farm employment on grain production is nonlinear, and with an increase in labor transfer rate, it shifts from benefiting grain production to adversely affecting grain production.

H2: The impact of rural labor transfer to off-farm employment on grain production and other agricultural land use varies, prompting an adjustment in agricultural land use structure toward non-grain production.

H3: Rural labor transfer to off-farm employment changes farmers' labor and capital constraints, affecting the input–output relationship (agricultural land use efficiency) in agriculture. This impact is nonlinear, and with an increase in rural labor transfer rate, it shifts from benefiting agricultural land use efficiency to adversely affecting it.

Model Specification and Variables and Data

Model Specification

This study utilizes panel data from 1961 counties in China for the years 2000–2014, which helps address issues related to omitted variables caused by unobserved individual differences that do not change over time. Given the large cross-sectional capacity, substantial individual heterogeneity, and potential autocorrelation of individual disturbances, this study incorporates different intercepts in the regression equations for each county to capture this heterogeneity. This forms the basis of the model for analyzing the impact of off-farm employment on agricultural land use transformation, which is specified as follows:

$$Y_{it} = a + bL_{it} + X_{it}g + m_i + e_{it} \tag{16.1}$$

where Y_{it} represents agricultural land use conditions in county i in year t, including grain yield per unit of land, agricultural yield per unit of land, and agricultural land use efficiency.[19] This study uses grain yield per unit of land and agricultural yield per unit of land as dependent variables to account for both the impact of labor loss on agricultural land use area and output. Due to the diversity of agricultural products, it is challenging to find a unified output indicator when assessing overall agricultural land use, so output value is used. Additionally, to analyze the comprehensive impact of off-farm employment on the input–output relationship in agriculture, this study employs the DEA (Data Envelopment Analysis) Hybrid model to measure agricultural land use efficiency.[20] L_{it} is the key explanatory variable, representing the rural labor off-farm employment transfer rate, which reflects the outflow of rural labor. X_{it} denotes a set of control variables affecting the agricultural input–output relationship (including economic development level, importance of agriculture, agricultural land use capital constraints, land management scale, and rural labor quality), reflecting other explanatory variables influencing agricultural land use factors in county i in year t. μ_i is the unobserved individual-specific random effect, representing individual heterogeneity, and ε_{it} is the random disturbance term. $\mu_i + \varepsilon_{it}$ is the composite disturbance term, while a 、 β 、 γ are the parameters to be estimated.

Based on the previous analysis, the impact of rural labor off-farm employment transfer on agricultural land use is constrained by whether the rural labor is surplus, resulting in a nonlinear relationship in its impact characteristics. Therefore, this study includes a quadratic term for the key explanatory variable to determine the relationship characteristics of off-farm employment impact on agricultural land use. The adjusted model is specified as follows:

$$Y_{it} = a + bL_{it} + b'L_{it}^2 + X_{it}g + m_i + e_{it} \tag{16.2}$$

Since one of the dependent variables, agricultural land use efficiency, is bounded within [0,1], but the data is continuous without truncation, regression analysis based on Equation 16.2 will still be conducted. To ensure the reliability of the regression results, this study calculates agricultural land use efficiency under variable and constant returns to scale for robustness analysis of the agricultural land use efficiency regression model.[21] The model for calculating agricultural land use efficiency is shown in Equation 16.3:[22]

$$\left\{\begin{array}{l} \min y = 1 - \frac{1}{w}(1 - eff) - \frac{1}{w}\sum_{i=1}^{w-1} S_i^{N-} / I_{ik}^{N} \\ s.t. I^{R}\lambda + S^{R-} - effI_k^{R} = 0 \\ \quad I^{N}\lambda + S^{N-} = I_k^{N} \\ \quad O\lambda \geq o_k \\ \quad \lambda, S^{-} \geq 0 \end{array}\right\} \tag{16.3}$$

where I represents input indicators, O represents output indicators, i denotes decision units (in this study, county-level administrative units). S^{-} represents the slack variable, indicating the redundancy rate of inputs. y is the maximum eigenvalue of the projection value correlation matrix of non-radial input indicators, eff represents agricultural land use efficiency, w is the number of input indicators, R denotes radial indicators, and N denotes non-radial indicators. This study selects the total power of agricultural machinery, agricultural fertilizer usage, number of agricultural workers, and farmland area as indicators for technological, capital, labor, and land inputs, respectively. Given the substitution relationship between capital and technology with labor, the first three indicators are set as non-radial indicators.

Data Sources and Variable Descriptions

County-level economies are fundamental units of China's national economy. Major issues in the current economy are concentrated at the county level. Research focused on county-level administrative units is therefore highly relevant for decision-making. According to data published by the National Bureau of Statistics, there were 2,854 county-level administrative units in China up to 2014.[23] Given the large cross-sectional span and extended time series involved in this study, administrative divisions adjustments and the special circumstances of some county-level administrative units in farmland use, such as urban areas with minimal or no farmland, have been considered, leading to the exclusion of such cases. Ultimately, this study selects 1,961 county-level administrative units. Data sources for this study include the *China County Statistical Yearbook* (2001–2015), provincial and municipal statistical yearbooks (2001–2015), and county statistical bulletins (2001–2015).

For the key explanatory variable, the rural labor transfer rate (*pRural*), this study adopts the method from Liu Xiaoguang (2015),[24] Xu Jianguo and Zhang Xun (2016),[25] calculated as (number of rural workers - number of agricultural workers)/number of rural workers. The number of agricultural workers is the labor force engaged in agricultural land use, reflecting the actual number of rural labor engaged in agriculture. Other control variables include economic development level (per capita GDP at the county level, *gdp_ca*); importance of agriculture (represented by two indicators: secondary industry added value/primary industry added value, *gdpratio_se*, and tertiary industry added value/primary industry added value, *gdpratio_te*); agricultural land use capital (per capita net income of farmers, *rIncome_ca*); land management scale (average cultivated land area per household, *aLand_hou*); and rural labor quality (proportion of ordinary middle school students in total population, *edu_ratio*).[26]

The dependent variables in this study are grain yield per unit of land (*grain_pro*), agricultural yield per unit of land (*agri_pro*), and agricultural land use efficiency (variable returns to scale efficiency, *eff_vrs*, and constant returns to scale efficiency, *eff_crs*). Descriptive statistics for these variables are shown in Table 16.1.

Model Estimation Results and Analysis

Given the presence of between-group heteroscedasticity, within-group autocorrelation, and cross-sectional dependence in the panel data, and with N>>T, this

Table 16.1 Descriptive statistics of variables

Variable		*Mean*	*Std. dev.*	*Unit*	*Time period*	*Obs*
Independent Variables, IV	*pRural*	0.352	0.172	%	2000~2014	29,415
	gdp_ca	1.315	1.625	10,000 RMB per capita	2000~2014	29,415
	gdpratio_se	3.253	7.931	–	2000~2014	29,415
	gdpratio_te	2.579	27.765	–	2000~2014	29,415
	rIncome_ca	4655	3259	RMB per capita	2000~2014	29,415
	aLand_hou	1.197	8.792	Mu per household	2000~2014	29,415
	edu_ratio	0.057	0.017	%	2000~2014	29,415
Dependent Variables, DV	*grain_pro*	0.284	0.175	Ton per mu	2000~2014	29,415
	agripro_pro	0.128	0.230	10,000 RMB per mu	2000~2014	29,415
	eff_vrs	0.235	0.124	–	2000~2014	29,415
	eff_crs	0.185	0.116	–	2000~2014	29,415

Note: Mu is a unit of area in China, 1 mu ≈ 0.0667 hectares, All value variables have been deflated using the consumer price index of 2000 as the base year. The variable rIncome_ca is transformed into its natural logarithm (lnrIncome_ca) before entering the model.

study employs the Driscoll and Kraay (1998) standard error method for estimation.[27] This method extends the White (1980) and Newey and West (1987) estimators to panel data, addressing heteroscedasticity, serial correlation, and cross-sectional dependence simultaneously. By using a mixed OLS/WLS or fixed-effects model to obtain robust standard errors, this method is referred to as Driscoll and Kraay Standard Errors (DK-SE).[28] This study further uses the fixed-effects model for estimation.

Impact of Off-Farm Employment Transfer on Grain Production

The impact of rural labor off-farm employment on grain production shows a robust "inverted U-shaped" relationship as shown in Table 16.2. This relationship can be explained from the following two aspects: on the one hand, regarding the demand for grain production, during the early stages of industrialization and urbanization, the reallocation of rural labor between urban and rural areas and between industries resulted in a significant increase in urban demand for grain due to population influx, which increased the demand to enhance grain production. On the other hand, regarding the supply of grain production, as some family members shift to off-farm employment, under the historical tradition of valuing grain production, households will adjust the distribution of family labor within agricultural land use. Therefore, in terms of grain production alone, the initial transfer of rural labor actually promoted an increase in grain yield per unit of land. However, this situation changed with the increase in the labor transfer rate. After rural labor transfer reaches a certain stage, from the perspective of demand for grain production, economic development and reform have increased grain trade, diversified dietary structures, and changed the consumption structure of agricultural products, reducing social demand for grain. From the supply perspective, as the cost of rural labor increases, the economic status of grain production in farm households declines. Even if domestic grain production is insufficient, farm households have sufficient funds to purchase grain, and the importance of relying on grain sales as a family economic source also decreases. Farmers lack the motivation to increase grain production, and the transfer of more family labor leads to insufficient labor, which negatively impacts grain yield per unit of land.

Based on the regression results and analysis, this study accepts hypothesis H1, which posits that the impact of off-farm employment transfer on grain production is nonlinear and changes from beneficial to detrimental to grain production as the rate of rural labor transfer increases. This finding has significant policy implications for China's current food security strategy. With the continued advancement of urbanization, industrialization, and modernization in China, the large-scale transfer of rural labor to off-farm employment will persist for a considerable period. Therefore, while promoting agricultural development and adjustments, it is crucial to unwaveringly continue the implementation of the food security strategy.

Table 16.2 Regression results of factors affecting grain yield per unit of land

FE(DK-SE)					
DV	*grain_pro*				
IV	(1)	(2)	(3)	(4)	(5)
pRural	0.2562***	0.2693***	0.164***	0.1627***	0.1578***
	(9.01)	(10.63)	(10.79)	(10.93)	(9.18)
pRural_sq		−0.1374*	−0.2192***	−0.215***	−0.2642***
		(−1.84)	(−3.14)	(−3.12)	(−3.67)
gdp_ca			0.0223***	0.025***	0.0206***
			(9.14)	(9.02)	(8.18)
gdp_casq			−0.0009***	−0.0008***	−0.0007***
			(−6.30)	(−6.05)	(−5.16)
gdpratio_se				−0.0016***	−0.0013***
				(−6.40)	(−5.25)
gdpratio_te				0.0001*	0.0001*
				(1.83)	(1.77)
aLand_hou				0.0001	
				(1.42)	
edu_ratio					−0.0001***
					(−6.31)
Constant	0.1940***	0.1918***	0.2058***	0.2077***	0.2569***
	(12.82)	(13.26)	(18.00)	(18.09)	(25.84)
Obs.	29,415	29,415	29,415	29,415	29,415
F	81.24***	64.50***	137.55***	123.00***	108.39***
Within−R2	0.1103	0.1125	0.1591	0.1630	0.1809

Note: 1) The numbers below the variable coefficients in the DK-SE model are t-statistics; 2) *, **, and *** denote significance levels of 0.1, 0.05, and 0.01, respectively; 3) The dependent variable lnagri_income interfered with the significance of other variables and thus was excluded from the final model results; 4) X_sq denotes the squared term of the variable, which was mean-centered when included in the model.

Analysis of the Impact of Off-Farm Employment Transfer on Agricultural Output

The impact of rural labor off-farm employment on agricultural output shows a robust "U-shaped trend" relationship as shown in Table 16.3. This relationship is distinct from the effect of off-farm employment on grain production. In the early stages of industrialization and urbanization, the transfer of family labor to off-farm employment led to adjustments in the allocation of family labor within agricultural land use. Consequently, in terms of grain production, there was no decrease in productivity due to labor transfer; instead, increased focus on grain production resulted in higher grain yield per unit of land. At the same time, due to the weakening of labor input into other agricultural land use by rural households, the increase in per capita agricultural land use value in the early stage of urbanization showed a decreasing trend. However, as the urbanization process advances and various socioeconomic changes occur, such as diversified dietary patterns and the development of agricultural product trade, the economic significance of grain production within households diminishes. This results in a more diversified agricultural land use structure, which benefits the increase in agricultural output value. Thus, while off-farm employment negatively affects grain yield per unit of land, it positively impacts the incremental agricultural yield per unit of land.

Implications of the differences in the impact of off-farm employment on grain production and agricultural output

Based on the analysis of the impact of off-farm employment on grain yield per unit of land and incremental agricultural yield per unit of land, we observe different trends in agricultural output and grain production as the rate of off-farm employment increases. This divergence prompts us to consider the following question: What causes the different characteristics of grain output and agricultural output with the increase in labor transfer? We believe the differences are due to the varying labor demands between grain production and other agricultural activities. This discrepancy has led to significant changes in China's agricultural land use structure as rural labor is transferred.

This inference can be further validated in Figure 16.1. In Figure 16.1, *grain_ratio* represents the proportion of grain planting area to total crop sowing area, and *nagripro_ratio* represents the proportion of narrow agriculture output value, that is the proportion of crop production value to total agricultural output value. It is evident that before 2004, both the proportion of grain planting area and the proportion of narrow agriculture output value showed an upward trend, indicating a tendency toward grain cultivation in agricultural land use during this period. Subsequently, both proportions exhibited a downward trend, and the timing closely aligns with the "Lewis turning point" inferred in numerous studies, further confirming the direction of China's agricultural land use structure adjustment away from grain.

Table 16.3 Model regression results of per-acre agricultural gross output

FE(DK-SE)						
DV	Δ*agri_pro*					
IV	(6)	(7)	(8)	(9)	(10)	(11)
pRural	0.0494***	0.0434***	0.0204***	0.0198***	0.0122***	0.0124***
	(4.44)	(4.28)	(4.08)	(4.10)	(3.45)	(3.31)
pRural_sq		0.0573***	0.0379**	0.0387**	0.0466**	0.0411**
		(3.10)	(2.75)	(2.74)	(2.68)	(2.57)
gdp_ca			0.005***	0.0061***	0.0046***	0.0044***
			(3.40)	(3.31)	(3.33)	(3.37)
gdp_casq			−0.0002**	−0.0002**	−0.0001**	−0.0001**
			(−2.93)	(−2.84)	(−2.79)	(−2.82)
gdpratio_se				−0.0007***	−0.0006***	−0.0006***
				(−4.33)	(−4.32)	(−4.31)
gdpratio_te				0.0005***	0.0005***	0.0001***
				(3.22)	(3.27)	(3.35)
lnrIncome_ca					0.0044**	0.0037**
					(2.65)	(2.21)
aLand_hou				0.0001	0.0001	0.0001
				(1.41)	(1.49)	(1.54)
edu_ratio						−0.0711**
						(−2.74)
Constant	−0.0055	−0.0051	−0.0026	−0.0019	−0.4323**	−0.0233
	(−1.27)	(−1.12)	(−0.74)	(−0.53)	(−2.26)	(−1.63)
Obs.	27,454	27,454	27,454	27,454	27,454	27,454
F	19.71***	9.37***	15.24***	54.49***	82.62***	75.54***
Within–R2	0.024	0.0262	0.041	0.0449	0.0464	0.0474

Note: 1) The numbers below the variable coefficients in the DK-SE model are t-statistics; 2) *, **, and *** denote significance at the 0.1, 0.05, and 0.01 levels, respectively; 3) *agri_pro* is AR(1) and is entered into the model in first-difference form.

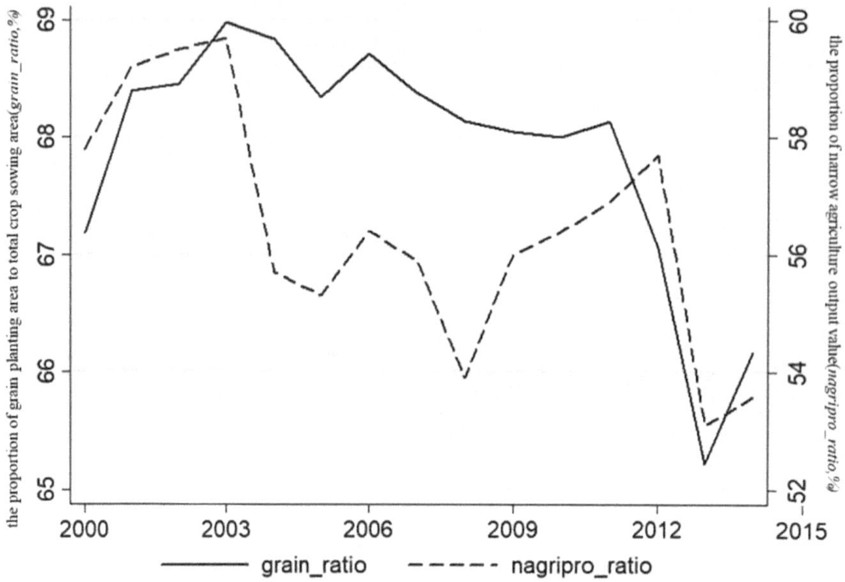

Figure 16.1 Proportion of grain sowing area and narrow agriculture output value.

Data source: China Economic and Social Development Statistical Database.

Based on the earlier analysis, this study accepts Hypothesis H2. The differences in labor demand between grain production and other agricultural activities have led to a subtle transformation in agricultural land use during the process of rural labor transfer to off-farm employment. This transformation resembles a "silent revolution," guiding the shift in agricultural land use structure away from grain production. However, this transformation also reflects that in the process of further deepening urbanization, industrialization, and modernization in China, appropriate measures should be taken to coordinate the relationship between food security strategy, diversified agricultural production needs, and agricultural production structure adjustment.

Analysis of the Impact of Off-Farm Employment on the Agricultural Input–Output Relationship

The impact of off-farm employment on the agricultural input–output relationship (agricultural land use efficiency) shows a "U-shaped trend" characteristic, as indicated in Table 16.4. This characteristic is similar to the impact of off-farm employment on agricultural output but differs from the "inverted U-shaped" relationship between labor transfer and agricultural land use efficiency anticipated by dual economy theories of urban and rural areas. The cause of this difference, we believe, is related to the dual economy theory's focus on the impact of labor transfer on output and the assumption of no capital and technology substitution effect for labor.

Table 16.4 Factors affecting agricultural land use efficiency – model regression results

FE(DK-SE)						
DV	*eff_vrs*					
IV	(12)	(13)	(14)	(15)	(16)	(17)
pRural	0.1274*** (5.67)	0.0939*** (5.76)	0.0446** (2.60)	0.0429** (2.58)	0.0557*** (4.09)	0.0605*** (4.94)
pRural_sq		0.3514*** (11.70)	0.3102*** (12.03)	0.3084*** (11.81)	0.2943*** (10.04)	0.2376*** (6.61)
gdp_ca			0.0103*** (9.27)	0.0144*** (13.30)	0.0168*** (7.56)	0.0147*** (6.51)
gdp_casq			−0.0003*** (−6.73)	−0.0003*** (−9.14)	−0.0004*** (−7.10)	−0.0003*** (−5.24)
gdpratio_se				−0.0023*** (−6.81)	−0.0024*** (−6.49)	−0.0023*** (−5.80)
gdpratio_te				0.0002*** (4.64)	0.0002*** (4.58)	0.0002*** (4.40)
lnrIncome_ca					0.0072 (1.29)	0.0128* (2.01)
aLand_hou				0.0007** (2.31)	0.0007** (2.19)	0.0006** (2.18)
edu_ratio						−0.0001*** (−5.33)
Constant	0.1904*** (21.14)	0.1918*** (25.41)	0.1977*** (25.14)	0.1992*** (27.42)	0.2506*** (5.66)	0.3359*** (6.49)
Obs.	29,415	29,415	29,415	29,415	29,415	29,415
F	32.14***	73.94***	45.86***	84.28***	157.32***	188.56***
Within–R2	0.0359	0.055	0.0692	0.0802	0.0811	0.101

Note: 1) The numbers below the variable coefficients in the DK-SE model represent t-statistics, while the numbers below the variable coefficients in the Tobit model represent z-statistics; 2)*,**, and *** denote significance levels at 0.1, 0.05, and 0.01, respectively.

These theoretical assumptions significantly differ from the actual agricultural land use situation in China.

China's land use system, based on the HCRS, has weak capital and technology substitution effects for labor. As a result, rural labor surplus is not absolute. Thus, the initial stage of labor migration exhibits a certain negative effect on agricultural land use efficiency. As the migration rate increases, the impact of rural labor migration on agricultural land use efficiency aligns with the expectations of the dual-economy theory of urban and rural areas (showing a significant and robust positive effect). The difference is that this impact direction does not change even when the labor migration rate reaches a certain level. This indicates that, on a national scale, the substitution effect of capital and technology for labor has mitigated the impact of large-scale rural labor migration, improving the agricultural input–output relationship and promoting agricultural land use efficiency. This is due to the influence of China's household registration and related systems, where rural labor migration is conducted on an individual basis rather than on a household basis. Under the HCRS, this migration characteristic ensures that the substitution effect of capital and technology for labor is carried out in a manner that considers the household as the production unit, and the flexibility of the small-scale farming economy is fully utilized. The reduction in labor input due to off-farm employment migration is effectively substituted by capital and technology, thus preventing a significant decline in agricultural land use efficiency.

Endogeneity analysis and robustness tests

Since labor off-farm employment decisions and agricultural land use decisions may be made simultaneously, leading to more general endogeneity issues arising from unobservable factors, it is necessary to further test the robustness of the parameters estimated based on panel data and fixed-effects models. Given the large sample size, heteroscedasticity, and serial correlation, this study uses the Three-Stage Least Squares (3SLS) method of simultaneous equation models to test for potential endogeneity issues.[29] Additionally, although the efficiency of land use does not suffer from truncation of values, the data is grouped into the interval (0,1], so a panel Tobit model can be used to analyze the robustness of the parameter estimation methods. Furthermore, this study also examines the sensitivity of the model parameters to the method of calculating land use efficiency (eff_crs) under constant returns to scale. The results of these robustness checks indicate that, whether considering potential endogeneity, model specification issues, or land use efficiency measurement methods, the parameters for both key explanatory variables and control variables exhibit strong robustness.

Based on the previous analysis, this study rejects hypothesis H3. Although the impact of off-farm employment on agricultural land use efficiency is nonlinear, it does not exhibit a "inverted U-shaped" pattern with increasing rural labor transfer rates. Instead, it shows a trend of initially declining and then increasing as the transfer rate rises. This study argues that, under the current land use system, considering the capital and technology's substitution effects for labor, rural labor remains relatively surplus in relation to agricultural land use in China. Therefore, its transfer has not negatively impacted land use efficiency.

Conclusions and Discussion

The progress of urbanization and industrialization in China has been accompanied by a large-scale transfer of rural labor to off-farm employment. This shift not only affects the labor input in agricultural land use but also influences other production factors through a series of chain and ancillary effects, thereby impacting the agricultural input–output relationship and leading to changes in grain production and adjustments in the agricultural land use structure. In this context, off-farm employment has subtly transformed agricultural land use in China, marking what can be described as a "silent revolution" in the agricultural sector. Understanding the characteristics of this transformation and analyzing the underlying reasons are crucial for scientifically assessing whether there is still surplus labor in China, whether to continue with food security strategies and implement the "three-rights separation" reform for agricultural land, and how to promote adjustments in the agricultural land use structure.

The study, based on panel data from 1,961 counties across China from 2000 to 2014 and framed by urban–rural dual economic theory and related analyses, presents the following main conclusions: 1) The relationship between off-farm employment and grain production is nonlinear. As the rate of rural labor transfer increases, it shows a significant and stable "inverted U-shaped" pattern. 2) The impact of off-farm employment on agricultural land use varies within agricultural land use itself. This variation is primarily due to the differing labor demands of grain production versus other agricultural activities, indicating a shift in agricultural land use structure toward non-grain conversion. 3) The impact of off-farm employment on agricultural land use efficiency is nonlinear but shows a distinct "U-shaped trend" pattern, which is different from its effect on grain production. This pattern suggests that, due to the substitution effect of capital and technology for labor, there is still some capacity for releasing rural labor with the reform of land transfer and use systems. The implications of these findings for policy are as follows:

1) In the processes of urbanization and industrialization, the application and promotion of capital and technology have a significant impact on agricultural land use. There is still surplus labor in rural areas relative to farmland use. Only with the advancement of capital and technology can this surplus labor continue to be extracted, thereby further optimizing labor allocation across industries. Additionally, the decline in agriculture does not necessarily mean a reduction in agricultural land use efficiency. Although the overall level of agricultural land use efficiency in China is still low, the transfer of rural labor has a significant positive impact on agricultural land use efficiency. Therefore, it is necessary to integrate the "Three Rights Separation" reform of agricultural land and land transfer to

further liberate rural labor from agricultural land use and enhance agricultural productivity.

2) The current transfer of rural labor has not yet impacted the pathways of farmland scale management. It is necessary to promote farmland transfer through relevant institutional reforms to increase the scale of land management by farmers. Empirical analysis indicates that, on a national average level, increasing the scale of family operated farmland is beneficial for boosting grain and agricultural land use and improving land use efficiency. Therefore, appropriate institutional and technical measures should be taken to integrate land scale management with rural labor transfer, in conjunction with China's household contract responsibility system and the "Three Rights Separation Reform" of agricultural land, to enhance agricultural land use efficiency.

This study also has some limitations. Due to constraints in obtaining county-level data, indicators such as rainfall and accumulated temperature could not be used to control variables. Additionally, due to the lag in updating county-level indicators across the country, data period could not be further extended. Furthermore, due to the content limitations, there was insufficient focus on regional differences, which could be further explored and optimized in future research.

Notes

1 Wang, Y. Q., & Chen, Y. Z. (2016). The impacts of labor migration on farm households' cropping structure. *Issues in Agricultural Economy, 37*(2), 41–48.

2 Tian, Y. J., Li, X. B., Ma, G. X., & Hao, H. G. (2010). Influences of labor emigration from agriculture on the production abandonment of cultivated land in ecological sensitive areas. *China Land Science*, *24*(7), 4–9; Zhang, Y., Li, X.-B., Song, W., & Zhang, L. (2016). Land abandonment under rural restructuring in China explained from a cost-benefit perspective. *Journal of Rural Studies*, *47*, 524–532.

3 Qian, W.-R., & Zheng, L.-Y. (2011). The impact of labor migration on farmers' agricultural production: Current research and future prospects. *China Rural Observation*, (1), 31–38, 95; Wang, Z.-C. (2015). Has rural labor migration reduced agricultural efficiency? *Statistical Research*, (3), 54–61.

4 Cai, F. (2010). Urbanization and the contribution of migrant workers: Reflections on China's economic growth potential in the post-crisis era. *China Population Science*, (1), 2–10, 111.

5 Rozelle, S., Taylor, J. E., & De Brauw, A. (1999). Migration, remittances and agricultural productivity in China. *American Economic Review*, *2*, 287–291;Taylor, E. J., Rozelle, S., & De Brauw, A. (2003). Migration and incomes in source communities: A new economics of migration perspective from China. *Economic Development of Cultural Change*, *1*, 75–101; Gai, Q. E., Zhu, X., & Shi, Q. H. (2014). Labor's migration and Chinese agricultural production. *China Economic Quarterly*, *13*(3), 1147–1170; Zou, B.-L., Mishra, A. K., & Luo, B.-L. (2018). Aging population, farm succession, and farmland usage: Evidence from rural China. *Land Use Policy*, *77*, 437–445.

6 Qian, W.-R., Wang, D.-H., & Zheng, L.-Y. (2016). The impact of migration on agricultural restructuring: Evidence from Jiangxi province in China. *Journal of Rural Studies*, *47*, 542–551.

7 Ji, Y.-Q., Yu, X.-H., & Zhong, F. (2012). Machinery investment decision and off-farm employment in rural China. *China Economic Review*, *23*, 71–80; Wang, X.-B., Huang,

J.-K., & Rozelle, S. (2017). Off-farm employment and agricultural specialization in China. *China Economic Review, 42*, 155–165.

8 Qin, H., & Liao, T. F. (2016). Labor out-migration and agricultural change in rural China: A systematic review and meta-analysis. *Journal of Rural Studies, 47*, 533–541.

9 Feng, S.-Y., Heerink, N., Ruben, R., & Qu, F. (2010). Land rental market, off-farm employment and agricultural production in Southeast China: A plot-level case study. *China Economic Review, 21*, 598–606.

10 Lin, J., & Li, D. (2013). Off-farm employment and grain production: Substitution or complementarity–an analysis based on the perspective of farmers in major grain-producing areas. *China Rural Economy*, (9), 54–62.

11 Wang, Y.-M., Yao, X.-G., & Zhou, M.-H. (2013). Rural labor outflow, regional differences, and grain production. *Management World*, (11), 67–76; Jiang, L., Deng, X.-Z., & Seto, K. C. (2013). The impact of urban expansion on agricultural land use intensity in China. *Land Use Policy, 35*, 33–39.

12 The dual economy theory, represented by Lewis (1954) and Fei and Rains (1964), posits that in developing countries, during the early stages of urbanization and industrialization, traditional agriculture is characterized by surplus rural labor with zero or negative marginal productivity. The transfer of this surplus labor does not affect agricultural land use but rather optimizes the land–labor relationship, which is beneficial for improving agricultural productivity. With the emergence of the "Lewis Turning Point," when the marginal productivity of rural labor becomes positive, the transfer of labor out of agriculture has adverse effects on land use, prompting a transformation in agricultural production. See Lewis, A. W. (1954). Economic development with unlimited supply of labor. *The Manchester School of Economic and Social Studies, 2*, 139–191; Fei, J. C. H., & Rains, G. (1964). *Development of the labor surplus economy: Theory and policy* (p. 35). Irwin.

13 According to the *China Statistical Yearbook* (2017), China's urbanization rate was 17.92% in 1978 and first exceeded the 30% threshold in 1996, entering a rapid urbanization phase. The 30% threshold is one of the two inflection points in the "S-curve" of urbanization growth proposed by American economic geographer R. M. Northam.

14 Wang, J., & Zhong, X.-H. (2011). Has China reached the Lewis turning point? – theoretical analysis and international experience. *Chinese Social Sciences*, (5), 22–37, 219.

15 Fei, J. C. H., & Rains, G. (1964). *Development of the labor surplus economy: Theory and policy* (p. 35). Irwin.

16 Wan, G. H., & Cheng, E.-J. (2001). Effects of land fragmentation and returns to scale in the Chinese farming sector. *Applied Economics, 33*, 183–194.

17 According to the China Economic and Social Development Statistical Database, the total power of agricultural machinery was 11,749.9 (10,000 kilowatts) in 1978 and increased to 111,728.1 (10,000 kilowatts) in 2015, representing an approximate ninefold increase.

18 Zhong, F., Lu, W.-Y., & Xu, Z.-G. (2016). Is rural labor migration detrimental to grain production?–An analysis of farmers' factor substitution, cropping structure adjustment behaviors, and constraints. *China Rural Economy*, (7), 36–47.

19 Grain is generally planted on cultivated land. The grain yield per unit of land is calculated as total grain production divided by the area of cultivated land, while the agricultural yield per unit of land is calculated as total agricultural output value divided by the area of farmland. Land factors are measured by actual input quantities, incorporating crop index and recycling conditions into the analysis. China conducted its second national land survey in 2007, and the area of land used for agriculture (including cultivated land) fluctuated significantly around 2009. This study uses data on increases and decreases in land types covered by agricultural land at the end of 2009 to infer the data for early 2009, which is then used as the data for the end of 2008. This data is compared with the original end-2008 data to infer the actual data for the period before 2008.

20 The reason for using DEA to measure agricultural production efficiency is that this method can assess the relationship between various types of input and output indicators (Fried, 2002). Fried, H. O., Lovell, C. A.-K., Schmidt, S. S., et al. (2002). Accounting for environmental effects and statistical noise in Data Envelopment Analysis. *Journal of Productivity Analysis*, *17*(1), 157–174. The Hybrid model considers the substitutability between factors by examining the relationship of input factors projected onto the optimal production frontier, allowing for the evaluation of how capital and technology substitute for the loss of rural labor. See Cheng, G. (2014). *Data envelopment analysis method and MaxDEA software* (p. 134). Intellectual Property Publishing House.

21 Due to the current inability to obtain a sufficient estimator for individual unobservable effects, the conditional maximum likelihood estimation of fixed effects cannot be estimated. Therefore, this study does not employ the Tobit model in the main regression model for agricultural production efficiency.

22 This study focuses on how to optimize the input–output relationship in agricultural production without reducing output, thus adopting an input-oriented goal programming approach.

23 Excluding Hong Kong, Macao and Taiwan, China.

24 Liu, X.-G., Zhang, X., & Fang, W.-Q. (2015). The urban-rural income distribution effect of infrastructure: A perspective based on labor transfer. *World Economy*, (3), 145–170.

25 Xu, J.-G., & Zhang, X. (2016). Agricultural productivity growth, labor transfer, and the linkage development between industry and agriculture. *Management World*, (7), 76–97.

26 Given the crucial role of cultivated land in agricultural production and its higher labor demand compared to other land types, the average cultivated land area per household is used as a substitute for the area of agricultural land per household as the measurement indicator. Due to the unavailability of data on the educational background of the existing labor force and the long-term nature of educational improvements, the proportion of students in ordinary secondary schools, which represents the local basic education situation, is used as a substitute indicator.

27 Due to space limitations, the results of the unit root tests for variables and the disturbance term tests are not included. Please contact the authors for these details. Refer to Driscoll, J. C., & Kraay, A. C. (1998). Consistent covariance matrix estimation with spatially dependent panel data. *Review of Economics and Statistics*, *80*(4), 549–560.

28 White, H. (1980). A heteroskedasticity-consistent covariance matrix estimator and a direct test for heteroskedasticity. *Econometrica*, *4*, 817–838; Newey, W. K., & West, K. D. (1987). A simple, positive semi-definite, heteroskedasticity and autocorrelationconsistent covariance matrix. *Econometrica*, *55*(3), 703–708.

29 Due to space limitations, the formulas for the simultaneous equations model, descriptive statistics for variables, and test results are not included. Please contact the authors if these details are needed.

17 Transitioning agricultural land to eco-estates in rural Gauteng

Farmers' perceptions and policy and planning implications

Ashley Gunter

Introduction

The transition of agricultural land to eco-estates is an emerging trend in rural South Africa, particularly in Gauteng, where the pressure for urban expansion intersects with the demand for environmentally sustainable living spaces (Alexander et al., 2019). Eco-estates, defined as residential developments that emphasize ecological conservation and sustainable living, are increasingly being developed on former agricultural lands. This trend reflects broader global shifts in land use, driven by urbanization, environmental concerns, and changing socio-economic priorities (Walker & Fortmann, 2003). However, the implications of such transitions for local farmers, communities, and broader land use policies remain complex and multifaceted, raising critical questions about sustainability, social equity, and agricultural security.

In South Africa, land use change has long been a contentious issue, rooted in the country's history of colonialism and apartheid, which left deep scars on the agricultural landscape and landownership patterns (Hall, 2014). The post-apartheid era has seen significant efforts to redress these historical injustices through land reform policies aimed at redistributing land to historically disadvantaged communities (Hall, 2011). However, the rapid urbanization and the rise of eco-estates present new challenges to these efforts, as agricultural lands, often the focus of land reform, are increasingly being converted into residential developments catering to middle- and upper-income groups (Sherriff-Shüping, 2015). This chapter explores the implications of this trend, focusing on the perspectives of farmers who are directly affected by the transition from agricultural land to eco-estates in rural Gauteng. The province of Gauteng is South Africa's smallest yet most urbanized province and serves as an ideal study site due to its rapid urban expansion and the increasing pressure on agricultural land for development.

The study is grounded in the recognition that farmers' perceptions and experiences are crucial to understanding the broader impacts of land use change. Farmers, who have traditionally been the stewards of agricultural land, are facing new pressures as eco-estates encroach on land that was once used for food production and other agricultural activities (Ballard & Jones, 2011). These pressures are not merely economic but also social and cultural, as farming communities are

DOI: 10.4324/9781032672779-21

displaced or forced to adapt to new realities imposed by eco-estate developments (Frayne et al., 2014; Jonah & May, 2020). Moreover, the transition raises important questions about the sustainability of agricultural practices in the face of increasing urbanization and the push towards more environmentally friendly land uses.

This chapter employs qualitative interviews with 15 farmers in Gauteng to explore their perceptions of eco-estate developments, their impact on livelihoods and communities, and their experiences with planning and policy processes related to agricultural land use. By focusing on these farmers' perspectives, the study aims to provide a nuanced understanding of the challenges and opportunities associated with the transition from agriculture to eco-estates. The analysis will highlight the intricate sociopolitical and economic factors that influence this transition, shedding light on the broader implications for landownership, agricultural security, and sustainable development in South Africa.

The findings of this study have significant implications for policymakers, developers, and communities in Gauteng and beyond. As South Africa continues to grapple with issues of land reform, rural development, and environmental sustainability, understanding the impacts of land use transitions on local farmers is essential (Akinola et al., 2021). This chapter argues that planning processes need to be more inclusive and participatory, taking into consideration the perspectives and needs of local farmers and communities. This could involve increased farmer participation in decision-making processes, enhanced support for alternative livelihoods, and the development of more sustainable land use policies that prioritize food security, environmental conservation, and social equity (Borras & Franco, 2010).

By contributing to the ongoing debates about sustainable rural development, land tenure, and community participation, this chapter aims to advance our understanding of the complex dynamics of agricultural land use transitions in South Africa. The transition from agricultural land to eco-estates presents both challenges and opportunities, and it is through careful consideration of all stakeholders' perspectives that sustainable and equitable land use strategies can be developed.

Agricultural Land Use and Rural Development in South Africa

The transition of agricultural land to eco-estates in rural Gauteng, South Africa, represents a significant shift in land use patterns, reflecting broader global trends in sustainable development, urbanization, and environmental conservation. This chapter explores the existing body of work on key themes relevant to this transition, including agricultural land use, eco-estates, land use planning and policy in South Africa, and farmers' perceptions of land use change. By examining these themes, the chapter aims to contextualize the transition from agriculture to eco-estates within the broader debates on rural development, land reform, and sustainable land use practices.

Evolving Land Dynamics and the Future of Rural South Africa

Agricultural land use in South Africa has historically been shaped by the country's colonial and apartheid past, which resulted in highly unequal land distribution and a bifurcated agricultural sector (Mukaddam, 2024). The post-apartheid era has seen efforts to redress these imbalances through land reform policies aimed at redistributing land to historically disadvantaged communities (Hall, 2011). However, these efforts have faced numerous challenges, including the slow pace of land reform, the difficulties of supporting emerging farmers, and the pressures of urbanization and economic development (Lahiff, 2007).

Rural development in South Africa has traditionally been linked to agriculture, with the sector playing a crucial role in providing livelihoods, ensuring food security, and maintaining rural communities (Baiphethi & Jacobs, 2009). However, recent decades have seen increasing pressures on agricultural land, driven by urban expansion, economic development, and environmental conservation efforts (Aliber & Hart, 2009). These pressures have led to significant changes in land use, including the conversion of agricultural land to residential, commercial, and industrial uses. The transition from agricultural land to eco-estates is one such change, reflecting broader shifts in land use patterns and raising important questions about the future of agriculture in South Africa.

Eco-Estates: Definitions, Characteristics, and Global Trends

Eco-estates, also known as green or sustainable developments, are residential communities designed to minimize their environmental impact and promote sustainable living. These developments typically incorporate features such as energy-efficient buildings, water conservation systems, waste recycling, and the preservation of natural habitats (Alexander, 2021). Eco-estates have become increasingly popular globally as a response to growing concerns about climate change, environmental degradation, and the need for sustainable urban development (Gałecka-Drozda et al., 2021; Agbajor & Mewomo, 2024).

In South Africa, the development of eco-estates has been driven by a combination of factors, including rising demand for environmentally sustainable living options, increasing awareness of environmental issues, and the desire to preserve natural landscapes and biodiversity (Sherriff-Shüping, 2015; Schoeman & Gunter, 2018). Eco-estates are often marketed as offering a high quality of life, with access to green spaces, recreational facilities, and a sense of community. However, they also raise important questions about social equity, land use, and the displacement of existing communities, particularly when they are developed on agricultural land (Soga et al., 2014).

The global trend towards eco-estates reflects broader shifts in urban planning and development, with increasing emphasis on sustainability, environmental conservation, and the need to create resilient communities in the face of climate change (Tezangi, 2014; Farr, 2018). However, the development of eco-estates is not without its challenges, including the need to balance environmental goals with social and economic considerations, the potential for social exclusion, and the impact on existing land uses and communities (Rogerson, 2023).

Land Use Planning and Policy in South Africa

Land use planning and policy in South Africa have historically been shaped by the country's dual goals of promoting economic development and addressing social inequalities (Nunn, 2020). The post-apartheid era has seen significant changes in land use policy, with a focus on promoting equitable development, sustainable land use, and the integration of previously excluded communities into the formal economy (Turok, 2012). However, these goals have often been in tension with the pressures of urbanization, economic growth, and environmental conservation.

The development of eco-estates on agricultural land raises important questions about the adequacy of existing land use policies and the extent to which they support sustainable development (Thinda et al., 2020). While eco-estates are often promoted as environmentally friendly developments, their impact on agricultural land and rural communities can be significant, particularly when they lead to the displacement of existing land uses and the exclusion of local communities from decision-making processes (Van Noorloos & Kloosterboer, 2018; Côté-Roy & Moser, 2019).

The South African government has recognized the need for more integrated and sustainable land use planning, with recent policy frameworks emphasizing the importance of balancing economic development, social equity, and environmental sustainability (Department of Rural Development and Land Reform, 2019). However, the implementation of these policies has been uneven, with significant challenges remaining in terms of coordinating land use planning across different levels of government, ensuring meaningful community participation, and addressing the complex sociopolitical dynamics of land use change.

Farmers' Perceptions and Attitudes Towards Land Use Changes

Farmers' perceptions and attitudes towards land use changes are crucial to understanding the broader impacts of transitions from agriculture to eco-estates. Farmers, as the primary stewards of agricultural land, are directly affected by such transitions, which can have significant implications for their livelihoods, social networks, and cultural practices (Frayne et al., 2014). Understanding farmers' perspectives is essential for developing more inclusive and participatory land use planning processes that take into account the needs and priorities of local communities.

Research on farmers' perceptions of land use change has highlighted the complex and often conflicting views that farmers hold about developments such as eco-estates. Some farmers may view eco-estates as an opportunity to diversify their income streams, sell their land at a higher value, or transition to less labour-intensive forms of land use (Mistry & Spocter, 2022). However, many farmers are concerned about the loss of agricultural land, the impact on food security, and the displacement of rural communities (Walker & Fortmann, 2003).

The transition from agricultural land to eco-estates can also raise important questions about landownership and tenure. In South Africa, where land reform remains a contentious issue, the conversion of agricultural land to residential developments can exacerbate existing inequalities in landownership and access, particularly when it leads to the exclusion of marginalized communities from land and resources (Hall, 2011). This underscores the importance of developing land use policies that prioritize social equity, food security, and the rights of local communities.

The literature on agricultural land use, eco-estates, land use planning and policy, and farmers' perceptions of land use change provides a critical foundation for understanding the transition from agricultural land to eco-estates in rural Gauteng (Hall, 2011; Sherriff-Shüping, 2015). This transition reflects broader global trends in sustainable development and urbanization, but it also raises significant challenges related to social equity, landownership, and the sustainability of agricultural practices. The review highlights the need for more inclusive and participatory land use planning processes that consider the perspectives of farmers and local communities, as well as the importance of balancing environmental goals with social and economic considerations.

Methodology

This study adopts a qualitative research design to explore the transition of agricultural land to eco-estates in rural Gauteng, South Africa. Qualitative research is particularly suited to this study's objectives, as it allows for an in-depth exploration of farmers' perceptions, experiences, and the sociopolitical dynamics surrounding land use changes (Creswell, 2014). The study employs a case study approach, focusing on a specific geographic region within Gauteng to provide a detailed understanding of the complexities involved in agricultural land transitions. The case study method is appropriate for investigating contemporary phenomena within real-life contexts, particularly when the boundaries between the phenomenon and context are not clearly evident (Yin, 2018).

Sampling Strategy

The study uses purposive sampling to select participants who can provide rich, relevant, and diverse insights into the phenomenon under investigation (Patton, 2015). The sample consists of 15 farmers in Gauteng who have been directly affected by the development of eco-estates on agricultural land. The selection criteria include farmers who own or manage agricultural land that has been converted or is in the process of being converted into eco-estates, as well as those who have been involved in the planning or decision-making processes related to these developments. This sampling strategy ensures that the participants are well-positioned to offer valuable perspectives on the research questions.

Data Collection

Data collection was conducted through semi-structured interviews, which are ideal for exploring participants' perceptions, experiences, and attitudes in depth (Kvale, 2007). The semi-structured format allows for flexibility, enabling the interviewer to probe further into specific areas of interest while ensuring that all key topics are covered. The interview guide was developed based on the research objectives and literature review, focusing on topics such as the impact of eco-estate developments on agricultural practices, farmers' experiences with land use planning and policy processes, and their views on the sustainability and social implications of the transition to eco-estates.

Interviews were conducted face-to-face, allowing for the establishment of rapport and a deeper understanding of the participants' contexts. Each interview lasted between 45 minutes and an hour and was audio-recorded with the participants' consent. Detailed field notes were also taken during and after the interviews to capture non-verbal cues and contextual information that could enrich the data analysis.

Data Analysis

The interview data were transcribed verbatim and analysed using thematic analysis, a method for identifying, analysing, and reporting patterns (themes) within data (Braun & Clarke, 2006). Thematic analysis was chosen because it provides a flexible approach to data analysis, allowing for both inductive and deductive coding processes. The analysis began with an initial reading of the transcripts to become familiar with the data, followed by the generation of initial codes that captured significant features of the data relevant to the research questions.

Themes were then developed by grouping related codes and refining them through a process of constant comparison, ensuring that they accurately reflected the data (Strauss & Corbin, 1998). The final themes were reviewed in relation to the entire data set and the study's research objectives, ensuring that they provided a coherent and compelling narrative of the findings.

Validity and Reliability

To ensure the validity and reliability of the study, several strategies were employed. Triangulation was used by comparing the interview data with relevant literature and policy documents to confirm the consistency of the findings. Member checking was also conducted by sharing the preliminary findings with a subset of participants to verify the accuracy of the interpretations (Miles & Jozefowicz-Simbeni, 2010). Additionally, an audit trail was maintained throughout the research process, documenting decisions made during data collection and analysis to provide transparency and enable replication.

Results Overview: the Dynamics of Land Use Transitions

This section presents a detailed exploration of how the transition from agricultural land to eco-estates in rural Gauteng has affected farmers and the broader rural community. It captures the range of experiences, from the emotional impact of losing ancestral land to the financial benefits some farmers gained. The analysis addresses the complexity of these changes, considering the challenges related to policy frameworks, community engagement, and long-term sustainability. Farmers' voices highlight the need for more inclusive and balanced approaches to land use planning that protect rural livelihoods while accommodating development pressures.

Overview of Farmers' Experiences and Reactions to the Emergence of Eco-Estates

The emergence of eco-estates in rural Gauteng has elicited mixed reactions from farmers, who have experienced both positive and negative impacts as their agricultural lands are transformed into residential developments. Many farmers expressed a sense of loss and uncertainty as the traditional uses of their land gave way to new and unfamiliar developments. One farmer, who has been cultivating maize for over two decades, remarked, "It feels like losing a part of our heritage. This land has been in our family for generations, and now it's being turned into something completely different" (Participant A).

However, some farmers acknowledged that the development of eco-estates also brought certain advantages, particularly in terms of financial gains. "The offer to sell my land was too good to refuse . . . with the money, I was able to pay off debts and invest in a smaller, more manageable plot of land elsewhere" (Participant B). Despite this, the same farmer admitted to feeling conflicted about the long-term implications: "It's a relief financially, but I do worry about what this means for the future of farming in this area" (Participant C).

Analysis of Perceived Benefits and Challenges Associated With the Transition

The perceived benefits of the transition from agricultural land to eco-estates were largely tied to the immediate financial incentives. Several farmers mentioned the substantial offers they received from developers, which often far exceeded what they could earn from farming. "The money was good," said Participant C. "I could never have made that much from farming alone. It allowed me to retire earlier than I had planned" (Participant C). This sentiment was echoed by others who viewed the sale of their land as an opportunity to escape the increasingly challenging economic conditions faced by small-scale farmers.

However, the transition also presented significant challenges, particularly in terms of the impact on the farming community and the local economy. "It's not just about the land, when the farms go, so do the jobs. There's a whole chain of people who rely on the farms for work, from the workers in the fields to those who supply us with seeds and equipment" (Participant D). The loss of agricultural land was also seen as a threat to food security and the sustainability of local food systems. "We're losing valuable farmland, and once it's gone, it's gone for good, that's land that could be used to feed our communities" (Participant E).

Farmers' Views on the Long-Term Sustainability and Viability of Eco-Estates

When asked about the long-term sustainability and viability of eco-estates, farmers expressed a range of views, often reflecting a balance between scepticism and cautious optimism.

"Eco-estates sound good on paper–sustainable living, green spaces–but I'm not sure how sustainable they really are . . . What happens when the novelty wears off?

Will people still be committed to the eco-friendly practices, or will it just become another suburban [sprawl]?" (Participant I).

Others were more hopeful, seeing the potential for eco-estates to set a new standard for environmentally conscious living.

"If done right, these eco-estates could be a model for [sustainable] development but it needs real commitment from developers and residents to maintain those standards over the long time" (Participant J).

However, there was also concern about the exclusivity of such developments and their accessibility to the broader population.

"These estates are marketed to the rich, I wonder if they're really making a difference in the bigger picture, or if it's just a way to make people feel good about living green without addressing the real issues of inequality and access to resources" (Participant K).

While the transition to eco-estates has provided certain economic benefits to individual farmers, the broader implications for the farming community and the sustainability of rural life in Gauteng remain uncertain. The farmers' experiences and views underscore the need for a more inclusive approach to land use planning that considers the long-term impacts on agriculture, food security, and community cohesion.

Policy and Planning Processes: Examination of Existing Agricultural Land Use Policies in Gauteng

Agricultural land use policies in Gauteng are embedded within a broader national framework aimed at balancing economic development with the preservation of agricultural resources. These policies are designed to protect prime agricultural land from encroaching urban development while also allowing for the strategic re-zoning of land to meet the growing demands for housing and infrastructure (Department of Agriculture, Forestry and Fisheries, 2018). Despite these provisions, the policies have been criticized for their lack of clarity and consistency, particularly in the face of increasing pressure to convert agricultural land into residential developments such as eco-estates. One farmer remarked, "The policies seem to change with every new government official that comes in. One moment we're being told to preserve our land, and the next, they're approving these big developments" (Participant L).

The existing policies also struggle to address the unique challenges faced by farmers in peri-urban areas, where the proximity to expanding cities often leads to conflicting land use priorities. "We're stuck between a rock and a hard place," said Participant M. "On one hand, the policy says we should keep farming, but on the other, the city keeps growing, and we're being squeezed out" (Participant M). This tension highlights the need for more nuanced and adaptable policies that can respond to the specific circumstances of regions like Gauteng.

Farmers' Engagement With Policy and Planning Processes

Farmers' engagement with policy and planning processes has been limited, with many feeling marginalized and excluded from key decisions that affect their land

and livelihoods. "We're the ones working the land every day, but when it comes to planning, we're often the last to know," said Participant N. "By the time we're involved, the decisions have already been made" (Participant N). This lack of meaningful engagement has left many farmers feeling disillusioned with the policy process, viewing it as something distant and disconnected from their realities.

Some farmers have attempted to engage more actively with local authorities and planning bodies, but they often encounter bureaucratic hurdles and a lack of responsiveness. "I've attended several meetings, but it feels like we're just ticking boxes. There's a lot of talk, but when it comes to actually listening to our concerns, nothing seems to change" (Participant O). This sentiment reflects a broader disconnect between policymakers and the farming community, where the latter's voices are often overshadowed by the interests of developers and urban planners.

Critical Assessment of Inclusivity and Participation in Decision-Making

The inclusivity of the decision-making process has been a major concern among farmers, who argue that their interests are not adequately represented. "These eco-estates are being planned by people who don't understand farming or the challenges we face. We need to be part of the conversation from the beginning, not just brought in at the end to sign off on something we had no say in" (Participant A). The current planning processes often lack mechanisms for genuine participation, where farmers can contribute their knowledge and insights in a meaningful way.

The exclusion of farmers from decision-making processes is further compounded by the power dynamics at play, where larger, more influential stakeholders, such as developers and local government officials, dominate the discussions. "It feels like the developers have all the power, they have the money, they know the right people, and they push through their projects while we're left scrambling to understand what's happening" (Participant C). This imbalance underscores the need for more equitable and inclusive planning processes that give farmers a stronger voice in decisions that directly impact their land and livelihoods.

Challenges Faced by Farmers in Navigating Policy Frameworks

Navigating the complex policy frameworks governing agricultural land use has proven to be a significant challenge for many farmers. The policies are often seen as overly complicated and difficult to interpret, particularly for those without legal or planning expertise. "The paperwork is overwhelming, because I'm a farmer, not a lawyer, and trying to make sense of these rules is almost impossible" (Participant F). This complexity has led to frustration and a sense of powerlessness among farmers, who feel ill-equipped to advocate for their rights and interests within the existing system.

Moreover, the bureaucratic nature of the policy processes has made it difficult for farmers to access the support and resources they need to navigate these transitions. "You have to go through so many hoops just to get a straight answer by the time you figure out what you need to do, it's often too late" (Participant G). This challenge is exacerbated by the lack of clear guidance and support from local authorities, leaving farmers to fend for themselves in a system that seems more geared towards developers than those working the land.

The Role of Local Government and Other Stakeholders in Facilitating or Hindering Transitions

Local government and other stakeholders play a crucial role in facilitating or hindering the transition of agricultural land to eco-estates. The support, or lack thereof, provided by these entities can significantly influence the outcomes for farmers.

"Local government can either be a friend or an enemy. Some officials are really helpful and try to support us, but others seem more interested in pushing through developments without considering the impact on farming" (Participant O).

The relationship between farmers and local government is often characterized by a lack of trust, with many farmers feeling that their concerns are secondary to the economic interests associated with new developments.

"We've been here for generations, but it feels like the government is more interested in pleasing the developers than in supporting us. They say they want sustainable development, but it's hard to see how that's possible when they're willing to sacrifice farmland for estates" (Participant M).

Other stakeholders, such as non-governmental organizations (NGOs) and community groups, have also been involved in the process, often advocating on behalf of farmers and pushing for more sustainable and inclusive development practices. However, their influence is often limited compared to that of developers and government bodies. "The NGOs try to help, but they don't have the same clout as the developers, it's a tough battle, and sometimes it feels like we're in on our own" (Participant I).

The policy and planning processes surrounding the transition of agricultural land to eco-estates in Gauteng are fraught with challenges. Farmers face significant hurdles in navigating the complex policy frameworks, and their engagement with these processes is often limited and frustrating. The role of local government and other stakeholders is crucial in shaping the outcomes of these transitions, but the current dynamics often favour development interests over the preservation of agricultural land and the well-being of farming communities. These findings highlight the need for more inclusive, transparent, and supportive planning processes that better balance the needs of all stakeholders involved.

Broader Implications for Rural Development in South Africa

The transition of agricultural land to eco-estates has broader implications for rural development in South Africa, particularly in terms of balancing development with the preservation of rural livelihoods and cultural heritage. Rural areas in South Africa are often characterized by a dependence on agriculture, both as an economic activity and as a way of life. The conversion of farmland into residential developments can undermine these rural economies and contribute to the decline of rural communities. "Rural development should be about supporting agriculture and the people, not building new housing, we need development that benefits everyone, not just a few" (Participant F).

Moreover, the rise of eco-estates raises questions about the inclusivity and equity of rural development. These developments are often targeted at wealthier, urban-based individuals seeking a rural lifestyle rather than addressing the needs of existing rural populations.

"These estates aren't for us; they're for people with money who want a piece of the countryside but what about the people who have been living and working here for generations? Where do we fit into this new vision of rural South Africa?" (Participant F).

The transition of agricultural land to eco-estates in Gauteng has far-reaching implications across sociopolitical, economic, environmental, and rural development dimensions. While these developments may offer certain benefits, such as financial gains for individual landowners and potential conservation opportunities, they also pose significant challenges. These include the disruption of traditional landownership patterns, the erosion of agricultural security, negative impacts on local economies, and environmental concerns. The broader implications for rural development in South Africa highlight the need for a more balanced approach that considers the long-term sustainability and inclusivity of these transitions, ensuring that rural communities are not left behind in the pursuit of new development paradigms.

Farmers' Perceptions, Policy Realities, and Practical Implications

The transition of agricultural land to eco-estates in rural Gauteng reveals a significant disconnect between farmers' perceptions and the realities of existing policy frameworks. Farmers view these transitions through a lens of personal and community impact, emphasizing the socio-economic and cultural losses associated with the conversion of agricultural land. As highlighted in the previous sections, many farmers feel that the policy processes governing land use change are distant, opaque, and largely driven by the interests of developers rather than the needs of rural communities. This perception is compounded by the complexities and inconsistencies in land use policies, which, while ostensibly designed to protect agricultural land, often fail to account for the nuanced challenges faced by farmers in peri-urban areas.

The policy realities, on the other hand, reflect a broader agenda focused on balancing economic development with environmental conservation, often at the expense of agricultural security and rural livelihoods. The policies guiding the development of eco-estates prioritize sustainable urbanization and the creation of environmentally friendly residential spaces, yet they do not adequately address the long-term implications for food security, land tenure, and the sociocultural fabric of rural communities. This disconnect underscores the need for more integrated and farmer-inclusive approaches to land use planning that align policy objectives with the lived experiences of those directly affected by these transitions.

The practical implications of this study are significant for policymakers, developers, and rural communities. For policymakers, the findings highlight the urgent need to revise land use policies to ensure they are more inclusive and responsive to the needs of farmers and rural communities. This could involve creating more transparent and participatory planning processes that actively involve farmers in decision-making, as well as developing policies that balance the demands of urban expansion with the preservation of agricultural land and rural livelihoods (Cousins, 2007; Beinart, 2012).

For developers, the study suggests that the success of eco-estates depends not only on their environmental credentials but also on their ability to integrate with and support the existing rural fabric. Developers should consider the broader social and economic impacts of their projects and work collaboratively with local communities to mitigate negative consequences. This could include investing in community development initiatives, providing support for displaced farmers, and ensuring that eco-estates genuinely contribute to sustainable rural development (Mancebo, 2015; Swilling, 2016; Zantsi, 2021).

For rural communities, the findings underscore the importance of advocacy and engagement in policy processes. Farmers and community members need to be proactive in asserting their rights and interests in the face of land use changes. Building alliances with NGOs, community groups, and other stakeholders can enhance their capacity to influence policy and planning decisions. Additionally, exploring alternative livelihood strategies and diversification can help rural communities adapt to the changing landscape while preserving their socio-economic and cultural integrity (Walker et al., 2010; Kepe & Tessaro, 2014).

The transition of agricultural land to eco-estates in Gauteng presents both challenges and opportunities. By synthesizing farmers' perceptions with policy realities, comparing with other regions, and drawing on theoretical frameworks, this study provides a comprehensive understanding of the implications of these transitions (Cadieux, 2008). The findings offer valuable insights for creating more equitable, sustainable, and inclusive land use strategies that benefit all stakeholders involved in the rural development process.

Discussion and Conclusion

The transition of agricultural land to eco-estates in rural Gauteng embodies broader global trends where urban expansion and the demand for sustainable living intersect with traditional rural land use. These transitions are not only complex but also carry multifaceted implications for sociopolitical dynamics, economic security, environmental sustainability, and the cohesion of rural communities. While 60% of farmers reported financial gains from selling their land, 80% expressed concerns about the long-term consequences, such as threats to food security and the erosion of community bonds (Sherriff-Shüping, 2015; Hall, 2014).

There is a significant disconnect between farmers' perceptions and the realities of existing policy frameworks. Although policies are intended to balance economic development with environmental conservation, they often fall short in addressing the specific challenges rural communities face, particularly in the areas of agricultural security, land tenure, and social cohesion. 73% of farmers interviewed felt excluded from meaningful engagement in land use planning, reinforcing the view that the policy landscape remains distant and unresponsive to their needs and lived experiences (Lahiff, 2007; Department of Rural Development and Land Reform, 2019).

Comparative analyses with other regions, both within South Africa and internationally, underscore that these challenges are not unique to Gauteng. Similar tensions between development and agricultural sustainability are evident in regions where urban expansion encroaches on farmland, leading to the displacement of farming communities and raising critical questions about the long-term viability of such transitions (Van Noorloos & Kloosterboer, 2018). These comparative insights reinforce the need for context-specific strategies that consider local conditions while also drawing lessons from global experiences to inform more effective and inclusive policy frameworks (Borras & Franco, 2010).

This chapter contributes to a deeper understanding of land use transitions through the lens of political ecology and sustainable rural development. These frameworks illuminate the power dynamics and socio-environmental interactions that shape land use decisions, highlighting the unequal distribution of benefits and burdens associated with these transitions (Walker & Fortmann, 2003). The study suggests that true sustainability in rural development requires a holistic approach that integrates economic, social, and environmental goals, ensuring that the benefits of development are equitably shared among all stakeholders (Farr, 2018).

Practically, the implications of this study are significant for policymakers, developers, and rural communities. Policymakers must urgently revise land use policies to ensure they are more inclusive and responsive, actively involving farmers in decision-making processes and balancing the demands of urban expansion with the preservation of agricultural land and rural livelihoods (Turok, 2012). Developers, too, must recognize that the success of eco-estates depends not only on their environmental credentials but also on their ability to integrate with and support the existing rural fabric (Sherriff-Shüping, 2015). For rural communities, the findings underscore the importance of advocacy and engagement in policy processes, as well as the need to explore alternative livelihood strategies to adapt to the changing landscape (Aliber & Hart, 2009).

The transition of agricultural land to eco-estates in Gauteng presents both challenges and opportunities. By synthesizing farmers' perceptions with policy realities, comparing with other regions, and drawing on theoretical frameworks, this chapter provides a comprehensive understanding of the implications of these transitions. The findings offer valuable insights for creating more equitable, sustainable, and inclusive land use strategies that benefit all stakeholders involved in the rural development process. As South Africa continues to grapple with issues of land reform, rural development, and environmental sustainability, understanding and addressing the impacts of land use transitions on local farmers and communities is essential for achieving long-term, sustainable development outcomes.

References

Agbajor, F. D., & Mewomo, M. C. (2024). Green building research in South Africa: A scoping review and future roadmaps. *Energy and Built Environment*, *5*(2), 316–335.

Akinola, A. O., Kaseeram, I., & Jili, N. N. (2021). Expropriation and the discontent of land reform in South Africa: An introduction. In *The new political economy of land reform in South Africa* (pp. 1–9), Springer.

Alexander, J., Smith, D. A. E., Smith, Y. C. E., & Downs, C. T. (2019). Eco-estates: Diversity hotspots or isolated developments? Connectivity of eco-estates in the Indian Ocean Coastal Belt, KwaZulu-Natal, South Africa. *Ecological Indicators*, *103*, 425–433.

Alexander, J., Smith, D. A. E., Smith, Y. C. E., & Downs, C. T. (2021). Urban land development for biodiversity: Suggested development and management guidelines for eco-estates using case studies from coastal KwaZulu-Natal, South Africa. *Urban Forestry and Urban Greening*, *65*, 127–347.

Aliber, M., & Hart, T. G. B. (2009). Should subsistence agriculture be supported as a strategy to address rural food insecurity? *Agrekon*, *48*(4), 434–458.

Baiphethi, M. N., & Jacobs, P. T. (2009). The contribution of subsistence farming to food security in South Africa. *Agrekon*, *48*(4), 459–482.

Ballard, R., & Jones, G. A. (2011). Natural neighbors: Indigenous landscapes and eco-estates in Durban, South Africa. *Annals of the Association of American Geographers*, *101*(1), 131–148.

Beinart, W. (2012). Beyond 'homelands': Some ideas about the history of African rural areas in South Africa. *South African Historical Journal*, *64*(1), 5–21.

Borras, S. M., & Franco, J. C. (2010). Contemporary discourses and contestations around pro-poor land policies and land governance. *Journal of Agrarian Change*, *10*(1), 1–32.

Braun, V., & Clarke, V. (2006). Using thematic analysis in psychology. *Qualitative Research in Psychology*, *3*(2), 77–101.

Cadieux, K. V. (2008). Political ecology of exurban "lifestyle" landscapes at Christchurch's contested urban fence. *Urban Forestry & Urban Greening*, *7*(3), 183–194.

Côté-Roy, L., & Moser, S. (2019). 'Does Africa not deserve shiny new cities?' The power of seductive rhetoric around new cities in Africa. *Urban Studies*, *56*(12), 2391–2407.

Cousins, B. (2007). Agrarian reform and the 'two economies': Transforming South Africa's countryside. In *The land question in South Africa: The challenge of transformation and redistribution* (pp. 220–245), HSRC Press.

Creswell, J. W. (2014). *Research design: Qualitative, quantitative, and mixed methods approaches*. Sage Publications.

Department of Agriculture, Forestry and Fisheries. (2018). *National policy on the preservation and development of agricultural land*. Government of South Africa, Pretoria.

Department of Rural Development and Land Reform. (2019*). National spatial development framework*. Government of South Africa.

Farr, D. (2018). *Sustainable urbanism: Urban design with nature*. John Wiley and Sons.

Frayne, B., Crush, J., & McLachlan, M. (2014). Urbanization, nutrition and development in Southern African cities. *Food Security*, *6*, 101–112.

Gałecka-Drozda, A., Wilkaniec, A., Szczepańska, M., & Świerk, D. (2021). Potential nature-based solutions and greenwashing to generate green spaces: Developers' claims versus reality in new housing offers. *Urban Forestry & Urban Greening*, *65*, 127345.

Hall, R. (2011). Revisiting unresolved questions: Land, food and agriculture. *Transformation: Critical Perspectives on Southern Africa*, *75*(1), 81–94.

Hall, R. (2014). *Land reform in South Africa: Troubled past, tenuous future. Institute for Poverty, Land and Agrarian Studies* (PLAAS). University of the Western Cape.

Jonah, C. M., & May, J. D. (2020). The nexus between urbanization and food insecurity in South Africa: Does the type of dwelling matter? *International Journal of Urban Sustainable Development*, *12*(1), 1–13.

Kepe, T., & Tessaro, D. (2014). Trading-off: Rural food security and land rights in South Africa. *Land Use Policy*, *36*, 267–274.

Kvale, S. (2007). *Doing interviews*. Sage Publications.

Lahiff, E. (2007). *Land reform in South Africa: Is it meeting the challenge?* (Policy Brief, 20).

Mancebo, F. (2015). Insights for a better future in an unfair world: Combining social justice with sustainability. *Transitions to Sustainability*, 105–116.

Miles, B. W., & Jozefowicz-Simbeni, D. M. H. (2010). Naturalistic inquiry. In B. Thyer (Ed.), *The handbook of social work research method* (pp. 415–425). Pearson.

Mistry, A., & Spocter, M. (2022). Exploring the "eco-ness" of South Africa's eco-estates. *Journal of Urbanism: International Research on Placemaking and Urban Sustainability*, 1–20.

Mukaddam, F. (2024). South Africa's history of colonialism and apartheid. In F. Mukaddam (Ed.), *Muslim women between community and individual rights: Legal pluralism and marriage in South Africa*. Springer Nature.

Nunn, N. (2020). The historical roots of economic development. *Science*, *367*(6485), 9986–9997.

Patton, M. Q. (2015). *Qualitative research & evaluation methods: Integrating theory and practice*. Sage Publications.

Rogerson, J. M. (2023). The limits of green infrastructure development in urban South Africa: The case of green roofs. *Analele Universității din Oradea, Seria Geografie*, *33*(1), 16–26.

Schoeman, R. C., & Gunter, A. W. (2018). Greenwashing: Exploring the role of green building rating agencies on the building industry, the case of Johannesburg. *South African Geographical Society Conference Proceeding*, *1*, 345.

Sherriff-Shüping, R. D. (2015). *Buying into residential eco-estates: Perception and reality of "green living" in eco-estates in Gauteng, South Africa* [Doctoral dissertation]. University of South Africa, Department of Agriculture and Environmental Sciences.

Soga, M., Yamaura, Y., Koike, S., & Gaston, K. J. (2014). Land sharing vs. land sparing: Does the compact city reconcile urban development and biodiversity conservation? *Journal of Applied Ecology*, *51*(5), 1378–1386.

Strauss, A., & Corbin, J. (1998). *Basics of qualitative research: Techniques and procedures for developing grounded theory*. Sage Publications.

Swilling, M. (2016). *Just transitions: Explorations of sustainability in an unfair world*. UCT Press.

Tezangi, N. R. (2014). Sustainable urbanism: Analysis of sustainable environment principles in practical urban form. *Journal of Engineering and Architecture*, *2*(2), 197–205.

Thinda, K. T., Ogundeji, A. A., Belle, J. A., & Ojo, T. O. (2020). Understanding the adoption of climate change adaptation strategies among smallholder farmers: Evidence from land reform beneficiaries in South Africa. *Land Use Policy*, *99*, 104858.

Turok, I. (2012). *Urbanisation and development in South Africa: Economic imperatives, spatial distortions and strategic responses* (Urbanisation and Emerging Population Issues Working Paper 8). United Nations Population Fund (UNFPA).

Van Noorloos, F., & Kloosterboer, M. (2018). Africa's new cities: The contested future of urban development. *Urban Studies*, *55*(6), 1223–1241.

Walker, C., Bohlin, A., Hall, R., & Keep, T. (2010). *Land, memory, reconstruction, and justice: Perspectives on land claims in South Africa*. Ohio University Press.

Walker, P. A., & Fortmann, L. P. (2003). Whose landscape? A political ecology of the 'Exurban' Sierra. *Cultural Geographies*, *10*(4), 469–491.

Yin, R. K. (2018). *Case study research and applications: Design and methods*. Sage Publications.

Zantsi, S. (2021). Explaining land size demand among potential emerging farmers in South Africa: What does it mean for land redistribution policy? *South African Geographical Journal*, *103*(4), 519–539.

18 Long-term agricultural land use change analysis and land use policy in Karamoja, Uganda, since 2000

Catherine Nakalembe, Diana B. Frimpong, and Gordon Y. Mwesigwa

Introduction

Karamoja, in northeastern Uganda, is an ecologically fragile area traditionally characterized by pastoralist and agropastoralist livelihood systems Krätli (2010). This combination enabled communities to adapt to seasonal rainfall patterns and environmental uncertainties through seasonal migration. However, over the past several decades, government policies, development initiatives, and demographic shifts have catalyzed a transition toward more sedentary agricultural practices Filipová and Johanisova (2017); Nakalembe et al. (2017). This shift has led to the encroachment on and fragmentation of grazing lands better suited to the region's arid and variable climate Osaliya et al. (2020).

Pastoralism is a significant economic activity along Uganda's cattle corridor, with pastoralists being the country's primary livestock producers Krätli (2010). Despite contributing substantially to livelihoods, pastoral systems have been consistently undervalued in development policies, often leading to interventions that undermine traditional livelihoods Krätli (2010); Murphy et al. (2017). Current policies and their impacts on Karamoja often have roots in colonial-era approaches to pastoralism and development Johnson and Anderson (1988). The legacy of system-blind pastoral development, embedded in infrastructure, institutions, and expertise, continues to influence present-day approaches Krätli (2010).

Land use in Karamoja is further complicated by a history of significant conflict and displacement Kumssa et al. (2009); Nakalembe et al. (2017). However, violence within Karamoja must be examined within the broader context of the region in Uganda. During the same period that disarmament operations were conducted in Karamoja, other parts of Uganda were experiencing significant displacement and conflict, with an estimated 8% of the Ugandan population internally displaced in 2006 Krätli (2010).

Agricultural Expansion and Land Cover Changes in Karamoja

Climate change, population growth, and socioeconomic factors have accelerated agricultural expansion, profoundly impacting food security and environmental sustainability Egeru et al. (2014a); Stavi et al. (2022). The shift from pastoral to agricultural land use has significant ecological implications for Karamoja's semi-arid

DOI: 10.4324/9781032672779-22

environment. Studies have shown that cropland expansion has led to a decline in perennial grasses and increased soil erosion, raising concerns about the long-term sustainability of these land use changes Baker (1975); Krätli (2010). Quantifying and mapping these rapid land use transitions, particularly the expansion of cropland, are crucial for understanding their implications. The challenges faced in Karamoja are not isolated, and throughout the cattle corridor, land tenure reforms and the shift toward formalized private property regimes have led to land access issues, conflicts, and marginalization of pastoralist communities Murphy et al. (2017). The cattle corridor is a rangeland ecosystem that stretches from the southwest to the northeast of the country and is home to a large concentration of livestock Hasahya et al. (2023). While land tenure reforms, such as the 1995 Constitution and the 1998 Land Act, aimed to recognize customary tenure, formalizing ownership has often favored settlers and town dwellers over pastoralist communities. The high costs and bureaucratic procedures for registering communal land have disadvantaged many pastoralists Krätli (2010).

Analyses of land cover changes across the Karamoja region over the past several decades reveal extensive cultivation and expansion of smallholder cropland (Figure 16.1). Nakalembe et al. (2017) documented a 299% increase in cropland area for the entire region from 2000 to 2011, with the most rapid expansion occurring around major towns. This trend is corroborated by more localized studies on the Lokere and Lokok catchments by Osaliya et al. (2020) and broader dynamics in forage availability by Egeru et al. (2014a). Future projections by Osaliya et al. (2020) suggest that cropland could potentially double from 13.8% of land area in 2003 to over 30% by 2050, accompanied by a decline in grassland. Concurrently, significant losses have occurred in woodland and bushland vegetation across the region. Egeru et al. (2014a) reported that woodland conversion to grassland was a major shift from 1986 to 2000, attributing this to increased livestock numbers and grazing pressure. However, Osaliya et al. (2020) observed an increase in bushland in later years, linking this to reduced burning and grazing due to periods of instability.

Driving Factors and Impacts: Multiple complex factors drive these land cover changes, including government policies and development programs encouraging settled cultivation (Krätli, 2010; Murphy et al., 2017), land tenure reforms favoring privatization and restriction of pastoralist mobility (Murphy et al., 2017), and climatic factors such as droughts (Abrahams, 2021). These changes have significantly transformed land use patterns, concentrating more on permanent settlements. Pastoralist communities have been particularly impacted, experiencing loss of grazing lands, increased conflicts over resources, and food insecurity (Abrahams, 2021; Murphy et al., 2017).

Drought and Food Insecurity – A Complex Interplay: Recent studies underscore the complex relationship between drought and food insecurity in Karamoja, highlighting the interplay of climate variability, land use change, conflict, and vulnerability to climatic shocks. Drought remains a major driver of food insecurity in the region, leading to decreased agricultural productivity and exacerbating food shortages and malnutrition. Egeru et al. (2014a) demonstrated how frequent droughts have led to declining pasture and water resources, affecting livestock productivity. Nakalembe (2018) identified extreme agricultural droughts in 2002, 2004, and 2009, revealing significant spatial variations in impacts across districts. While their model linked vegetation data to food aid needs, the study emphasizes that food insecurity is driven by multiple factors beyond drought. Recurrent droughts force changes in land use, including the increased movement of pastoralists seeking grazing lands, leading to resource conflicts (McGuirk & Nunn, 2024; Muluken, 2020). Chaplin (2017) corroborated this multidimensional nature of food insecurity, while Mkutu (2003) explored the link between climate-induced resource scarcity and conflict. Datzberger (2022) highlighted how the shift from pastoralism to agro-pastoralism has introduced new vulnerabilities. These studies collectively emphasize the need for a holistic approach to address food insecurity in Karamoja, considering climatic, environmental, socioeconomic, and political factors.

Environmental Resilience and Policy Implications: The transition from traditional pastoral systems to settled agriculture in Karamoja has profound implications for environmental resilience and sustainability. Traditional pastoral practices evolved over centuries to optimize resource use in semi-arid environments through mobility and flexible grazing patterns (Krätli, 2010). Current agricultural policies, however, often fail to account for the region's inherent ecological constraints and climate variability, potentially undermining long-term environmental sustainability. Land use changes have triggered cascading effects on local ecosystems. The expansion of cropland into traditional grazing areas has led to soil degradation, reduced vegetation cover, and altered hydrological patterns (Egeru et al., 2014a). These changes diminish the landscape's natural resilience to climate shocks, particularly during drought periods. The fragmentation of grazing corridors further compounds these issues by limiting the ability of pastoral communities to implement traditional drought-coping strategies (Murphy et al., 2017). The impact on biodiversity is equally significant. The conversion of natural vegetation to cropland has reduced habitat connectivity and altered species composition. While some studies indicate increased bushland in certain areas (Osaliya et al., 2020), this change often represents ecological degradation rather than recovery, as it frequently involves the proliferation of less desirable species at the expense of native vegetation communities. Local communities bear the brunt of these environmental changes. Reduced access to diverse natural resources undermines traditional livelihood strategies and food security options. The degradation of ecosystem services, particularly those related to water retention and soil fertility, creates a feedback loop that further threatens agricultural productivity (Stavi et al., 2022). This situation is particularly acute for marginalized groups who lack the resources to invest in agricultural inputs or alternative livelihood strategies. Current agricultural policies often prioritize short-term productivity gains over long-term environmental resilience. The push for intensive cultivation in areas better suited to pastoral production risks creating a "poverty trap" where environmental degradation and economic vulnerability reinforce each other. This suggests a need for policy approaches that better integrate traditional ecological knowledge with modern agricultural practices, acknowledging the vital role of mobility and flexible resource use in maintaining environmental resilience (McGuirk & Nunn, 2024).

Recent Advances in Remote Sensing for Land Use Mapping in Drylands

Recent satellite remote sensing and cloud computing advancements have revolutionized large-scale land use mapping in dryland regions like Karamoja. Google Earth Engine (GEE) has emerged as a powerful platform for this study, offering cloud-based access to extensive satellite imagery archives and substantial computational resources (Nakalembe et al., 2024). These advancements facilitate the efficient processing of vast datasets needed for comprehensive land cover analysis (Gorelick et al., 2017). The Harmonized Landsat-Sentinel (HLS) dataset, which combines Landsat 8 and Sentinel-2 imagery, provides high temporal frequency observations with a 2–3-day revisit time at 30-meter resolution (Claverie et al., 2018), making it particularly valuable for monitoring seasonal changes in agricultural landscapes. For vegetation analysis in dryland regions, this study employs the Generalized Difference Vegetation Index (GDVI) (Wu, 2014), which offers particular advantages over traditional indices in areas with low vegetation cover. GDVI shows increased sensitivity in rangelands and sparsely vegetated areas, with values magnified by 190%–867% for rangeland compared to conventional indices. The versatility of GDVI's adjustable power *n* makes it especially suitable for Karamoja's mixed landscape of cropland and pastoral areas, with $GDVI^2$ providing optimal performance across the region's diverse land cover types (Wu, 2014). These technological advances enable more detailed analysis of land use changes in Karamoja, where data availability and resolution have historically limited comprehensive assessment (Nakalembe et al., 2017). By leveraging these tools, this study examines the linkages between agricultural expansion, land policy shifts, and climate impacts since 2000, providing new evidence to inform sustainable land governance aligned with the region's ecological and social context.

Study Area

The Karamoja region in northeastern Uganda comprises nine districts: Karenga, Kaabong, Kotido, Abim, Moroto, Napak, Nabilatuk, Amudat, and Nakapiripirit. It covers an area of 28,000 km^2 located between 1°and 4° N and 33°–35° E. This semi-arid region is situated on a large plateau at an average elevation of 1,000 m above sea level, characterized by undulating plains dominated by savanna grasslands, shrublands, and scattered trees, interspersed with a few prominent mountains (Nakalembe et al., 2017 & Kagan et al., 2009).

Karamoja encompasses three distinct livelihood zones running north to south, reflecting varying soils and rainfall patterns (USAID, 2017):

(i) The pastoral zone: A semi-arid area with annual rainfall below 500 mm, focused on livestock production with limited crop cultivation.
(ii) The agropastoral zone: Receives 500–800 mm of erratically distributed annual rainfall, supporting both livestock and crop farming.
(iii) The agricultural zone or "green belt": Located in the south and west, benefits from 800 mm to 1,200 mm of annual rainfall and fertile soils, enabling diverse crop production (Famine Early Warning Systems Network, 2016).

The region experiences a single rainy season from April to November, with significant spatial variability in annual precipitation (USAID, 2017). Average annual temperatures range from 16°C in highlands to 24°C in lowlands. Climate trends since the 1960s indicate warming and drying, with average temperature increases of 1.3°C (1975–2009), and a 15–20% reduction in annual rainfall (Mubiru, 2010). Projections for 2040–2069 suggest further temperature increases of 1.2–1.5°C on average, while rainfall projections remain uncertain (Le Groupe-conseil Baastel, 2015).

Agriculture faces significant challenges across all zones in Karamoja, including poor soil quality (predominantly sandy clay loams with low productivity), heavy reliance on rainfall, and limited resources (Gayfer & Shanahan, 2012). The region's main crops are sorghum and maize, supplemented by beans, groundnuts, cassava, and sweet potatoes, with crop choices varying by livelihood zone (FAO, 2014). Climate risks to agriculture include rising temperatures, more extreme weather events, potential crop failures, increased pest incidence, and shorter sporadic growing periods (Famine Early Warning Systems Network, 2016).

Karamoja faces significant development challenges, with the lowest socio economic indicators in Uganda. According to the Uganda Bureau of Statistics (UBOS), the population of Karamoja stands at 1.45 million in 2024, an increase from 0.97 million recorded in the 2014 Uganda National Household Survey of Statistics (UBOS, 2024). Despite representing only 2.8% of Uganda's population, the region contributes 65.7% to the total national poverty estimate. The population is young, with an average age of 15 years, and half of the total population being female (UBOS, 2016). Karamoja has the highest total fertility rate in Uganda, with women of reproductive age (15–49 years) giving birth to an average of eight children, compared to the national average of five children (UBOS, 2018).

The National Panel Survey 2019/2020 reveals that Karamoja had the lowest Primary School Net Enrollment Ratio and Gross Enrollment Ratio at 42%, compared to the rural average of 79%. Infrastructure and living conditions lag behind national standards, with only 17% of households having iron sheet roofs and nearly 70% lacking proper toilets. Government health facilities are the primary choice for health care services, utilized by 70% of the sick population. Economic indicators show a high proportion of the working population in employment (70%), primarily in subsistence agriculture (64%), and also high unemployment rates (12%).

Livelihoods in Karamoja are primarily agro-pastoralist, engaging in a combination of livestock rearing and subsistence farming (Nakalembe et al., 2017; USAID, 2017). Agriculture is rainfed, low-input, and vulnerable to frequent droughts that vary in severity and impact (Nakalembe, 2018). Historical factors such as conflict, environmental degradation, poor infrastructure, and limited education and health access constrain development (UBOS, 2017; USAID, 2017). Land use has shifted since the 1960s due to policies promoting sedentary agriculture and mining over pastoralism (Rugadya & Kamusiime, 2013).

Food insecurity remains high, with communities relying on food aid (Nakalembe et al., 2017; World Food Programme and UNICEF, 2016). A land use study could quantify agricultural expansion, its drivers, and its implications for livelihoods and sustainability in this climate-vulnerable region, considering the complex socioeconomic context and historical dynamics that have shaped the region's development.

Research Overview and Methodological Framework

This study investigates agricultural land use changes and climate trends in Karamoja, Uganda from 1981 to 2020. Through an integrated analysis combining remote sensing, climate data, and policy assessment, we examine three key research questions:

1) How have land use policies and development initiatives since 2000 shaped Karamoja's agricultural landscape?
2) What are the spatiotemporal patterns of cropland expansion in Karamoja, and how are these patterns influenced by policy implementation and correlate with climate trends?
3) How can recent advances in remote sensing technologies improve the quantification and understanding of land use changes in this complex pastoral-agricultural landscape?

By addressing these questions through multiple analytical approaches, our study aims to inform evidence-based policy decisions for sustainable land governance aligned with Karamoja's unique ecological and social context.

Materials and Methods

Policy Analysis

The timeline of land use changes and policies in Karamoja (1960–2020) was developed through a literature review of peer-reviewed articles, governmental and nongovernmental reports, and academic studies, with key sources including Nakalembe et al. (2017), Egeru et al. (2014b), and Krätli (2010). A systematic analysis of policy documents, focusing on national agricultural policies, land use regulations, and Karamoja-specific development plans, was conducted to identify objectives, implementation strategies, and potential impacts on land use. Estimates of cropland area were derived from multiple studies, primarily Nakalembe et al. (2017) and Egeru et al. (2014b), and cross-referenced with policy events for consistency. The timeline was chronologically arranged with key policy events, development initiatives, and land use changes, integrating cropland area estimates to illustrate the relationship between policy interventions and land use changes. Data synthesis involved correlating policy implementations with observed changes in cropland areas to identify potential causal relationships and major trends.

Remote Sensing Datasets

This study examines cropland changes from 2000 to 2019 using a combination of datasets, including GLAD cropland masks, a Very High Resolution (VHR) dataset, and the Uganda Land Cover Scheme II. These datasets collectively provide a multi-resolution view of cropland extent across the study area over time, as summarized in Table 18.1.

The analysis starts with the Uganda Land Cover Scheme II data from 2000, which establishes a baseline for cropland extent at a 30m resolution RCMRD (2018). For subsequent years, we utilized GLAD cropland masks from 2003, 2011, 2015, and 2019, derived from Landsat data, to maintain consistency in temporal coverage (Potapov et al., 2022). Additionally, to obtain a more granular perspective, we integrated a VHR dataset from 2012, providing 2m resolution data specifically for the Karamoja region Nakalembe et al. (2017).

Before employing the GLAD dataset in our analysis, we evaluated its performance using an independent sample of 100 randomly selected points across our region of interest. We manually classified these points using high-resolution satellite imagery, historical Google Earth imagery, Harmonized Landsat Sentinel-2 mosaics, and vegetation indices. These manually labeled points were then compared to the GLAD cropland classifications for the corresponding years to validate the dataset's accuracy and assess its applicability to our specific study area.

Change detection analysis directly compared the cropland masks from consecutive periods (Table 18.1), subtracting the earlier period's mask from the later period's mask, resulting in maps highlighting areas of cropland expansion. The total expansion area was calculated by summing the pixel areas classified as new cropland within the study area.

It's important to note that the 2012 VHR dataset, with its 2m resolution, provides a significantly more detailed view of cropland extent compared to the 30m resolution of other datasets. This higher resolution allows for the detection of smaller agricultural plots and more precise field boundary delineation, potentially leading to a higher overall cropland area estimate for that year. When interpreting results, special consideration should be given to this methodological difference.

Table 18.1 Summary of data used in the study

Period	*Dataset/Source*	*Data ID/Resolution*
2000	Uganda Land Cover Scheme II	30m resolution
1995–2003	Landsat 5 Surface Reflectance	LANDSAT/LT05/C02/T1 L2
2000–2003	GLAD Cropland Mask (2003)	users/potapovpeter/Global cropland 2003
2008–2011	GLAD Cropland Mask (2011)	users/potapovpeter/Global cropland 2011
2011–2012	High Resolution Cropland Extent, Zenodo for Karamoja Uganda 2010/11	2m resolution
2013–2015	GLAD Cropland Mask (2015)	users/potapovpeter/Global cropland 2015
2016–2019	GLAD Cropland Mask (2019)	users/potapovpeter/Global cropland 2019
2020–2023	HLS (Annual Mosaics)	HLSL30 v002 (2000–2019)
Auxiliary Data: GHSL Population Data (2000, 2005, 2010, 2020)		

To provide context for our cropland change analysis, we incorporated population data from the Global Human Settlement Layer (GHSL) for 2000, 2005, 2010, and 2020. This dataset helps understand potential relationships between cropland expansion and changes in human settlement patterns.

Climate Trends Analysis

We analyzed precipitation and temperature patterns in Karamoja, Uganda from 1981 to 2020 using CHIRPS rainfall and ERA5-Land temperature data.

Data Processing and Analysis: Data were extracted and processed using GEE. Monthly and seasonal statistics were calculated using ee.Reducer functions, and regional statistics were computed with ee.Image.reduceRegions() to analyze spatial variability across the entire region of Karamoja.

Trend Analysis: We quantified trends using the non-parametric Mann-Kendall test for significance and Sen's slope estimator for magnitude (Sen, 1968; Mann, 1945). These methods are robust for assessing monotonic trends in climate time series without requiring assumptions about data distribution (Wang et al., 2020; Banerjee et al., 2020; Zeng et al., 2019). The Mann-Kendall test was applied to each pixel time series using ee.Reducer.mannKendall() in GEE. Pixels with an absolute score value greater than 1.96 (p-value ≤ 0.05) were considered to have a significant trend at the 95% confidence level. Sen's slope was calculated using ee.Reducer.sensSlope() to estimate the magnitude of significant trends. For each pixel, the median slope was calculated if the Mann-Kendall test indicated a significant trend ($p \leq 0.05$). To summarize trends, we calculated spatial averages of annual time series within the entire region of Karamoja and exported these to R for further analysis using the "trend" package.

Results

Karamoja Land Use Change Timeline (1960–2020)

Since the 1960s, the Karamoja region has undergone significant changes in land use and policy. Figure 18.1 illustrates this evolution over six decades, highlighting major policy interventions, development projects, and cropland expansion. The timeline's left axis chronicles key events and policies that have shaped land use in the region. The right axis and accompanying color gradient represent the increasing percentage of land under crop cultivation. This visual representation demonstrates a clear trend of cropland expansion, from less than 5% of the total land area in the 1960s to between 35% and 40% by 2020 (Nakalembe et al., 2017; Egeru et al., 2014b). This dramatic increase reflects Karamoja's transition from a predominantly pastoral system to a more diversified land use model, influenced by a complex interplay of policy interventions, development projects, demographic changes, and socioeconomic factors.

Key interventions include introducing water development with dams and valley tanks in 1938. In the 1950s, the successful campaign against East Coast Fever led to increased cattle populations (Baker, 1975). From 1955 to 1965, a significant shift in land use policy occurred, with about 94.6% of Karamoja's land being gazetted for conservation purposes (Krätli, 2010). In 1963, the Ranch Scheme was introduced, aiming to commercialize pastoralism (Ocan & Ocan, 1994). The 1970s witnessed the establishment of protected areas and increased insecurity (Egeru et al., 2014b; Krätli, 2010). The 1980s and 1990s were marked by forced disarmament campaigns and the promotion of crop cultivation (Filipova & Johansson, 2017; Levine, 2010). Significant policy changes occurred in the late 1990s and early 2000s. The New Constitution of Uganda was established in 1995, followed by the Local Governments Act in 1997. The Land Act implementation began in 1999, with an amendment in 2001 that recognized customary land tenure (Filipova and Johansson, 2017). In 2002, over half of the previously gazetted protected areas were degazetted (Krätli, 2010). The 2000s saw the introduction of several development initiatives. The first phase of the Northern Uganda Social Action Fund (NUSAF 1) ran from 2003 to 2009, focusing on community empowerment and infrastructure development. In 2006, the Disarmament, Demobilization and Reintegration (DDR) program was implemented. By 2009, about 50% of nonconservation land was under 38 mining licenses (Krätli, 2010). The 2010s saw increased mining and infrastructure development (Egeru et al., 2014b), further impacting traditional land use patterns.

The second phase of NUSAF (2009–2015) emphasized livelihood support and institutional development. Recent years have seen further policy developments. The Karamoja Integrated Development Plan (KIDP) was introduced in 2011 (Filipova & Johansson, 2017). NUSAF 3 (2015–2020) focused on public works and introduced a Disaster Risk Financing component to enhance climate resilience (World Bank's Disaster Risk Finance and Insurance Program, 2023). Finally, in 2018, the National Land Policy was implemented (Filipova & Johansson, 2017), marking the region's most recent significant land policy change.

While NUSAF projects aimed to support development in northern Uganda, including Karamoja, their implementation often reflected broader national priorities that did not always align with the needs of pastoral communities (Krätli, 2010). In contrast to some national policies, the African Union adopted a Policy Framework for Pastoralism in 2010, recognizing the value of mobile pastoralism and the need to support existing pastoral production systems (Krätli, 2010).

Cropland Change

Karamoja Cropland Changes (2000–2019)

The analysis of cropland area changes in the region reveals a significant expansion over the observed period from 2003 to 2019. The data shows a steady increase in cropland area, with the most dramatic growth as shown in Figure 18.2. In 2011, the baseline year, the cropland area was approximately 71,430 hectares. By 2015, this had increased by 18.85%, reaching 84,895 hectares. The most substantial growth was observed in the four years that followed, with the cropland area reaching

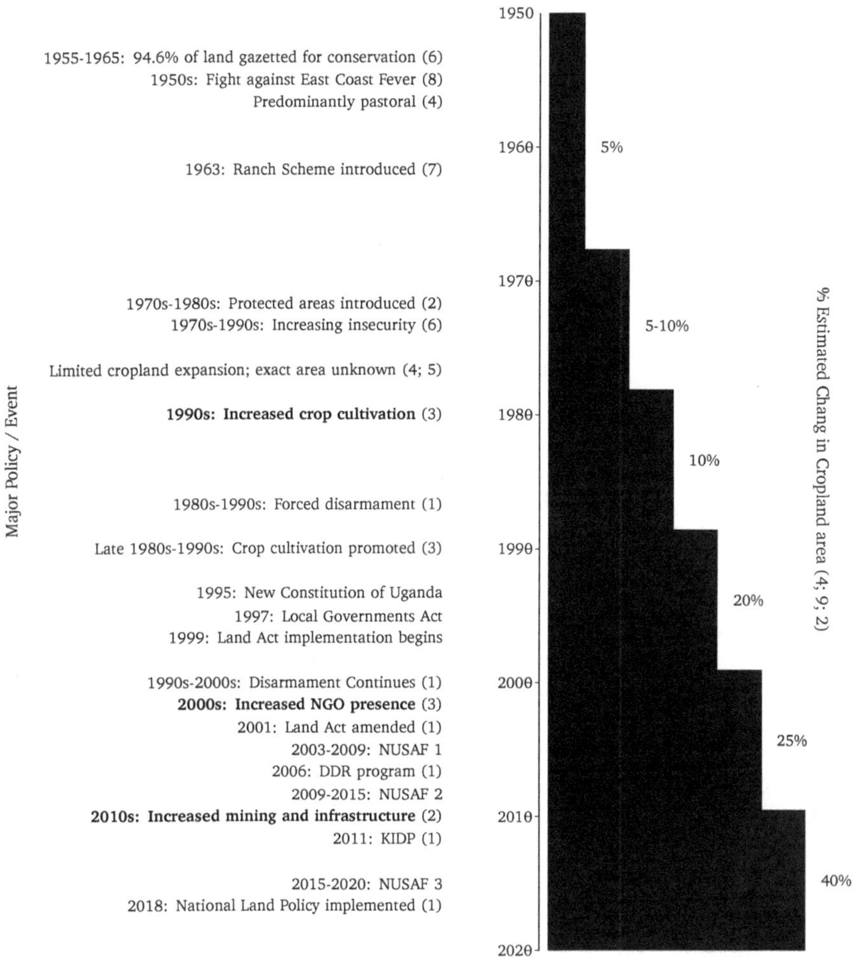

Figure 18.1 Karamoja land use change timeline (1960–2020). Major policy events (left) and cropland expansion (right, orange gradient) show the region's transition from pastoral to mixed land use systems. Cropland increased from < 5% (1960s) to 35–40% (2020) of the total land area.

130,347 hectares by 2019, representing an 82.48% increase from the 2011 baseline. This rapid expansion, particularly in the latter half of the decade, suggests significant agricultural development or land use changes in the region. Table 18.2 provides a detailed breakdown of these changes, showing both the absolute areas and the percentage changes for each period.

Analysis of cropland area changes in Karamoja from 2000 to 2019 reveals apparent fluctuations and an overall increase in agricultural land use (Table 18.2). In 2000, cropland based on RCMRD (2018) was estimated at 90,650.70 hectares (3.29% of the total regional area). Data from Potapov et al. (2022) indicates that by

Table 18.2 Changes in cropland area from 2003 to 2019 based on Potapov et al. (2022)

Period	*Cropland area (ha)*	*Change in cropland area (%)*
2003	64,270.93	baseline
2011	71,429.83	11.14%
2015	84,894.58	32.08%
2019	130,347.04	102.79%
Change 2011 vs 2003	7,158.90	11.14%
Change 2015 vs 2003	20,623.65	32.08%
Change 2019 vs 2003	66,076.11	102.79%

2011, cropland had decreased to 71,429.83 hectares (2.59%), seemingly representing a 21.20% reduction.

However, it is crucial to note that this apparent decrease may be largely attributable to methodological differences and varying definitions of cropland between the two assessments rather than an actual reduction in cultivated area. The 2000 and 2011 data points come from different sources, likely employing distinct methodologies, imagery resolutions, and potentially different cropland classifications, making direct comparisons challenging.

Focusing on the consistent dataset from Potapov et al. (2022), we observe a steady increase in cropland area from 2011 onwards. Cropland expanded from 71,429.83 hectares (2.59%) in 2011 to 84,894.58 hectares (3.08%) in 2015, and further to 130,347.04 hectares (4.73%) by 2019. The period from 2015 to 2019 witnessed the most rapid expansion, with a 53.54% increase in cropland area. Overall, between 2011 and 2019, cropland in Karamoja increased by 58,917.21 hectares, or 82.48%, according to this consistent dataset.

The 2012 data from Nakalembe et al. (2017) shows a significantly higher cropland area of 360,999.26 hectares (13.11% of the total regional area) compared to other years in the dataset, a discrepancy likely attributable to several factors. Methodological differences play a key role, as the 2012 data was developed using Very High Resolution (VHR) WorldView-1 and WorldView-2 satellite imagery, offering superior spatial resolution compared to other commonly used satellite data sources. This 2m resolution VHR dataset provides a markedly different representation of cropland extent compared to more recent 30m resolution datasets. While it captures substantial cropland areas often missed by coarser resolutions, including small plots and precise field boundaries, it may also misclassify some non-cropland pixels as cropland, particularly in complex landscapes. The resulting cropland extent estimate differs significantly from newer 30m data-derived estimates due to a combination of detection of cropland areas not visible in coarser resolutions, occasional misclassification, and actual land use changes that occurred after 2012. When interpreting these results, it's crucial to consider that apparent differences in cropland extent between the older VHR and newer 30m datasets reflect this complex interplay of resolution-based detection capabilities, potential classification errors, and genuine landscape transformations over time rather than solely representing contemporary land use patterns.

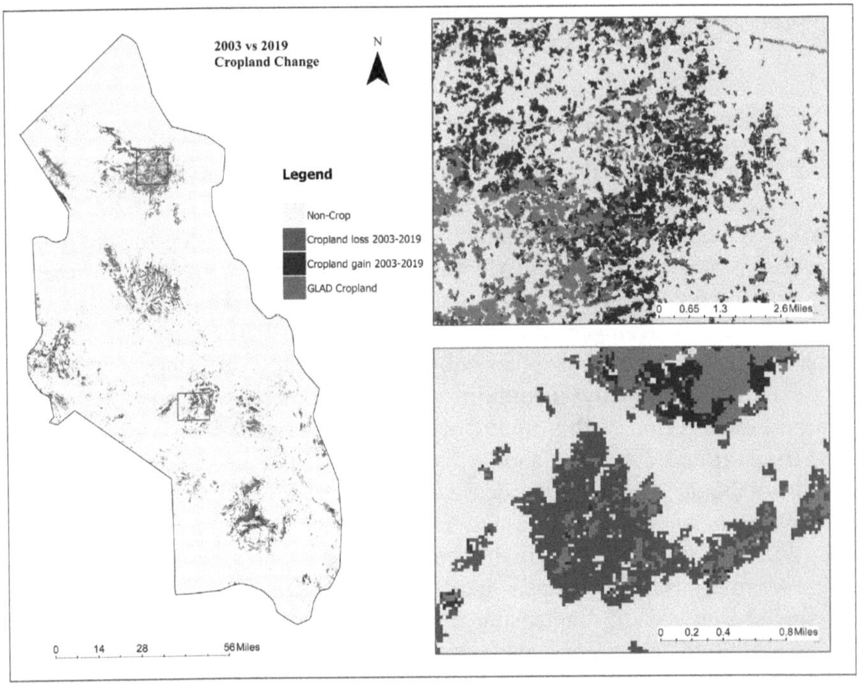

Figure 18.2 Change maps based on GLAD cropland data.

Source: (Potapov et al., 2022).

Additionally, agricultural land use can fluctuate due to factors such as weather patterns, economic conditions, or policy changes, and the 2012 data might reflect a peak in agricultural expansion. Given these potential sources of variation, direct comparisons between the 2012 data point and other years should be approached with caution. The stark difference in estimates also highlights the need for higher-resolution mapping in this region, which is dominated by small fields less than 2 hectares in size. Such small-scale agriculture is often underrepresented in lower-resolution assessments, potentially leading to significant underestimations of total cropland area. Importantly, even with the observed expansion, cropland occupied only 4.73% of Karamoja's total land area (27,528 km^2 or 2,752,800 hectares) by 2019, according to the Potapov et al. (2022) dataset. This suggests that while agricultural land use has increased significantly, there remains substantial potential for further agricultural expansion or conservation of nonagricultural lands in the region.

These findings underscore the complexity of tracking land use changes in regions like Karamoja, where small-scale agriculture predominates. They highlight the critical need for consistent, high-resolution monitoring approaches to assess and understand land use dynamics over time accurately. Future studies should aim to employ methodologies that can capture the nuances of small-scale agriculture while maintaining consistency for reliable trend analysis.

Climate Trend Analysis – Precipitation

Figure 18.3b illustrates the spatial distribution of precipitation magnitude and statistical confidence across the Karamoja region from 1981 to 2020. The map reveals significant annual precipitation increases (at the 95% confidence interval) in Amudat, Nakapiripirit, Napak, Moroto, Abim, and parts of Kotido districts, with magnitudes ranging from 3.3mm to 16mm per decade. Trend data shows the region experienced above-average precipitation from 2010 to 2020, except for 2017. This recent wet period likely contributes substantially to the overall upward precipitation trend. Statistical analysis confirms a significant increase in precipitation over the four-decade period, with a confidence level exceeding 95%.

Analysis of decadal trends shows that in some locations of Karamoja, mainly around Kaabong, and Kotido, decadal averages have consistently increased annually from approximately 10mm in 1981/1990 to 16mm in three decades (2011/2020). At the same time, in the same geography and other surrounding areas around Abim and Napak, we observe the same increasing rainfall intensities during the rainy season, steadily increasing from approximately 18mm per decade during 1981–1990 to 28mm per decade during 2011–2020. This pattern is consistent across seasons in each decade. However, rainfall intensities in the dry season are significantly less, ranging between 4mm and 10mm each decade.

Temperature

Figure 18.3a illustrates the spatial distribution of temperature trends across the Karamoja region. Areas exhibiting statistically insignificant trends are characterized by z-score values ≤ 1.96 ($\alpha \geq 0.05$). Significant positive trends in annual average temperatures were observed in Abim, Napak, central Kaabong, and Nakapiripirit districts. In contrast, northern Kaabong, Kotido, Amudat, and the areas surrounding Moroto experienced temperature increases that, while present, did not reach statistical significance (z-score ≤ 1.96).

Most of Karamoja's sub-districts experienced their highest decadal average temperatures during the drier months, ranging from 25°C to 28°C. Conversely, during the rainy season, these areas exhibited significantly lower temperatures, typically between 19°C and 23°C. Long-term analysis reveals a statistically significant increase in overall temperature across the entire district from 1981 to 2020. This warming trend is characterized by an R^2 value of 0.11 and a p-value ≤ 0.05, indicating significant change over the four-decade period.

Climate Anomalies and Shifts

Our analysis reveals significant climatic variability in the Karamoja region over the past four decades (1981–2020). This period is characterized by severe droughts

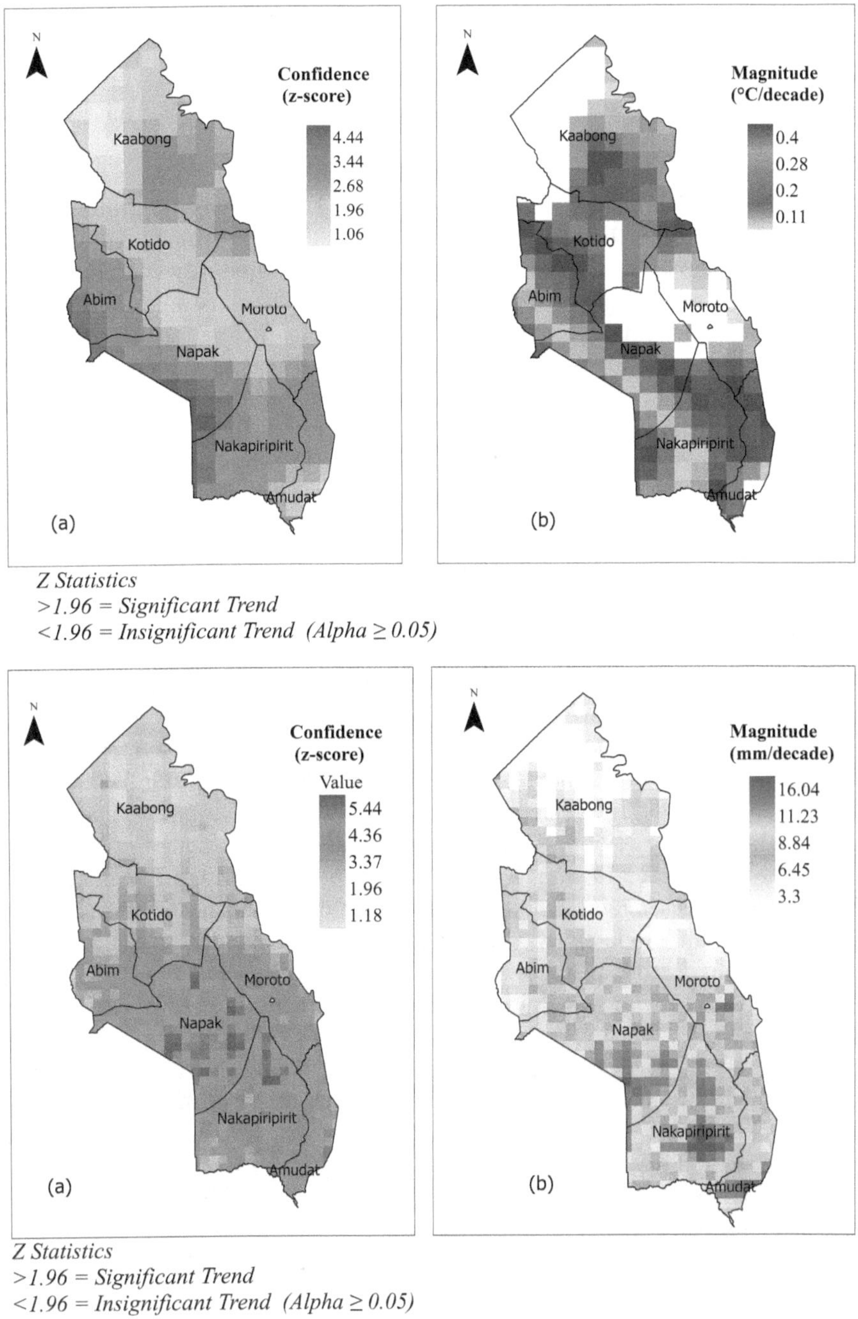

Figure 18.3 Spatial distribution of temperature and precipitation trends.

(a) Spatial distribution of temperature trends

(b) Spatial distribution of precipitation trends

and extreme deviations from decadal averages, interspersed with years of heavy rainfall. Drought years include 1984, 1993, 2002, 2005, and 2009. These years exhibited precipitation anomalies ranging from -15mm to -20mm, representing deviations of approximately -34%, -23%, -20%, -15%, and -17% from long-term averages, respectively.

In contrast, the most recent decade (2011–2020) exhibited a marked increase in precipitation. Our results indicate an estimated 17% increase in rainfall during this period, providing clear evidence of the region's erratic climate patterns. This recent trend underscores the complex nature of climate variability in Karamoja and highlights the need for adaptive management strategies.

Temperature patterns in Karamoja reveal significant fluctuations over the study period. Notable warming events occurred in 1987, 1993, 2002, 2005, 2009, 2015, 2016, and 2017, with temperatures exceeding 0.5°C to 1.0°C above the 40-year average. The most recent decade (2011–2020) exhibits a slight cooling trend, with temperatures decreasing by 0.44% compared to the previous decade. This cooling trend likely relates to the concurrent increase in precipitation observed during this period, highlighting the complex interplay between temperature and precipitation patterns in the region.

Discussion

Karamoja experienced substantial cropland expansion between 2011 and 2019, driven by agricultural development initiatives, including the Northern Uganda Social Action Fund (NUSAF) projects, the Karamoja Integrated Development Programme (KIDP), and disarmament policies that undermined pastoralism in favor of sedentary agriculture (Nakalembe et al., 2017). Climate change has dual impacts on agriculture in the region, affecting crop yields and land suitability for farming. While areas around Abim, Napak, Nakapiripirit, Moroto, and Amudat have experienced increased rainfall and warmer temperatures, this hasn't translated into higher agricultural productivity. Our analysis reveals severe droughts and extreme deviations from decadal averages, with notable drought years in 1984, 1993, 2002, 2005, and 2009, showing precipitation anomalies ranging from -15mm to -20mm. Although the most recent decade (2011–2020) exhibited a marked 17% increase in rainfall, erratic patterns and poor seasonal distribution have limited agricultural productivity. Despite the cropland expansion, crop production in Karamoja remains low by over 70%, according to the 2014 food security assessment conducted by WFP, FAO, MAAIF, and UBOS (Government of Uganda and World Food Programme (WFP), 2014). The region's predominantly smallholder farmers and pastoralists face recurring challenges as periodic droughts lead to crop failures and livestock mortality, pushing them further into economic hardship. These erratic rainfall patterns and prolonged dry seasons undermine agricultural viability and food security interventions, intensifying poverty as communities lose their primary sources of food and income.

While our analysis highlights the complex relationship between climate patterns and agricultural productivity in Karamoja, accurately quantifying the observed cropland expansion remains challenging due to limitations in data resolutions and classification accuracy. Though readily accessible 30m GLAD and LULC maps are available, evaluation against the 2012 Very High-Resolution (2m) map reveals a significant underestimation of cropland, particularly for small-scale agriculture plots interspersed with other land cover types. For instance, Potapov et al. (2022) reported a 102.8% increase in cropland between 2003 and 2019, but comparison with high-resolution data from Nakalembe et al. (2017) shows substantial discrepancies, with the 2012 map estimating 360,999.26 hectares of cropland – three times higher than the 130,347.04 hectares reported for 2019. These inconsistencies align with recent findings by Kerner et al. (2024) showing very low consensus across different land cover maps in sub-Saharan Africa, where unanimous agreement on cropland prediction occurred in fewer than 0.5% of pixels. These inconsistencies highlight fundamental challenges in accurately measuring agricultural expansion through remote sensing in complex landscapes like Karamoja.

While our study identifies agricultural expansion alongside climate and mapping patterns using readily available satellite data, a comprehensive assessment of policy impacts on Karamoja's land use transformation requires more in-depth analysis beyond our scope. This would need to examine how specific clauses within NUSAF, KIDP, and disarmament policies have influenced individual and community land use decisions, evaluate the implementation effectiveness of these policies, and assess their interaction with traditional land management systems. Such analysis would also benefit from extensive stakeholder interviews and historical records to understand the complex socioeconomic drivers behind the observed land use changes. Though crucial for informing future policy direction, these aspects were not addressed in our satellite-based analysis.

Conclusion

The complex ecological and socioeconomic landscape of Karamoja necessitates an integrated approach to address food insecurity, with traditional pastoralism potentially playing a crucial role (Levine, 2010). Central to this approach is comprehensive, very-high resolution (VHR) mapping of both croplands and rangelands, capturing their extent, status, seasonal variations, and interactions. This holistic mapping is essential for balancing agricultural and pastoral needs and designing sustainable land management interventions.

However, our study's reliance on 30m resolution products for change analysis presents limitations, including the potential to miss fine-scale changes, mixed pixel effects, and underestimation of small-scale land conversions. Additionally, climate trend analysis using satellite-based datasets faces challenges such as potential biases in complex topographies and the smoothing of fine-scale variability (Macharia et al., 2022; Stewart & Nitschke, 2017; Jhajharia et al., 2014).

To address these challenges, future research should focus on implementing VHR mapping initiatives, developing advanced multi-resolution change detection methods, and integrating cropland and rangeland mapping. Climate-resilient practices informed by fine-scale land use patterns, balanced agricultural expansion with sustainable pastoralism, and addressing land tenure issues are crucial. Integrated food security programs underpinned by comprehensive spatial analysis and improved climate monitoring systems are also essential.

By adopting this holistic approach that leverages scale-appropriate mapping and acknowledges methodological limitations, stakeholders can work toward improving food security and sustainable land use in Karamoja. The integration of VHR techniques for both cropland and rangeland analysis is crucial for capturing real change and informing evidence-based decision-making in this dynamic landscape. Addressing current methodological limitations and embracing comprehensive mapping approaches will enable a nuanced understanding of Karamoja's land use dynamics, facilitating more effective strategies for sustainable development and food security in the region.

References

Abrahams, D. (2021). Land is now the biggest gun: Climate change and conflict in Karamoja, Uganda. *Climate and Development*, *13*(8), 748–760.

Baker, R. (1975). "Development" and the pastoral people of Karamoja, North-Eastern Uganda: An example of the treatment of symptoms. In T. Monod (Ed.), *Pastoralism in tropical Africa* (Chapter 9, pp. 187–205). Oxford University Press.

Banerjee, A., Chen, R., E. Meadows, M., Singh, R., Mal, S., & Sengupta, D. (2020). An analysis of long-term rainfall trends and variability in the Uttarakhand Himalaya using Google earth engine. *Remote Sensing*, *12*(4), 709.

Chaplin, D. R. (2017). *Improving information uptake for climate change adaptation by integrating Indigenous knowledge systems with climate information services* [Doctoral dissertation]. Division of Risk Management and Societal Safety, Lund University, Sweden.

Claverie, M., Ju, J., Masek, J. G., Dungan, J. L., Vermote, E. F., Roger, J.-C., Skakun, S. V., & Justice, C. (2018). The harmonized landsat and sentinel-2 surface reflectance data set. *Remote Sensing of Environment*, *219*, 145–161.

Datzberger, S. (2022). Lost in transition? Modernisation, formal education and violence in Karamoja. *World Development*, *158*, 106013.

Egeru, A., Osaliya, R., MacOpiyo, L., Mburu, J., Wasonga, O., Barasa, B., Said, M., Aleper, D., & Majaliwa Mwanjalolo, G.-J. (2014a). Assessing the spatio-temporal climate variability in Semi-Arid Karamoja sub-region in North-Eastern Uganda. *International Journal of Environmental Studies*, *71*(4), 490–509.

Egeru, A., Wasonga, O., Kyagulanyi, J., Majaliwa, G. J., MacOpiyo, L., & Mburu, J. (2014b). Spatio-temporal dynamics of forage and land cover changes in Karamoja sub-region, Uganda. *Pastoralism*, *4*(1), 6.

Famine Early Warning Systems Network. (2016). *Uganda food security outlook*. Technical report, FEWS NET.

FAO. (2014). *Uganda food security assessment*. Technical report, Food and Agriculture Organisation of the United Nations.

Filipová, Z., & Johanisova, N. (2017). Changes in pastoralist commons management and their implications in Karamoja (Uganda). *Journal of Political Ecology*, *24*(1), 881–900.

Filipova, Z., & Johansson, E. (2017). *Karamoja systems analysis*. Technical report, Karamoja Resilience Support Unit (KRSU).

Gayfer, J., & Shanahan, M. (2012). *How climate change altered the face of hunger in Africa*. Technical report, CARE International.

Gorelick, N., Hancher, M., Dixon, M., Ilyushchenko, S., Thau, D., & Moore, R. (2017). Google earth engine: Planetary-scale geospatial analysis for everyone. *Remote Sensing of Environment*, *202*, 18–27.

Government of Uganda and World Food Programme (WFP). (2014). *Uganda: Karamoja food security assessment March 2014*. Posted on ReliefWeb on July 6, 2014.
Hasahya, E., Thakur, K., Dione, M. M., Kerfua, S. D., Mugezi, I., & Lee, H. S. (2023). Analysis of patterns of livestock movements in the cattle corridor of Uganda for risk-based surveillance of infectious diseases. *Frontiers in Veterinary Science*, *10*, 1095293.
Jhajharia, D., Dinpashoh, Y., Kahya, E., Choudhary, R. R., & Singh, V. P. (2014). Trends in temperature over Godavari River Basin in Southern Peninsular India. *International Journal of Climatology*, *34*(5), 1369–1384.
Johnson, D., & Anderson, D. M. (1988/2019). *The ecology of survival. Case Studies from Northeast African History* (pp. 1–54), Routledge (Original work published 1988 by Westview Press).
Kagan, S., Pedersen, L., Ollech, S., & Knautc, D. (2009). *The karamoja syndrome: Transdisciplinary systems research informing policy and advocacy*. Technical report, Cultura21.
Kerner, H., Nakalembe, C., Yang, A., Zvonkov, I., McWeeny, R., Tseng, G., & Becker-Reshef, I. (2024). How accurate are existing land cover maps for agriculture in Sub-Saharan Africa? *Scientific Data*, *11*(1), 486.
Krätli, S. (2010). Karamoja with the rest of 'the rest of uganda'. *Nomadic Peoples*, *14*(2), 3–23.
Kumssa, A., Jones, J. F., & Herbert Williams, J. (2009). Conflict and human security in the north rift and North Eastern Kenya. *International Journal of Social Economics*, *36*(10), 1008–1020.
Le Groupe-conseil Baastel. (2015). *Economic assessment of the impacts of climate change in Uganda*. Technical report, Climate Change Department, Ministry of Water and Environment.
Levine, S. (2010). *What to do about Karamoja? Why pastoralism is not the problem but the solution. A food security analysis of Karamoja*. Technical report, FAO/ECHO.
Macharia, D., Fankhauser, K., Selker, J. S., Neff, J. C., & Thomas, E. A. (2022). Validation and intercomparison of satellite-based rainfall products over Africa with Tahmo in Situ rainfall observations. *Journal of Hydrometeorology*, *23*(7), 1131–1154.
Mann, H. B. (1945). Nonparametric tests against trend. *Econometrica: Journal of the Econometric Society*, 245–259.
McGuirk, E. F., & Nunn, N. (2024). Transhumant pastoralism, climate change, and conflict in Africa. *Review of Economic Studies*, rdae027.
Mkutu, K. (2003). Pastoral conflict and small arms. In *Small arms and security in the Great Lakes region and the Horn of Africa* (vol. 48). Saferworld.
Mubiru, D. N. (2010). *Climate change and adaptation options in Karamoja*. FAO.
Muluken, T. K. (2020). The role of indigenous conflict resolution mechanisms in the pastoral community: An implication for social solidarity in Somali Region, Shineli Woreda. *Open Access Library Journal*, *7*(2), 1–16.
Murphy, S., Carmody, P., & Okawakol, J. (2017). When rights collide: Land grabbing, force and injustice in Uganda. *The Journal of Peasant Studies*, *44*(3), 677–696.
Nakalembe, C. (2018). Characterising agricultural drought in the Karamoja subregion of Uganda with meteorological and satellite-based indices. *Natural Hazards*, *91*(3), 837–862.
Nakalembe, C., Dempewolf, J., & Justice, C. (2017). Agricultural land use change in Karamoja region, Uganda. *Land Use Policy*, *62*, 2–12.
Nakalembe, C., Kerner, H., Zvonkov, I., Humber, M., Galvez, A., Makabe, E., Venturini, S., & Becker-Reshef, I. (2024). A framework for eo-based national agricultural monitoring (eo-nam)-for the African context. *Research Square*. Preprint.
Ocan, C. E., & Ocan, C. (1994). Pastoral resources and conflicts in North-Eastern Uganda: The Karimojong case. *Nomadic Peoples*, 123–135.
Osaliya, R., Wasonga, O., Mwanjalolo, J. M., MacOpiyo, L., Kironchi, G., & Adipala, E. (2020). Predicted land use and land cover outlook for Semi-Arid Lokere and Lokok catchments in Karamoja Region, Uganda. *African Crop Science Journal*, *28*(4), 595–616.
Potapov, P., Turubanova, S., Hansen, M. C., Tyukavina, A., Zalles, V., Khan, A., Song, X.-P., Pickens, A., Shen, Q., & Cortez, J. (2022). Global maps of cropland extent and

change show accelerated cropland expansion in the twenty-first century. *Nature Food*, *3*(1), 19–28.
RCMRD. (2018). *Uganda landcover 2000 scheme i*. GHG Land Cover Map for Uganda. https://data.rcmrd.org
Rugadya, M. A., & Kamusiime, H. (2013). Tenure in mystery: The status of land under wildlife, forestry and mining concessions in Karamoja region, Uganda. *Nomadic Peoples*, *17*(1), 33–65.
Sen, P. K. (1968). Estimates of the regression coefficient based on Kendall's Tau. *Journal of the American Statistical Association*, *63*(324), 1379–1389.
Stavi, I., Roque de Pinho, J., Paschalidou, A. K., Adamo, S. B., Galvin, K., de Sherbinin, A., Even, T., Heaviside, C., & van der Geest, K. (2022). Food security among dryland pastoralists and agropastoralists: The climate, land-use change, and population dynamics nexus. *The Anthropocene Review*, *9*(3), 299–323.
Stewart, S. B., & Nitschke, C. R. (2017). Improving temperature interpolation using modis lst and local topography: A comparison of methods in South East Australia. *International Journal of Climatology*, *37*(7), 3098–3110.
Uganda Bureau of Statistics (UBOS). (2016). *National population and housing census 2014 main report*. Technical report, UBOS.
Uganda Bureau of Statistics (UBOS). (2017). *Uganda national household survey 2016/17*. Technical report, UBOS.
Uganda Bureau of Statistics (UBOS). (2018). *Uganda demographic and health survey 2016*. Technical report, UBOS and ICF.
Uganda Bureau of Statistics (UBOS). (2024). *The national population and housing census 2024 – preliminary report*. UBOS, Kampala, Uganda.
USAID. (2017). *Climate risks in food for peace geographies: Karamoja region, Uganda*. Technical report, USAID.
Wang, F., Shao, W., Yu, H., Kan, G., He, X., Zhang, D., Ren, M., & Wang, G. (2020). Re-evaluation of the power of the mann-kendall test for detecting monotonic trends in hydrometeorological time series. *Frontiers in Earth Science*, *8*, 14.
World Bank's Disaster Risk Finance and Insurance Program. (2023). *Disaster risk finance diagnostic in Uganda*. World Bank, Washington, DC.
World Food Programme and UNICEF. (2016). *Food security and nutrition assessment*. Technical report, WFP.
Wu, W. (2014). The generalized difference vegetation index (gdvi) for dryland characterisation. *Remote Sensing*, *6*(2), 1211–1233.
Zeng, H., Wu, B., Zhang, N., Tian, F., Phiri, E., Musakwa, W., Zhang, M., Zhu, L., & Mashonjowa, E. (2019). Spatiotemporal analysis of precipitation in the sparsely gauged Zambezi River Basin using remote sensing and Google earth engine. *Remote Sensing*, *11*(24), 2977.

19 Rural idyll or forgotten backwaters

What and who determines the transition of rural areas in Denmark? A case study from Central Jutland, Denmark

Søren Bech Pilgaard Kristensen, Anne Gravsholt Busck, and Michael Helt Knudsen

Introduction

Many rural areas across Europe are undergoing transition. Some villages and areas suffer from decline resulting from a combination of depopulation, aging population, loss of economic activity, and services (Mohr et al., 2023; Dimopoulos et al., 2023; Newsham & Rowe, 2023; Eistrup et al., 2019). Other areas remain viable and even increase in population, as new residents seek a green environment and a rural lifestyle (Kristensen et al., 2019; Jończy et al., 2021). Many studies have pointed out how the combination of external and internal drivers and factors influences specific trends in a given area (Georgios et al., 2021; Gkartzios & Scott, 2014). For example, rich landscape and amenity values, vibrant local communities, reliable public transport, and accessibility to nearby jobs and services may attract "urbanites" and reinforce a positive trend, while the opposite may lead to rural decline (Mohr et al., 2023). Furthermore, agricultural production remains an important factor in many rural landscapes, as for example Denmark, where agriculture covers 60% of the area (Statistics Denmark, 2024) and local as well as external conditions influence the type and scale of agriculture (Busck & Kristensen, 2014).

The local landscape is an important element in the transformation experienced in rural areas. In several Mediterranean countries, rural decline in recent decades has led to abandonment of fields and pastures and ultimately spontaneous regrowth of shrubs (Van der Sluis et al., 2019; Dimopoulos et al., 2023). The impact on the aesthetic and amenity values is crucial for the attractiveness of rural areas, and abandonment may lead to further decline. Indeed, rich landscape and amenity values may be important criteria for potential residents, looking for attractive landscapes which offer recreational opportunities (e.g. hunting, hiking, biking) (Jończy et al., 2021; Aner, 2016). Indeed, some rural residents are not only consumers but also co-create and co-manage the landscape they live in through their landscape management activities (e.g. as full-time or hobby farmers) (Busck, 2002; Busck & Kristensen, 2014). The local landscape, therefore, serves as a multidimensional entry point to understand the type of transition rural areas undergo.

DOI: 10.4324/9781032672779-23

Rural areas in Denmark are experiencing both processes described here. In some areas, rural decline may even occur close to areas experiencing growth and rural development (Kristensen et al., 2009; Busck & Kristensen, 2014; Kristensen et al., 2019). Due to the traditional importance of agricultural activity as the main land use, agricultural production is one of the keys to understanding the type of transition experienced in different areas. This is true both in areas where agricultural production continues to be an important economic factor and in areas where it is stagnating, since the decline leaves a vacuum in terms of economic activity and land use (Primdahl, 2010).

In this chapter we explore the multiple dimensions and roles that rural landscapes play in rural development. Key questions concern 1) What is the relationship between agricultural production and landscapes? 2) What are the landscape management strategies of different types of rural residents? 3) What is the role of public policies and instruments (subsidies) in this process? The research was undertaken as a case study of a marginal agricultural landscape in central Jutland, Denmark, which has been strongly affected by both processes of productivism and post-productivism (Kristensen, 1999; Reenberg et al., 2010; Kristensen et al., 2009).

Agricultural systems in transition

Overall, agricultural land use in much of Europe has been strongly influenced by a structural adjustment of farms characterized by economy of scale, continued specialization, and concentration of production since 1950 (Latacz-Lohmann & Hodge, 2003; Primdahl & Swaffield, 2010). Assisted by the financial support provided under the EU Common Agricultural Policy (CAP), in particular the Pillar 1 subsidies, the industrialization of the agricultural sector has led to a polarization between fewer and larger farms on the one hand and a large number of small farms on the other hand (Guiomar et al., 2018; Kristensen et al., 2009; Reenberg et al., 2010; Almstedt, 2013). In broad terms, this corresponds to much of the period 1970–2000 in the (northern) EU countries, where the industrialization of agriculture was accompanied by a Fordist food regime. Alarmed by the environmental consequences of continued industrialized production and WTO concerns over the distorted free trade conditions, the EU shifted the CAP in a more comprehensive and green direction, leading to the term "greening of the CAP" (Almstedt, 2013). The latest CAP (2023–2027) is designed to deliver important elements in EU's overarching biodiversity and climate change agenda, that is by increasing compliance with environmental conditions for accessing CAP subsidies (Agriculture and Rural Development, 2022). However, recent events in 2024 have reversed some of these strategies, showing how politically volatile environmental and climate issues are in the agricultural sector (Arboleas, 2024).

The increased focus on other outputs than "food, feed and fibre", such as amenity values, rural vitality, natural resources, are often grouped together under the heading of public goods (Baldock et al., 2010). Primarily in NW Europe, this shift was described as a transition from a productivist to a post-productivist phase of agriculture (Almstedt, 2013). Central characteristics of post-productivism (or the post-productivist transition) consisted of a reversal of the concentration, intensification, and specialization of production, which had characterized much of NW European agriculture after 1950 (Bjørkhaug & Richards, 2008).

Critics of these concepts pointed to the limited empirical evidence that a major shift actually had occurred (Evans et al., 2002). The critique led to the emphasis on multifunctionality as a (new) and more encompassing term (Almstedt, 2013). Multifunctionality has also been adopted as the official "mantra" of the EU CAP when negotiating the free-trade arrangements during the WTO-conflict era. These public goods were described as vital for keeping the countryside alive and therefore should be eligible for support (Baldock et al., 2010; Pavlis et al., 2016). The agri-environmental (AE) schemes have been a central instrument in promoting environmentally friendly practices through subsidies to farmers (Pavlis et al., 2016). They also include nonagricultural uses of land and buildings, such as agro-tourism, farm shops, care farming, or nonagricultural enterprises (Vejre et al., 2007; Kristensen et al., 2019).

The specific production system and land use which evolves in an area is a result of multi-level drivers and factors (Kristensen et al., 2009). At a local level, Primdahl and Swaffield (2010) suggested that levels of urbanization (i.e. distance to urban centers) and agricultural production potential (i.e. soil fertility, market conditions) might explain the variation in agricultural production systems. National and international agricultural policy and support schemes also play an important role in favoring certain types of production (Latacz-Lohmann & Hodge, 2003; Pavlis et al., 2016). In conclusion, the drivers affecting the agricultural sector and through that the rural landscapes in Europe have become more diverse and complex.

The distribution of different farm and farmer types is closely associated with the relative advantages of different types of production in different areas. Thus, farming communities are seldom homogenous but consist of farms ranging in size (area, economic output, livestock numbers) and level of engagement in farming (ranging from full-time to no involvement (rural residents). Primdahl (2010) illustrates how local variation even within countries results in different landscape trajectories. Landscapes undergoing extensification of agriculture might experience gradual or complete abandonment, as seen in several Eastern European countries or in parts of southern Europe (Helfenstein et al., 2024; Primdahl, 2010; Van der Sluis et al., 2019).

Rural development – conceptualization of drivers

Changes in the agricultural sector have an impact on rural areas. When the primary sector has less economic importance, a broader array of drivers of rural development becomes evident. In the 1960s, especially rural areas far from urban centres experienced population decline, because people moved to urban areas seeking jobs. Declining rural areas were pictured as lacking economic or human resources, preventing them from embarking on a new positive rural development (Georgios et al., 2021). The presentation of ways forward was twofold. Related to globalization and the need for rural areas to be competitive, investments should seek to bring rural economies into the global market (Georgios et al., 2021).

However, in some declining rural areas the external investments either did not come or did not solve the challenges. In a European context, these areas were named "peripheral" (as opposed to "core" areas) (Georgios et al., 2021). In the late 1990s, endogenous rural development approaches started emerging, emphasizing the potential of local assets and resources. Citizen engagement and networking (Dargan & Shucksmith, 2008), community capacity building (Shucksmith, 2010), as well as inclusive planning (Healey, 2004) were seen as essential means of promoting locally driven development for rural areas. The EU LEADER programmes initiated in 1991 is an example of this focus on local assets and resources and fostered growth and revival of rural areas in the EU. Neo-endogenous drivers of rural development are based on a combination of exogenous and endogenous drivers. Thus, emphasizing that the two approaches should go hand in hand by, for example, state-led subsidies for local initiatives based on local assets and resources (Ray, 2000).

In this chapter, we focus on a neo-endogenous approach. Thus, we investigate how local actors use local assets to drive new initiatives but also how the overall context has an impact on rural development as a combination of location (e.g. relations to urban areas) and public initiatives (subsidies, planning, and regulation).

A central aspect of rural development is how to counteract population decline, as a decreasing population may lead to less opportunities (service, jobs etc.) for the remaining population (Newsham & Rowe, 2023). In addition, in some rural areas the demography becomes skewed, for example, with a high percentage of elderly people and the young people leaving for education and job possibilities (Maciej et al., 2020; Eistrup et al., 2019). This may add to a decline in the development momentum (Newsham & Rowe, 2023). Therefore, rural development strategies often have a focus on retaining and increasing the number of young people and families with kids.

The desire to attract new citizens and retain existing ones has fostered exploration of motives for living in (marginal) rural areas. This has been named "migration factors" or "attractors and push" factors. (Jończy et al., 2021) identified a number of such factors related to economic, environmental, and amenity motives. In a Danish context, Aner (2016) focused on the urban-to-rural migration of middle-aged families with kids. Here, two overall factors were identified, namely relatively cheap housing (attractor) and non-urban environments for kids (push factor). Oxford Research (2014) identified good public transport, public institutions for kids, cultural amenities (e.g. theaters), and more generally public services as important attractors in a rural area in Denmark.

Amenities related to the local landscape are also mentioned as attractors. This may be because of attractive elements in the landscapes (e.g. forests, beaches, and nature areas in general) or the fact that there are few people in the landscape and thus possibilities of experiencing silence and remoteness (Danish Outdoor Council, 2014).

In this chapter, we give emphasis to the landscape-based amenities as drivers of rural development, including as attractors for newcomers. In this perspective, the private landowners are central actors, as most rural areas in Denmark are in private ownership.

We are inspired by the concept of 'Landscape management styles' when we examine the landscape management implemented by private landowners. Landscape management styles are a way to categorize and analyse the combined pattern of motivation and practice of a certain type of landscape management (Busck, 2002). Thus, different styles have different landscape practices and different landscape implications. Busck (2002) describes five styles including 'production enthrone', 'conserve and enhance', and 'create a (green) paradise' which coexist on different properties in a small village in Denmark and thus create a mosaic of strategies and impact on the landscape at community level.

Analytical framework – rural landscape transitions

The previous sections have outlined the diversity and dynamics of processes and drivers of change in rural communities and landscapes. Evidently, they operate at different spatial levels and interact in complex ways to influence a given local landscape in a certain way. In this section, we synthesize the different dimensions into an analytical framework, which will be used for the analysis.

Holmes (2008) identified different landowner values concerning land use and management strategies. The three extreme positions were 'Production landscape' (e.g. landscapes dominated by mixed and specialized farming), 'Consumption landscapes' (e.g. landscapes dominated by residential use and hobby farming), and 'Protection landscapes' (e.g. landscapes dominated by conservation and abandoned marginalized agriculture). The most complex landscape type is the 'complex multifunctional' landscapes, which accommodate several (opposing) landscape trends (e.g. both productivist and post-productivist characteristics). Van der Sluis et al. (2019) and Pinto-Correia et al. (2016) applied the model in multi-country case studies of European rural landscapes, based on integration of agricultural and socio-economic criteria. Their work highlights the need for a comprehensive and multi-sector approach to the study of rural landscape transitions.

Based on these insights, we operationalize the analysis of landscape transition in Rostrup by integrating elements derived from the three main themes of the research. First, we study the agricultural transition in Rostrup by looking at landowner type, agricultural land use, and livestock. Second, we investigate rural development through the parameters attractors and push factors, non-agricultural, and economic activity (other gainful activities). Finally, we study planning perspectives of the transition by looking at landscape and recreational use, landscape management strategies, and the role of subsidies in landscape management.

Policy context: key physical planning and nature management instruments in Denmark

At the national level, Denmark is divided into three planning zones: urban, rural, and summer house zones, with differentiated regulation. In the rural zone, structural urbanization (e.g. construction of buildings for residential use) is restricted, and the areas are mostly used for primary production (agriculture, forestry, fishery), nature, and nature-related recreational purposes. The Summer house zone is, with a few exceptions, the only area where summer cottages may be located. Likewise, new urban development is supposed to take place in urban zones only to avoid urban sprawl. This zonation system is credited with maintaining a sharp distinction between rural land and built-up areas (Ministry of Industry, 2019).

The Nature Protection Law includes two fundamental regulations. One is a general regulation by which a number of nature and semi-natural nature types (heathlands, freshwater and salt meadows, ponds etc.) are protected from change and the landowner will not be compensated for the restriction on land use. Second, the Law contains provisions for Conservation Orders, which are specific protections of specified nature and semi-natural areas. The landowner will be compensated for the effort. Apart from the Law-based nature management, many natural and semi-natural areas are established and managed by private landowners. Some of these initiatives are supported by EU subsidies – for example through rural development programs under the CAP, which may co-finance, for example afforestation, hedgerow planting, or extensification of agricultural land uses.

Methodology

The case area is identical to an area researched in 1995 which encompasses Rostrup village and was selected for its unique opportunity to undertake a longitudinal study in an area with marginal potential for agriculture (Figure 19.1).

Whereas the 1995 study concentrated on agricultural transformation between 1973 and 1995, the 2018 research focused more broadly on landowner–landscape interaction and local rural development challenges in areas experiencing a decline of agriculture but at the same time endowed with considerable landscape amenities. The aerial photo in Figure 19.2 shows Rostrup village surrounded by agricultural fields, thickets, and forests.

Rostrup covers roughly 500 hectares and consisted of 24 farms in the 1995 survey. Out of the 24 farms, three had no owner in 2018, as the owner was deceased, or the property had been merged with another farm. Eight of the remaining 21 properties were used as hunting areas (forests, no agricultural production) and/or owned by an outsider (not residing in Rostrup), and only one of these was able to participate. Another five landowners were unavailable for an interview or did not want to participate. Thus, the research is based on face-to-face interviews with nine landowners in November 2018. Each interview lasted for 45–60 minutes and covered the following topics:

- Dynamics of agriculture (crops and livestock)
- Nature management
- Local community and local development

Figure 19.1 Location of Rostrup village.

Source: Authors

- Recreational use of area
- Role of planning and subsidy schemes

In addition, we obtained some additional information from neighbors and agricultural registers for some of the farms that were unavailable during the survey.

Landowners were classified according to their economic contribution of farming activity to the household economy (based on the proportion of income from agriculture for the person most involved in agriculture (see Table 19.1).

Table 19.1 Landowner types 1995 and 2018 (numbers in brackets show distribution of landowners interviewed in 2018)

Landowner type[1]	*Definition*	*1995*	*2018 (interviewed)*
Full-time farmers	100% of income from agriculture	9	1 (1)
Part-time/ hobby	0–99% of income from agriculture	15	1 (1)
Pensioners	Age > 67 or public welfare recipient	0	5 (4)
Not farmer or resident	As 'resident', but owner not residing on the property	0	8 (1)
Resident	Owner residing on the property, land may be used for agriculture (rented by others)	0	6 (2)
Total		24	21 (9)

1 The categories 'Part-time farmers' and 'Hobby farmers' have been merged to allow comparison with the categories in Kristensen (1999).*Source*: Kristensen and Busck (2024).

Results

Agriculture and landscape in Rostrup

The physical landscape in Rostrup posed some limitations on the area's agricultural potential. The sandy soils, undulating terrain with a mix of small depressions and hills was better suited for agricultural production 50 years ago than in 2018. Old hedgerows and stone walls, demarcating field and property boundaries, follow the complex landscape patterns and add to the challenges faced by modern farms (Figure 19.2). Numerous small forests and thickets have been planted over the last century in the most difficult areas. Together with many bogs and wetlands in the depressions, they contribute to a mosaic of landscape patches within the agricultural landscape.

Agricultural structure in Rostrup: 1995–2018

In 1995, there were 38% full-time farmers (see Table 19.1) among the 24 landowners, out of which 78% (seven out of nine full-time farmers) were dairy farmers. The remaining 15 landowners (62%) were part-time or hobby farmers and therefore not engaged in full-time agriculture (Table 19.1). In 2018, the number of full-time farmers had been drastically reduced to only one full-time mink farmer (who was one of two non-dairy farmers in 1995). In other words, only 5% (1 property out of 21) were managed as a full-time farm in 2018. In addition to these landowner categories, several new categories had appeared in 2018: residents, absent (non-agricultural) owners, and pensioners, none of whom were involved in agriculture, except for renting out land to tenant farmers. These three categories constituted the majority of landowners in 2018 (95%).

Figure 19.2 Rostrup village and landscape.

Source: Photo by authors.

Based on the interviews, the reasons for this change appear to be a mix of different issues and can be synthesized as follows:

- Present owners were once full-time farmers but are now part-time or hobby farmers.
- Old and run-down farms, requiring substantial renovations.
- Present owners are retired and too old to be motived to enlarge/modernize production.
- Present owners prefer to continue living on the farm, as mortgage and other costs are low but cannot afford investments required for renovations.
- Present owners are residents who have moved recently to Rostrup and rent out land to full-time farmers,
- Small farms, making it necessary to enlarge production (rent or buy land) to run a profitable operation, but general economic conditions may not allow this (see text for details).
- Small fields and heterogenous landscape, making it difficult to use modern farm machines to obtain economy of scale.

Not only are farms and fields small and expensive to cultivate, Rostrup village is also hilly and contains many small nature areas and cultural structures (stone and earthwalls) which make fields less uniform and rectangular shaped (Figure 19.2).

Land use changes between 1995 and 2018

Land use in Rostrup in 2018 was dominated by traditional crops, as 61.8% of the farmed area was sown with crops in rotation (cereals, grass in rotation, sowing seeds etc.) (Table 19.2). This distribution has only changed slightly since 1995, when 63.5% of the area was sown with these crops. However, crop composition changed between 1995 and 2018, as certain crops (e.g. fodder beets, alfalfa, and peas) are no longer cultivated in Rostrup, and others have appeared (e.g. sowing seeds and green fodder). The cultivation of Christmas trees, as five-year rotation crops, covered small areas both years (3% in 1995 and 1.5% in 2018). This crop is commonly found in areas of marginal agricultural potential. Part of the Christmas tree planting has taken place in an area previously used for gravel extraction.

Livestock

Livestock production is an important land use factor in Denmark, and the different types of animals have different landscape impacts. For example, full-time dairy cattle require large areas devoted to fodder production or grazing areas. The dynamics of livestock production is therefore an important dimension to understand the land use changes of rural landscapes.

In 2018, only four out of 21 properties kept livestock, about one-quarter the number found in 1995.

Table 19.2 Land use in Rostrup, 1995 and 2018

Land use type	*Land use 1995 (area in ha and %)*		*Land use 2018 (area in ha and %)*	
	ha	*%*	*ha*	*%*
Cereals	166.9	36.0	144.2	29.6
Oats	0.0	0.0	45.6	9.4
Maize	15.7	3.4	21.5	4.4
Rapeseed	1.4	0.3	0.0	0.0
Potato	0.0	0.0	0.0	0.0
Fodder beets	9.9	2.1	0.0	0.0
Pea	1.8	0.4	0.0	0.0
Alfalfa	13.3	2.9	0.0	0.0
Sowing seeds	0	0	8.2	1.7
Green fodder	0	0	22.3	4.6
Grass in rotation	85.2	18.4	59.2	12.2
Permanent grass	42.0	9.1	43.8	9.0
Christmas trees	14.0	3.0	7.3	1.5
Other set-aside, forest, bog etc.)	113	24.3	134.7	27.7
Total	463.2	100	486.8	100

The only full-time farmer in the village in 2018 was engaged in mink production, and there was thus no 'traditional' livestock production (dairy and pig) at all. In 1995, the dairy farmers dominated the livestock production, but they had completely disappeared by 2018. In contrast, farms with beef cattle and sheep dominated in 2018.

All in all, agricultural land use has remained quite stable between 1995 and 2018, in terms of area used for arable agriculture. However, minor crops had disappeared from Rostrup in this period, partly due to changes in livestock composition, as there were no longer any dairy cattle in 2018. In contrast, nearly one-third of the land use consisted of grassland, which was used as pasture for beef cattle. The large proportion of forest reflects the poor conditions for agriculture as well as an interest in hunting and nature conservation, as discussed in section 4.3.

Rural development

Rostrup is a small village, originally one of five hamlets, in Gadbjerg parish. The community depended on neighboring villages for most services, with Gadbjerg village, located 4 km to the north and Billund and Vejle as the closest urban centers at 15 km and 20 km distance respectively (Figure 19.1). The village is composed of 5–10 farms and a few single houses in a small village center and a similar number of farms scattered beyond the compact village center (Figures 19.2 and 19.3).

Attractors and push factors influencing purchase decisions

When asked about motives for the purchase of the property, most full-time and former full-time farmers mentioned the production potential of the farm. Other landowners who depend on other sources of income than agriculture mentioned relatively cheap housing and the location (relative proximity to towns, allowing easy commuting). Others pointed to the qualities of the property and the area for their hobby (horseback riding, hunting, nature) as main attractors.

Many landowners referred to a lack of services (schools, shops) in the village which increases transport demands and reduces the vitality of Rostrup. The distance to services means that "most people need at least 2 cars to live here" (Interview, R44).

Other gainful activities

As mentioned earlier, only 5% of the landowners were managing their property as a full-time farm in 2018 and had no other sources of income. Other income-generating activities can either be on the property (so-called On Farm Business Structure Diversification (OFBSD)) or off the farm (e.g. wage labor income). Most of the properties had rented out leftover buildings to other full-time farmers to raise animals (cattle to full-time farmers or horses for urban dwellers) or storage of slurry (slurry container). One landowner used farm equipment for contractual labor (snowploughing). We encountered one example of OFBSD in Rostrup in 2018 which was more of an entrepreneurial activity, disconnected from traditional agriculture, as a tailor had converted part of the dwelling house to her workshop.

Planning perspectives

A few planning perspectives were found to add extra challenges and uncertainties about the future, which may deter people from buying property in the village. For the last 5–10 years, it has been debated to construct a railway line from Vejle to Billund (Figure 19.2 for location). One of the 'lines' could pass through the area. The discussion of the possible location of a new railway has a big effect on the local community. Many see it as a source of uncertainty which makes it difficult to make long-term plans and even to put property on the market.

Nature and recreation in Rostrup

Rostrup represents a heterogenous landscape with several valuable habitats. The main biodiversity interests are linked to semi-natural landscape elements, which are protected by the general regulation in the Nature Conservation Act and private forests (see Table 19.3 and Figure 19.3). In addition to these, hedgerows are important linear landscape elements but are not protected by the Nature Conservation Act.

Dry grassland: Dry grasslands ('overdrev' in Danish) are located on well-drained soils (often slopes) dominated by grass species and were previously maintained through cattle grazing. Most have disappeared due to cultivation or increased nutrient loads, which have created oligotrophic conditions and reduced the nutrient-sensitive flora and fauna.

Heathland: Tiny fragments of the heathland vegetation which covered most of central Jutland until the last half of the 19th century are still found as small, uncultivated areas within a matrix of cultivated fields. The few remaining areas are protected by Nature Conservation Act, but many have been negatively affected by increased nutrient loads from surrounding areas.

Bogs and meadows: Numerous small bogs are found in natural depressions in the area. The number decreased greatly in the 20th century due to drainage and cultivation. In addition, a large contiguous bog/wetland area is located in the northern part of Rostrup village. The biodiversity potential of the wetland areas is high, and the location next to a large state forest makes Rostrup an attractive hunting area.

Table 19.3 Semi-natural landscape elements in Rostrup, 2018 (ha)

Dry grassland[1]	*Heathland*	*Bog*	*Meadow*	*Forest*	*All*
14.2	13.8	7.9	12.1	133.9	181.8

1 'Overdrev' in Danish.*Source*: GeoDanmark, 2024.

Figure 19.3 Land use in Rostrup. White: agriculture, Gray: forest.

Source: Danmarks Arealinformation (2024).

Woodlots/small forests: Almost 30 small and large woodlots (ranging in size from 0.3 hectares to 55 hectares) are scattered in Rostrup (Figure 19.3). Most of them are planted woodlots with evergreen vegetation. Although the biodiversity value of many evergreen forests and woodlots is limited, compared to deciduous forests, they nevertheless represent important habitat values. They have been left unmanaged for decades, and natural succession has led greater variation of species composition. The interviews also showed that woodlots and forests have a high value for hunting. None of the areas found in Rostrup are protected by nature conservation and could in principle be removed anytime.

Hedgerows: Rostrup village area is criss-crossed by numerous hedgerows, most of which were originally planted as shelterbelts. This is another indication of the relatively sandy and marginal soil quality for crop production. The hedgerows serve as important linear corridors in the landscape, allowing certain species to migrate through the area. They are also appreciated by hunters, as they add to the network of wooded landscape elements in Rostrup village.

A comparison of the semi-natural landscape elements in Rostrup in 1995 and 2018 (data not shown) reveals a large increase in forested areas, whereas bog and heathland areas declined, and the area of permanent grass stagnated. The forest area increased by almost 10% since 1995 (from 17.7% to 27.4%), which is a very high proportion compared with the national average of 14% (Statistics Denmark, 2024).

Recreational use of property

For many rural residents, the attraction of living in a rural landscape is linked to the outdoor opportunities the property (and surroundings) offers. Two of the most popular activities in the countryside in Denmark are horseback riding and hunting. In Rostrup, sufficient space for horseback riding was mentioned as important motives for purchase of two of the properties (Interview, R98, R350). This included sufficient stable space as well as small paddock for grazing and/or hay cutting. Both properties were mainly used for residential purposes, and the owners had no previous or current involvement in agriculture. They could therefore be described as rural residents who were attracted by the opportunity for horseback riding.

Recreational hunting is a very common activity in rural Denmark (Lundhede et al., 2015). It is also an activity which polarizes rural communities (Gamborg & Jensen, 2017). These opposing viewpoints and practices were also found in Rostrup village, where six of the nine landowners use the property for hunting. Two out of the remaining three properties where no hunting takes place are too small for this activity, while the last one is strongly against hunting (Interview, R184).

Landscape management

During interviews, the landowners were asked questions about their landscape management practice and motivation in this respect. Based on the findings, we identified four groups of landowners.

1) Retired or still active full-time farmers, who would often own a large area of permanent grassland (sometimes heathland and meadow), which was grazed by suckling cows or sheep. This activity was a way to connect with their past career, by keeping a small herd of livestock. At the same time, this would often bring a small income and would frequently be subsidized through national or EU AE schemes. (N=5).
2) At the other end of the landscape management spectrum, we found landowners with limited land resources who use the property mainly as a residence. The very small property was the result of selling the arable part of the farm while keeping just enough land to cater for a few horses. (N=2).
3) A third group of landscape management was represented by one hobby farmer who specialized in Christmas trees production. This took place in a hilly area with a high concentration of woodlots. This was regarded as 'mixed' motives, with both nature management and commercial production playing a role. (N=1).

4) A fourth group was represented by a former full-time farmer, who in 2010 engaged with a 'landscape subsidy consultant' to implement a very ambitious landscape design. The consultant had prepared a project consisting of 5 hectares woodland and 5 hectares recreated wetland, which should have benefited from various subsidy schemes. At a very late stage, the plan was not approved by the authorities and thus never materialized. The uniqueness for this group consists of the very ambitious and proactive landscape management strategy, effectively piecing together enough funds from different subsidy schemes (N=1).

Such a diversity of interests does not come without controversy. The dislike for public interference in landscape management led one landowner to accuse two public employees inspecting land for the big rewetting project of trespassing private property. A neighbor describes the situation as "He (the aggressive neighbor) wants to be the sole decision maker, not just for his own land but the neighbors' land as well, as any change may impact his hunting interests" (Interview, R139). One landowner expressed equally strong feelings against hunting in the area and had "not allowed trespassers on his property for 50 years" (Interview, R184). Thus, the clashes are not only between "outsiders and insiders" but may also occur between landowners as "insider versus insiders" conflicts.

The role of public support for nature conservation (often termed 'public goods') was found to be very significant for landscape management in Rostrup. Most landowners who were eligible for subsidies, in the form of CAP support (AE schemes) or national funds (hedgerow planting, rewetting project), had participated in such programs, even if they came with restrictions and conditions.

Discussion

This chapter has explored the different drivers behind landscape changes in a marginal agricultural area and the nexus between agriculture, rural development (or decline), landscape management, and the landscape characteristics. It addressed these themes through a case study of agricultural changes and landscape activities in the period 1995–2018 in Rostrup village, located in Vejle municipality in Jutland, Denmark. The discussion is structured into the three main topics investigated in the research: agriculture, rural development, and landscape management.

Agriculture: the polarization of farm structure

The results revealed that the agricultural sector had changed tremendously between 1995 and 2018, as the full-time farms have virtually disappeared. A few of the former full-time farmers have reduced their activity level and are now either retired or part-time farmers, partly since this allows them certain tax benefits as long as the property is recognized as having a certain minimum production level and partly because they don't have a lot of expenses, in some cases leading to decay of buildings, as there is limited budget for upkeep of buildings. The outlook for full-time farms in Rostrup village does not look promising, as the minimum size to be profitable is larger than the present farm size (of farmland and buildings), according to statements by most interviewees. It would require a massive investment, which deters newcomers from taking over the properties. The present owners are past the point of realistically planning to expand and invest in the farm. This is also compounded by the financial situation of many farms. Due to the investment needs and current low profitability, farmers responded that financial institutions are not inclined to lend capital to either the current or potential new owners in the area. The low profitability was a combination of the outdated farm structure and the relatively low productivity of the land (sandy soils, undulating terrain) (Kristensen, 1999).

The increase in the number of small hobby farms between 1995 and 2018 was accompanied by an increase in the size of the remaining full-time 'professional' farms, as these have bought neighboring farms to benefit from economy of scale, thus leading to a further polarization of the farm structure. This mirrors a transformation seen in Denmark in general (Statistics Denmark, 2024; Kristensen et al., 2009). Also, we found that most hobby farms had rented out farmland (and buildings in one case) to the remaining full-time farms, some of which were located at quite a distance from Rostrup. This means that decision-making on land use in areas dominated (by share) by small hobby farms may actually be concentrated on a relatively limited number of farms, and some may not even be part of the local community.

Conditions for the customary practice of renting out land (and buildings) to full-time farmers were not very positive in 2018. Due to a general decline in the economy of agriculture (high prices on fodder and low prices on dairy and pork products), full-time farmers with livestock were generally struggling in this period. In fact, some of the interviewees reported that their tenants might not be able to pay the rent that they had agreed upon. This left the landowners with a difficult dilemma, whether to accept a lower income from renting out land or risk not having any income at all if they decided to terminate the contract and look for a new tenant.

Despite the changes in farm structure, the share of agricultural land use remained similar, occupying 76% in 1995 and 72% in 2018. We did therefore not see widespread abandonment of agricultural land and spontaneous afforestation as seen in Southern and Eastern EU countries (Dimopoulos et al., 2023). Nevertheless, the large forest area, which covered more than a quarter of the area in 2018, reflects a general trend in increased interest in hunting and nature conservation in areas where the profitability of agricultural production is declining (Lundhede et al., 2015; Lund & Jensen, 2017).

The reduction in livestock numbers also has consequences for the grazing of nature areas. Some landowners have entered AE schemes to graze designated areas with a small herd of suckling cows/beef cattle. The decline in the number of properties with grazers (sheep and cattle) and their numbers as such limits this important nature management practice.

Rural development – future of Rostrup

The amenity values found in the heterogenous landscape in Rostrup is an attractor for newcomers and visitors interested in using the landscape for recreational activities (in particular hunting and horseback riding) and was mentioned by new residents as important for their decision to move to Rostrup. The importance of a green environment and recreational opportunities aligns with findings by Aner (2016) and Georgios et al. (2021), who also stress the role of local amenity values of the landscape as an important attractor.

Despite the qualities of the property and Rostrup village that once attracted the owners of the properties, most respondents expressed anxiety about the future of the area. In terms of farming potential, most expressed that the farms are too small and not fit for modern farming, as discussed in section 7.1. From a community and 'liveability' perspective, residents refer to the lack of services in the village which increases transport demands. The negative trend started by such challenges, and with few (young) newcomers in recent years, means that the population becomes increasingly old (and small). Often, they have limited resources to engage in community affairs, the kind of self-help spirit which is otherwise high in demand in rural communities and often seen as fundamental to revitalize rural communities (Alvarez-Montoya & Ruiz-Ballesteros, 2024).

Landscape management: public goods or private Eden?

The interviews revealed a diversity of motives, values, and practices, which could in some cases even lead to conflicts internally between landowners or with public authorities. Inspired by the concept of landscape management styles (Busck, 2002), we identified four types of landscape managers based on a combination of their practice and motivation.

Seen together, the four types illustrate what a challenging societal task landscape management on private land is, straddling private and public ownerships and interests (public goods versus private ownership), motives (ranging from utilitarian to idealistic/moral), levels of capacity (limited/ordinary to plentiful/extraordinary), and landscape resources and structural resources (availability of land, labor, and capital). The resulting landscape management situation in Rostrup village can best be described as a mosaic of styles and practices, leading to some properties managed mainly for nature management goals intermixed with other properties. As described in section 4.3, such a diversity of interests does not come without issues and conflicts over, for example, access and user rights and hunting options. These often boil down to long and strong ties to the area, and conflicts were found within the community as well as between 'outsiders and insiders'.

The importance of public support for landscape management, in this case in the form of different types of EU or national subsidy programs, is clear from the findings. In addition, various designations and conservation orders had a clear impact on landowners' management strategies. Thus, a combination of 'stick and carrot' instruments has been crucial to guide landscape management activities to conserve and even enhance the semi-natural areas in Rostrup. Another noticeable finding concerns the role of 'landscape subsidy consultants', who specialize in the niche market of landscape designs aimed at tapping into the variety of subsidy schemes. The appearance of these consultants or brokers could be seen as a symptom of a "rural commodification" (Holmes, 2008) in a multifunctional countryside (Helfenstein et al., 2024).

Concluding remarks

Over the past three decades, rural landscapes in Europe have become increasingly polarized under the influence of endogenous and exogenous drivers. In Rostrup village, a marginal agricultural area in central Jutland, Denmark, this transformation has led to the virtual disappearance of full-time farmers, and most landowners are not involved in agriculture themselves, even if agricultural land use continues to dominate the landscape.

What does the future look like for Rostrup village? The outlook for traditional agricultural production is bleak, and it is more likely that the amenity values of the landscape could play a key role in local development, for example, by attracting new residents. The focus on recreational hunting, as seen by the large number of hunting areas (often with absent owners) and many of the landscape activities undertaken by landowners, indicates that the area may be transforming into a consumption landscape (Holmes, 2008; Van der Sluis et al., 2019) or a complex landscape where traditional agriculture (by tenant farmers) is intermixed with recreational use of semi-natural areas in a multifunctional landscape (Helfenstein et al., 2024).

As mentioned earlier, policies and subsidies played an important role in the landscape transitions in Rostrup – both related to agricultural production (e.g. the changes in use of set-aside land) and development of semi-natural landscape elements (e.g. forests). Thus, landscape management may be promoted by policies and subsidies. However, it is also important to acknowledge that the aspirations of the landowners are important for how landscapes are managed by landowners – regardless of potential subsidies. Therefore, informing about the aims of a given policy or subsidy is relevant. In addition, looking into how policies and subsidies align with the aspirations of different types of landowners may be important to better direct initiatives toward the relevant types of landowners. As far as rural development is concerned, more attention to the challenges facing rural communities in general (i.e. less restrictive mortgage financing requirements or tax subsidies) could compensate for some of the adverse conditions.

References

Agriculture and Rural Development. (2022). *Common agricultural policy for 2023–2027*. European Union.

Almstedt, Å. (2013). Post-productivism in rural areas: A contested concept. In Camilla Sandström Linda Lundmark (Ed.), *Natural resources and regional development theory*. Institutionen för geografi och ekonomisk historia, Umeå universitet.

Alvarez-Montoya, J. M., & Ruiz-Ballesteros, E. (2024). Newcomers and rural crisis: Beyond the demographic challenge. A case study in Andalusia (Spain). *Journal of Rural Studies*, *108*, 103292.

Aner, L. G. (2016). Dwelling habitus and urban out-migration in Denmark. *European Urban and Regional Studies*, *23*, 662–676.

Arboleas, M. S. (2024). EU countries seal relaxation of CAP green requirements until 2027. *Euractiv*.

Baldock, D., Hart, K., & Scheele, M. (2010). *Public goods and public intervention in agriculture*. European Network for Rural Development & European Commission.

Bjørkhaug, H., & Richards, C. A. (2008). Multifunctional agriculture in policy and practice? A comparative analysis of Norway and Australia. *Journal of Rural Studies*, *24*, 98–111.

Busck, A. (2002). Farmers' landscape decisions: Relationships between farmers' values and landscape practices. *Sociologia Ruralis*, *42*, 233–249.

Busck, A. G., & Kristensen, S. B. P. (2014). From agriculture to nature – a study of drivers of land use change in a peri-urban landscape. *Geografisk Tidsskrift-Danish Journal of Geography*, *114*, 41–58.

Danish Outdoor Council. (2014). *Danskernes brug af naturen – og omfanget af generende oplevelser i mødet med andre brugere*. Friluftsrådet, Copenhagen.

Danmarks Arealinformation. (2024). Retrieved March 3, 2025, from https://support.miljoeportal.dk/hc/da/articles/207560605-Brugerstyring-Adgang-til-fagsystemer-gennem-Danmarks-Miljoeportal

Dargan, L., & Shucksmith, M. (2008). LEADER and innovation. *Sociologia Ruralis*, *48*, 274–291.

Dimopoulos, T., Helfenstein, J., Kreuzer, A., Mohr, F., Sentas, S., Giannelis, R., & Kizos, T. (2023). Different responses to mega-trends in less favorable farming systems. Continuation and abandonment of farming land on the islands of Lesvos and Lemnos, Greece. *Land Use Policy*, *124*, 106435.

Eistrup, M., Sanches, A. R., Muñoz-Rojas, J., & Correia, T. P. (2019). A "young farmer problem"? Opportunities and constraints for generational renewal in farm management: An example from Southern Europe. *Land*, *8*(4), 70. https://doi.org/10.3390/land8040070

Evans, N., Morris, C., & Winter, M. (2002). Conceptualizing agriculture: A critique of post-productivism as the new orthodoxy. *Progress in Human Geography*, *26*, 313–332.

Gamborg, C., & Jensen, F. S. (2017). Attitudes towards recreational hunting: A quantitative survey of the general public in Denmark. *Journal of Outdoor Recreation and Tourism*, *17*, 20–28.

GeoDanmark, 2024. Retrieved March 3, 2025, from https://www.geodanmark.dk/home/vejledninger/vilkaar-for-data-anvendelse/

Georgios, C., Nikolaos, N., & Michalis, P. (2021). Neo-endogenous rural development: A path toward reviving rural Europe. *Rural Sociology*, *86*, 911–937.

Gkartzios, M., & Scott, M. (2014). Placing housing in rural development: Exogenous, endogenous and neo-endogenous approaches. *Sociologia Ruralis*, *54*, 241–265.

Guiomar, N., Godinho, S., Pinto-Correia, T., Almeida, M., Bartolini, F., Bezák, P., Biró, M., Bjørkhaug, H., Bojnec, Š., Brunori, G., Corazzin, M., Czekaj, M., Davidova, S., Kania, J., Kristensen, S., Marraccini, E., Molnár, Zs., Niedermayr, J., O'Rourke, E., . . . Wästfelt, A. (2018). 'Typology and distribution of small farms in Europe: Towards a better picture. *Land Use Policy*, *75*, 784–798.

Healey, P. (2004). The treatment of space and place in the new strategic spatial planning in Europe. *International Journal of Urban and Regional Research*, *28*, 45–67.

Helfenstein, J., Hepner, S., Kreuzer, A., Achermann, G., Williams, T., Bürgi, M., Debonne, N., Dimopoulos, T., Diogo, V., Fjellstad, W., Garcia-Martin, M., Hernik, J., Kizos, T., Lausch, A., Levers, C., Liira, J., Mohr, F., Moreno, G., Pazur, R., . . . Herzog, F. (2024). Divergent agricultural development pathways across farm and landscape scales in Europe: Implications for sustainability and farmer satisfaction. *Global Environmental Change*, *86*, 102855.

Holmes, J. (2008). Impulses towards a multifunctional transition in rural Australia: Interpreting regional dynamics in landscapes, lifestyles and livelihoods. *Landscape Research*, *33*, 211–223.

Jończy, R., Śleszyński, P., Dolińska, A., Ptak, M., Rokitowska-Malcher, J., & Rokita-Poskart, D. (2021). Environmental and economic factors of migration from urban to rural areas: Evidence from Poland. *Energies*, *14*(24), 8467. https://doi.org/10.3390/en14248467

Kristensen, S., Reenberg, A., & Peña, J. (2009). Exploring local rural landscape changes in Denmark: A human-environmental timeline perspective. *Danish Journal of Geography*, *109*, 47–67.

Kristensen, S. B. P., Præstholm, S., Busck, A. G., Winther, L., Fertner, C., Vesterager, J. P., & Vejre, H. (2019). On-farm business structure diversification in Greater Copenhagen–Farmers in an urban landscape or entrepreneurs in a rural landscape? *Land Use Policy*, *88*, 104093.

Kristensen, S. B. P., & Busck, A. G. (2024). *Agriculture and landscape in Gadbjerg and Givskud parishes 2008. Results from a questionnaire survey* (IGN Working Paper). University of Copenhagen.

Kristensen, S. P. (1999). Agricultural land use and landscape changes in Rostrup, Denmark: Processes of intensification and extensification. *Landscape and Urban Planning*, *46*, 117–123.

Latacz-Lohmann, U., & Hodge, I. (2003). European agri-environmental policy for the 21st century. *Australian Journal of Agricultural and Resource Economics*, *47*, 123–139.

Lund, J. F., & Jensen, F. S. (2017). Is recreational hunting important for landscape multi-functionality? Evidence from Denmark. *Land Use Policy*, *61*, 389–397.

Lundhede, T. H., Jacobsen, J. B., & Thorsen, B. J. (2015). A hedonic analysis of the complex hunting experience. *Journal of Forest Economics*, *21*, 51–66.

Maciej, K., Jennifer-Ellen, R., Vera, W., & Anne-Katrin, B. (2020). *Farmers of the future*. European Commission.

Ministry of Industry, Business and Financial Affairs (MIBF). (2019). *Landsplanredegørelse 2019*. Ministry of Industry, Business and Financial Affairs, Copenhagen.

Mohr, F., Diogo, V., Helfenstein, J., Debonne, N., Dimopoulos, T., Wenche, D., García-Martín, M., Hernik, J., Herzog, F., Kizos, T., Lausch, A., Lehmann, L., Levers, C., Pazur, R., Ruiz-Aragón, V., Swart, R., Thenail, C., Ulfeng, H., Verburg, P., Williams, T., Zarina, A., & Bürgi, M. (2023). Why has farming in Europe changed? A farmers' perspective on the development since the 1960s. *Regional Environmental Change*, *23*, 156.

Newsham, N., & Rowe, F. (2023). Understanding trajectories of population decline across rural and urban Europe: A sequence analysis. *Population, Space and Place*, *29*, e2630.

Oxford Research. (2014). *Til- og fraflytningsanalyse, motiver for til- og fraflytning på Lolland-Falster*. Frederiksberg.

Pavlis, E. S., Terkenli, T. S., Kristensen, S. B. P., Busck, A. G., & Cosor, G. L. (2016). Patterns of agri-environmental scheme participation in Europe: Indicative trends from selected case studies,. *Land Use Policy*, *57*, 800–812.

Pinto-Correia, T., Guiomar, N., Guerra, C. A., & Carvalho-Ribeiro, S. (2016). Assessing the ability of rural areas to fulfil multiple societal demands. *Land Use Policy*, *53*, 86–96.

Primdahl, J. (2010). Globalisation and the local agricultural landscape: Current change patterns and public policy interventions. In J. Primdahl & S. Swaffield (Eds.), *Globalisation and agricultural landscapes: Change patterns and policy trends in developed countries*. Cambridge University Press.

Primdahl, J., & Swaffield, S. (2010). Globalisation and the sustainability of agricultural landscapes. In J. Primdahl & S. Swaffield (Eds.), *Globalisation and agricultural landscapes: Change patterns and policy trends in developed countries*. Cambridge University Press.

Ray, C. (2000). Editorial. The EU leader programme: Rural development laboratory. *Sociologia Ruralis*, *40*, 163–171.

Reenberg, A., Langanke, T., Kristensen, S. B. P., & Colding, T. S. (2010). Globalisation of agricultural landscapes: A land systems approach. In J. Primdahl & S. Swaffield (Eds.), *Globalisation and agricultural landscapes: Change patterns and policy trends in developed countries*. Cambridge University Press.

Shucksmith, M. (2010). Disintegrated rural development? Neo-endogenous rural development, planning and place-shaping in diffused power contexts. *Sociologia Ruralis*, *50*, 1–14.

Statistics Denmark. (2024). *AFG5: Cultivated area by region, unit and crop*. Copenhagen.

Van der Sluis, T., Pedroli, B., Frederiksen, P., Kristensen, S. B. P., Busck, A. G., Pavlis, V., & Cosor, G. L. (2019). The impact of European landscape transitions on the provision of landscape services: An explorative study using six cases of rural land change. *Landscape Ecology*, *34*, 307–323.

Vejre, H., Abildtrup, J., Andersen, E., Andersen, P. S., Brandt, J., Busck, A., Dalgaard, T., Hasler, B., Huusom, H., Kristensen, L. S., Kristensen, S. P., & Præstholm, S. (2007). Multifunctional agriculture and multifunctional landscapes – land use as an interface. In *Multifunctional land use: Meeting future demands for landscape goods and services*. Springer.

Index

Note: numbers in **bold** indicate a table. Numbers in *italics* indicate a figure on the corresponding page.

FSC
www.fsc.org
MIX
Papier | Fördert
gute Waldnutzung
FSC® C083411

Kolibri 360 GmbH
Ferdinand-Jühlke-Straße 7
99095 Erfurt, Deutschland
produktsicherheit@kolibri360.de